EAI/Springer Innovations in Communication and Computing

Series Editor
Imrich Chlamtac, European Alliance for Innovation, Ghent, Belgium

Editor's Note

The impact of information technologies is creating a new world yet not fully understood. The extent and speed of economic, life style and social changes already perceived in everyday life is hard to estimate without understanding the technological driving forces behind it. This series presents contributed volumes featuring the latest research and development in the various information engineering technologies that play a key role in this process.

The range of topics, focusing primarily on communications and computing engineering include, but are not limited to, wireless networks; mobile communication; design and learning; gaming; interaction; e-health and pervasive healthcare; energy management; smart grids; internet of things; cognitive radio networks; computation; cloud computing; ubiquitous connectivity, and in mode general smart living, smart cities, Internet of Things and more. The series publishes a combination of expanded papers selected from hosted and sponsored European Alliance for Innovation (EAI) conferences that present cutting edge, global research as well as provide new perspectives on traditional related engineering fields. This content, complemented with open calls for contribution of book titles and individual chapters, together maintain Springer's and EAI's high standards of academic excellence. The audience for the books consists of researchers, industry professionals, advanced level students as well as practitioners in related fields of activity include information and communication specialists, security experts, economists, urban planners, doctors, and in general representatives in all those walks of life affected ad contributing to the information revolution.

Indexing: This series is indexed in Scopus, Ei Compendex, and zbMATH

About EAI

EAI is a grassroots member organization initiated through cooperation between businesses, public, private and government organizations to address the global challenges of Europe's future competitiveness and link the European Research community with its counterparts around the globe. EAI reaches out to hundreds of thousands of individual subscribers on all continents and collaborates with an institutional member base including Fortune 500 companies, government organizations, and educational institutions, provide a free research and innovation platform.

Through its open free membership model EAI promotes a new research and innovation culture based on collaboration, connectivity and recognition of excellence by community.

More information about this series at http://www.springer.com/series/15427

Khairol Amali Bin Ahmad • Khaleel Ahmad
Uma N. Dulhare
Editors

Functional Encryption

Editors
Khairol Amali Bin Ahmad
National Defence University of Malaysia
Kem, Kuala Lumpur, Malaysia

Khaleel Ahmad
Department of Computer Science
and Information Technology
Maulana Azad National Urdu University
Hyderabad, Telangana, India

Uma N. Dulhare
Muffakham Jah College of Engineering and
Technology
Hyderabad, Telangana, India

ISSN 2522-8595 ISSN 2522-8609 (electronic)
EAI/Springer Innovations in Communication and Computing
ISBN 978-3-030-60892-7 ISBN 978-3-030-60890-3 (eBook)
https://doi.org/10.1007/978-3-030-60890-3

This Springer imprint is published by the registered company Springer Nature Switzerland AG
The registered company address is: Gewerbestrasse 11, 6330 Cham, Switzerland

Foreword

Digital security is of utmost importance since half of the world's population uses smartphones. Thus, it touches the lives of common people worldwide. In this digital age, hiding and encrypting important data is one of the major challenges. Functional encryption schemes enhance the security, confidentiality and access control by using the function which allows the sharing of information with authorized people. For newcomers, students from another branch, or researchers who are new and want to understand and learn the basics of functional encryption, the search for any source material that fulfills these demands is often unsuccessful. Most books written on this topic are targeted at those who already possess knowledge of the basics. This can lead to the novice losing interest in this field because they cannot find a book written for them. Thus, this book will satisfy these readers, as it has been written with them in mind.

This is perhaps the first book about functional encryption written specifically for the novice. This book covers functional encryption algorithms and its modern applications in developing secure systems. The latest functional encryption algorithms are explained in a simple and precise manner. Examples are given to solidify the concepts and increase understanding.

This book helps professionals, researchers, scientists, faculty members, research scholars, graduate students, and software developers in the domain of Cryptography/Cybersecurity/Information Security/Software Security/Database Security/Web Security/Wireless Network Security/Cloud Security/Online Transactions/E-Commerce Security/M-Commerce Security for better understanding of the basic concepts and techniques to build functional encryption and various encryption mechanisms such as identity-based encryption (IBE) and attribute-based encryption (ABE) into real-world systems. Those who seek to understand these concepts and techniques will find this book a valuable asset. The editors have edited this book to provide awareness of the methods used for functional encryption in the academic and professional communities.

Riverside, USA   Mohammad Sufian Badar

Preface

Information security is the protection of information systems, hardware, software, and information from damages as well as theft, interruption, or misdirection to any of these resources. In other words, cybersecurity focuses on protecting computers, networks, programs, and data (in use, in rest, in motion) from unauthorized or unintended access, change, or destruction (all aimed for exploitation). It is estimated that 3300 million people are using smart mobile phones globally, which is more than half of the world's population. Hence, digital security is no longer limited to the scholarly community but is now the concern of all users of computers worldwide.

In acknowledging such expansion and needs of information security, this book is aimed to provide awareness of methods used for functional encryption in the academic and professional community. While this book would dwell on the foundations of functional encryption as part of security, it will also focus on contemporary topics for Research and Development.

The chapters cover functional encryption algorithms and its modern applications in developing secure systems, viz. entity authentication, message authentication, software security, cybersecurity, hardware security, Internet of Thing (IoT), cloud security, smart card technology, CAPTCHA, digital signature, and digital watermarking. This book is organized into 15 chapters, i.e., Foundations of functional encryption, Impact of Group Theory in Cryptosystem, Elliptic Curve Cryptography, Hyper Elliptic Curve Cryptography (HECC), XTR algorithm: Efficient and Compact Subgroup Trace Representation, Pairing-based cryptography, NTRU Algorithm: Nth Degree Truncated Polynomial Ring Units, Cocks IBE scheme, Boneh-Franklin IBE, Boneh-Boyen IBE, Sakai-Kasahara IBE, Hierarchical Identity-Based Encryption, Attribute-based Encryption, Extensions of IBE and Related Primitives, Digital Signatures.

Finally, it gives us great pleasure to acknowledge the contributions and supports of many individuals. Indeed, we would like to express our gratitude to all the authors who had contributed in the forms of the submitted chapters without which, the production of this book is not possible. We are also thankful to the team from

Springer for the meticulous service in timely publication of this book. We would like also to thank our college/University for their encouragement and last but not least, we gratefully appreciate the support, encouragement, and patience of our families.

Kuala Lumpur, Malaysia   Khairol Amali Bin Ahmad

Hyderabad, Telangana, India   Khaleel Ahmad

Hyderabad, Telangana, India   Uma N. Dulhare

Contents

1   **Foundations of Functional Encryption** 1
     Md. Sharif Hossen

2   **Impact of Group Theory in Cryptosystem** 19
     Priyanka Singh, Manju Khari, and Nikhil S. Kaundanya

3   **XTR Algorithm: Efficient and Compact Subgroup Trace
     Representation** 37
     Pinkimani Goswami, Madan Mohan Singh, and Dimpi Biswas

4   **HECC (Hyperelliptic Curve Cryptography)** 59
     Taspia Salam and Md. Sharif Hossen

5   **Pairing-Based Cryptography** 79
     Ansh Riyal, Geetansh Kumar, and Deepak Kumar Sharma

6   **NTRU Algorithm: N^{th} Degree Truncated Polynomial Ring Units** 103
     Afsar Kamal, Khaleel Ahmad, Rosilah Hassan,
     and Khujamatov Khalim

7   **Cocks IBE Scheme** 117
     Deepak Kumar Sharma, Bhanu Tokas, Venkata Rohit Jakkinapalli,
     and Ritvik Nagpal

8   **Boneh-Franklin IBE** 137
     Deepak Kumar Sharma, Bhanu Tokas, Venkata Rohit Jakkinapalli,
     and Ritvik Nagpal

9   **Boneh-Boyen IBE** 151
     Ankita Karale, Vladimir Poulkov, Milena Lazarova,
     and Pavlina Koleva

10   **Sakai-Kasahara IBE** 171
     Hamza Mutaher and Mahmoud E. Hodeish

11 HIBE: Hierarchical Identity-Based Encryption 187
     Tawseef Ahmed Teli, Faheem Syeed Masoodi, and Alwi M. Bahmdi

12 Extensions of IBE and Related Primitives 205
     Syed Taqi Ali

13 Attribute-Based Encryption 225
     Ankita Karale, Vladimir Poulkov, and Milena Lazarova

14 Digital Signatures 243
     Pinkimani Goswami, Madan Mohan Singh,
     and Khandakar Tahidur Rahman

15 QUIET: Quatro-Inverse Exponential Cipher Technique 279
     Harshit Bhatia, Rahul Johari, and Kalpana Gupta

Index 303

Chapter 1
Foundations of Functional Encryption

Md. Sharif Hossen

Abstract Functional encryption is an emerging paradigm of public-key encryption that permits the enormous flexibility in accessing the ciphertext and helps to compute the function of the ciphertext using a decryption key. Nowadays, it contributes to protect the more large and complex data in the cloud. In this chapter, we will discuss the functional encryption, its relationship with fully homomorphic encryption, challenges, and future directions.

Keywords Functional encryption · Obfuscation · Prime numbers · Greatest common divisor · Linear congruence · Multiplicative inverse

1 Introduction

With the fast advancement of technology, the number of mobile or laptop devices is growing rapidly. The people in this modern world are daily sharing millions of files via email and other media. Nowadays, most people are storing their resources in the cloud. The cloud provides us many facilities free, e.g., hardware, software, use of power. That means we can easily access our data without buying or controlling any hardware or software. We do not need to pay for storing the data and using the power to maintain the data in a remote high-powerful server computer. The main problem here is the security of data we store in the remote computer as the shared data can be any of the credential information like password, credit card details, and personal information. This information can be controlled by third parties because we are storing the data in the outside location where we do not have any idea about security issues. So, we are at risk of our data [1].

The data security can be assured by encrypting the information with a common shared secret key used by both sender and receiver, which is called encryption where the encrypted text instead of the credential information is sent through the media

Md. S. Hossen (✉)
Dept. of Information and Communication Technology, Comilla University, Comilla, Bangladesh

© Springer Nature Switzerland AG 2021
K. A. B. Ahmad et al. (eds.), *Functional Encryption*, EAI/Springer Innovations in Communication and Computing, https://doi.org/10.1007/978-3-030-60890-3_1

so that the third parties could not be able to break the original information. This encrypted text is read by breaking it by the receiver using the common key. This process is termed as decryption. This overall technique of sharing the common key to both parties is called the symmetric encryption. In two-way communication, the conversation is compromised as we do not know the way of sharing the shared key. So, anyone knowing the shared key can decrypt the message [2, 3]. Another approach called public-key encryption or asymmetric encryption uses two keys, public and private. Using this approach, someone getting the private key can retrieve the messages sent to you, but would not be able to see what you send to others as different key pairs are used. In an asymmetric paradigm, on the sender side, the sender uses the receiver's public key to encrypt the original messages, while on the receiver side, the receiver decrypts the encrypted messages using its own private key to get the original plaintext. In both cases, i.e., symmetric and asymmetric, if someone does not have any idea about the key, one can learn only the message length not the plaintext from the encrypted version of the original message [4, 5]. Different encryption techniques, e.g., triple DES, Blowfish, AES, RSA, Diffie-Hellman, and ECC, have been developed and implemented to ensure data security.

Accessing to the encrypted version (i.e., ciphertext) of the plaintext, traditional asymmetric encryption can provide the details of the original message or expose nothing. This introduces the idea of functional encryption (FE). FE is an emerging paradigm that permits a person enormous flexibility in accessing the ciphertext and helps to learn the function of the ciphertext using a decryption key. In FE scheme, a trusted authority uses a master secret key (k_{ms}) and produces a secret key $k_s[g]$ embedded with a function g. Getting $k_s[g]$, a user can determine $g(m)$ from the ciphertext of message m [6, 7]. Consider k_{pu} and k_{pr} indicate the receiver's public and private keys, respectively. Then, the encryption and decryption functions using functional encryption approach are expressed as follows:

$$\text{Encryption} : ciphertext, c = Enc\left(k_{pu}, m\right) \tag{1.1}$$

$$\text{Decryption} : plaintext, p = Dec\left(k_s\left[g\right], c\right) = g(m) \tag{1.2}$$

From Eq (1.2) of the decryption process, plaintext, $p = g(x)$. So, the k_{ms} provider can learn only about $g(m)$ not m. Original plaintext should be m, which is derived by $p = Dec(k_{pr}, c) = m$. So, only the receiver decrypts the message.

The rest of this chapter covers the following. Section 2 discusses the syntax of functional encryption. Section 3 shows the relationship between functional encryption, obfuscation, and fully homomorphic encryption. Challenges and applications are elaborated in Sects. 4 and 5, respectively. Present and future directions in functional encryption are included in Sect. 6. Sections 7 and 8 represent the details of additive and multiplicative inverse. Sections 9 and 10 describe the matrices and linear congruence, respectively. Prime and relatively prime numbers are mentioned in Sec. 11. Finally, in Sect. 12, we provide the discussion of the greatest common divisor using Euclid, Bezout, and Extended Euclid algorithms.

2   Functional Encryption Syntax

Here, we mention the syntactic discussion of FE. The functionality G explains how the information of plaintext can be obtained from the encrypted data.

2.1   Functionality

Assume that K and M refer to the set of all possible keys and messages, respectively. Then, the function G is defined as $G : K \times M \rightarrow \{0, 1\}$, which is a deterministic Turing Machine. In K, there is an empty key ϵ, which purposely tries to store the details of the original messages from the encrypted data, e.g., the length of it. Then, a user can get the leaked message using the following equations:

$$c \xleftarrow{R} Enc\left(k_{pu}, m\right) \tag{1.3}$$

$$Dec\left(\epsilon, c\right) \tag{1.4}$$

Let $k \in K$, $m \in M$. Given the ciphertext of m is c, and secret key of k is k_s. Then, the decryption function will be

$$z = G\left(k_s, c\right)$$

2.2   Functional Encryption Scheme

For all $k \in K$, $m \in M$, FE scheme requires four functions, setup (*Set*), key generation (*KeyGen*), encryption (*Enc*), decryption (*Dec*) to get the messages leaked from c. Here, the public key is k_{pu}, the master secret key is k_{ms}, the secret key for k is k_s

$$\left(k_{pu}, k_{ms}\right) \leftarrow Set\left(1^{\gamma}\right)$$

$$k_s \leftarrow KeyGen\left(k_{ms}, k\right)$$

$$c \leftarrow Enc\left(k_{pu}, m\right)$$

$$z \leftarrow Dec\left(k_s, c\right)$$

We need to fulfill $z = G(k, m)$ with probability 1.

In the above FE scheme, at first k_{pu} and k_{ms} are generated using the algorithm *Set*. Then, k_s is found by *KeyGen* algorithm. After that m is encrypted and we found the ciphertext, c. Finally, *Dec* algorithm with k_s is called to find the leaked message $G(k, m)$ from c.

Let $K : \{1, \epsilon\}$. For $m = 1$, encrypted data is completely decrypted using k_s. For $k = \epsilon$, it will return the total number of bits having in the message [5].

$$G(k, m) := \begin{cases} m & if \ m = 1 \\ length(m) & if \ k = \epsilon \end{cases}$$

3   Relationship Between Functional Encryption, Obfuscation, Fully Homomorphic Encryption

Homomorphic encryption is a type of encryption that permits the computation over encrypted text without any need of a secret key. It ensures the data to be encrypted even it is processed and computed. Anyone can apply the function on the ciphertext, but the plaintext will not be revealed. Using this approach, the employee in an organization can access and process the encrypted data and can get the text, but it is not the original plaintext. Only, the authenticated user can understand the text using its own secret key. That is, this approach ensures the integrity of the data as only the key holder can access or realize the decrypted versions of the encrypted text. This type of encryption is used in the computation of preserving privacy in the cloud. It can be of three types, namely partially, somewhat, and fully homomorphic encryption. In partially homomorphic encryption (which is used in RSA), a certain operation using a selected function is applied on ciphertext many times to build a secure connection. In somewhat homomorphic encryption, few operations (either multiplication or addition) are applied with a limited number of times.

Fully homomorphic encryption (FHE) is an encryption technique that allows us to perform arbitrary computations on the ciphertext with privacy where we will be able to access our data without any permission of the service provider. It is still under the development phase. For example, we are storing our personal data in the cloud without any guarantee that the cloud provider can access the data. So, there is a risk of our data. FHE can be used to prevent the cloud provider to access our data. Using functional encryption (FE), someone can access the encrypted data to compute the function on it but cannot reach to break the data to understand the original text. Obfuscation is the process of blurring the original message to sense it someone to be confused, ambiguous, and difficult to understand [6, 8].

It is unknown to understand the construction of FE from FHE. The functionalities used in FHE are also not known clearly because the output of it is also a result of the encryption of the encrypted data while the output of FE is understandable. FE is an openly discussed problem while FHE is not. FHE and FE can ensure the confidentiality of the data, i.e., the unauthorized people cannot access the data to use while obfuscation cannot guarantee the confidentiality of the data.

Using obfuscation, one can hide his email address (i.e., avoid as a spam email) to be marked as a spam email by spammers or spam filters. There are many methods of obfuscation. Some of these are reversing the text direction, adding null text, encoding "@" and ".", replacing "@" with "AT" and "." with "DOT," and using "urlencode" method. The best appropriate scheme is to reverse the text direction. Let us consider the email "mshossen@cou.ac.bd" to be obfuscated. At first, we will write the email backward on the page of the website and then include it in tag as follows:

db.ca.uoc@nessohsm

To see the email correctly to the visitors, the following code of CSS is written as follows:

.obfuscate {unicode-bidi: bidi-override; direction: rtl;}

On the other hand, FE and FHE ensure data confidentiality. Using FE, the spam filter can block the email sent from the sender to the receiver if it looks like spam where FE does not know about the contents of email. While FHE does not block the spam email but indicates it as spam. Hence, the importance of FE is superior to FHE [9].

4 Challenges

An attacker tries to use different keys to compute the function on the ciphertext. An FE system will be secure if the attacker cannot get the original message using its keys but only read the ciphertext. There are several challenges to ensure secure functional encryption (FE) system. Some of the challenges are mentioned below [5, 10]:

- It is a challenge to generate an efficient FE system for supporting various general functionalities.
- The main challenge in functional encryption is the generation of public and private keys to compute any function on the encrypted data. The creation of such keys could help the integration of anything from a random spam filter to the image to be used for recognition.
- Different secret keys are used to attack the FE. These attacks are called Collusion attacks which are another challenge to build a secure FE system because they

have the ability to merge different keys to generate a new key which tends to break the functionality of the FE system.

- Constructing a secure FE for all the polynomial-time functionalities is also a challenge. Hence, the predicate encryption is more realistic now. At the moment, the construction of inner products and fully homomorphic encryption can help to derive the functionalities of predicates.

5 Applications

Ensuring data security is a fast consideration of the system. There are lots of applications for the FE system. Some of the areas where FE is used are mentioned below [5, 11].

- Spam filtering.
- Increase the security of the cloud storage system.
- Privacy-preserving data mining system.
- Fine-grained access control.
- Access to the encrypted text using arbitrary policy over the sender's personal details.

6 Present and Future Directions in Functional Encryption

At present, the researchers are trying to develop a secure functional encryption system. Researchers from the University of California in Los Angeles already have built a technique of obfuscation using the FE to prevent hackers from attacking the programs. Their developed method is very useful to make the data to be unintelligible for the attacker [12].

The main challenge in the functional encryption is the generation of public and private keys to compute the function on the encrypted data. The creation of such keys could help the integration of anything from a random spam filter to the image to be used for recognition. Hence, the certain image of a person can be used to encrypt the text, and only the face of that person can be used to decrypt the text. More research investigations are needed to build a more flexible real-time system so that different accessing policies should be developed. These policies will help to maintain the access of data by the authorized users, i.e., who are the valid users or not to access the data. Signature verification in FE can also be taken into count to ensure the confidentiality of data [5].

7 Additive Inverse

In public-key encryption, we know two keys, private and public are used. In most of the cases, these keys are calculated using modular arithmetic [13].

7.1 Modular Arithmetic

Suppose we divide the integer X by the integer Y. Then, we will get a quotient Q and a remainder R. For cryptographic analysis, we are interested only in the value of R. Using the modulo operator (mod), the remainder R can be calculated easily as $X \, mod \, Y = R$, where Y is the modulus. For example, $18/10 = 1$ remainder 8, i.e., 18 mod $10 = 8$.

7.2 Addition and Subtraction Property of Modular Arithmetic

$$(X + Y) \bmod Z = (X \bmod Z + Y \bmod Z) \bmod Z$$

$$(X - Y) \bmod Z = (X \bmod Z - Y \bmod Z) \bmod Z$$

For example, for addition mod 10,

$$(8 + 9) \bmod 10 = (8 \bmod 10 + 9 \bmod 10) \bmod 10 = 17 \bmod 10 = 7.$$

Figure 1.1 shows the addition modulo 10. Plaintext can be encrypted using this addition operation. For example, for secret key $k = 4$, the encrypted text of plaintext 4592 is 8936.

Plaintext can be retrieved from the encrypted text, i.e., ciphertext using the inverse of k, called the additive inverse of k. The additive inverse of a number (k) refers to the negative $(-k)$ of that number, and when added to the number yields zero, $k + (-k) = 0$. The inverse of the number is also termed as opposite number, negation, and sign change. There is no effect of double additive inverse of a number, e.g., $-(-k) = k$. For $k = 4$, the inverse of k is $-k = 6$ since $4 + 6 = 0$, $(4 + 6 = 10 \bmod 10 = 0)$. Hence, the retrieved plaintext from ciphertext 8936 using $-k = 6$ is 4592.

In Fig. 1.1, the intersection of a number and its inverse will be zero, e.g., the intersection of 6 and 4 is zero.

+	0	1	2	3	4	5	6	7	8	9
0	0	1	2	3	4	5	6	7	8	9
1	1	2	3	4	5	6	7	8	9	0
2	2	3	4	5	6	7	8	9	0	1
3	3	4	5	6	7	8	9	0	1	2
4	4	5	6	7	8	9	0	1	2	3
5	5	6	7	8	9	0	1	2	3	4
6	6	7	8	9	0	1	2	3	4	5
7	7	8	9	0	1	2	3	4	5	6
8	8	9	0	1	2	3	4	5	6	7
9	9	0	1	2	3	4	5	6	7	8

Fig. 1.1 Addition mod 10

8 Multiplicative Inverse

Multiplicative inverse refers to the reciprocal of a number (Z) denoted by $1/Z$ or Z^{-1}. When a number is multiplied by its multiplicative inverse yields 1, e.g., $Z \times \left(1/Z\right) = 1$, but except the number zero (0) as $1/0$ is undefined [13].

8.1 Multiplication Property of Modular Arithmetic

$$(X \times Y) \bmod Z = (X \bmod Z \times Y \bmod Z) \bmod Z$$

For example, for multiplicative mod 10,

$$(8 \times 9) \bmod 10 = (8 \bmod 10 \times 9 \bmod 10) \bmod 10 = 72 \bmod 10 = 2$$

Figure 1.2 shows the multiplication modulo 10. Plaintext can be encrypted using this multiplication mod operation. For example, for secret key $k = 3$, the encrypted text of plaintext 4592 is 2576.

The plaintext can be retrieved from the encrypted text using the reciprocal of k, called the multiplicative inverse of k, i.e., k^{-1}. For $k = 3$, the multiplicative inverse of k is $k^{-1} = 7$ since $3 \times 7 = 1$, $(21 \bmod 10 = 1)$. Hence, the retrieved plaintext from the ciphertext 2576 using $k^{-1} = 7$ is 4592. In Fig. 1.2, the intersection of k and k^{-1} is 1 where only the numbers $\{1, 3, 7, 9\}$ satisfy this.

×	0	1	2	3	4	5	6	7	8	9
0	0	0	0	0	0	0	0	0	0	0
1	0	1	2	3	4	5	6	7	8	9
2	0	2	4	6	8	0	2	4	6	8
3	0	3	6	9	2	5	8	1	4	7
4	0	4	8	2	6	0	4	8	2	6
5	0	5	0	5	0	5	0	5	0	5
6	0	6	2	8	4	0	6	2	8	4
7	0	7	4	1	8	5	2	9	6	3
8	0	8	6	4	2	0	8	6	4	2
9	0	9	8	7	6	5	4	3	2	1

Fig. 1.2 Multiplication mod 10

9 Matrices

Different approaches are used to prevent attackers so that they could not be able to break the encoded data sent from the sender to access the original messages. Large matrices are extremely difficult to break, so the encryption and the decryption using matrices can be applied to ensure the data security of different organizations, governments, or non-governments [13].

For simplicity, we consider the assignment of each letter, i.e., A to Z, by a number sequentially.

A	B	C	D	E	F	G	H	I	J	K	L	M	N	O
1	2	3	4	5	6	7	8	9	10	11	12	13	14	15

P	Q	R	S	T	U	V	W	X	Y	Z
16	17	18	19	20	21	22	23	24	25	26

For the encryption and the decryption operation, we need to choose the plaintext and the encoding matrix in the sender side. The encrypted code is decrypted using a decoding matrix, which is also called the inverse of encoding matrix. We have to choose such a matrix that has the inverse of it.

Consider the plaintext is $P = $ "*ICT DEPARTMENTS*" and the 3×3 encoding matrix A is as follows:

$$A = \begin{vmatrix} 1 & 2 & -1 \\ -2 & 0 & 1 \\ 1 & -1 & 0 \end{vmatrix}$$

The inverse of A is called the inverse of encoding matrix, i.e., the decoding matrix, A^{-1}.

$$A^{-1} = \begin{vmatrix} 1 & 1 & 2 \\ 1 & 1 & 1 \\ 2 & 3 & 4 \end{vmatrix}$$

The plaintext "*ICT DEPARTMENTS*" corresponds to the following numbers. For whitespace "" in the plaintext, we consider the number 27.

I	C	T	*	D	E	P	A	R	T	M	E	N	T	S
9	3	20	27	4	5	16	1	18	20	13	5	14	20	19

Now, we form a 3 by 1 vector, v, for the above corresponding numbers of the plaintext as follows:

$$\begin{vmatrix} 9 \\ 3 \\ 20 \end{vmatrix} \quad \begin{vmatrix} 27 \\ 4 \\ 5 \end{vmatrix} \quad \begin{vmatrix} 16 \\ 1 \\ 18 \end{vmatrix} \quad \begin{vmatrix} 20 \\ 13 \\ 5 \end{vmatrix} \quad \begin{vmatrix} 14 \\ 20 \\ 19 \end{vmatrix}$$

$$v = \begin{vmatrix} 9 & 27 & 16 & 20 & 14 \\ 3 & 4 & 1 & 13 & 20 \\ 20 & 5 & 18 & 5 & 19 \end{vmatrix}$$

To form the encrypted version of the plaintext P, we multiply the encoding matrix by the above vector as follows:

$$E = A * v = \begin{vmatrix} 1 & 2 & -1 \\ -2 & 0 & 1 \\ 1 & -1 & 0 \end{vmatrix} \begin{vmatrix} 9 & 27 & 16 & 20 & 14 \\ 3 & 4 & 1 & 13 & 20 \\ 20 & 5 & 18 & 5 & 19 \end{vmatrix}$$

After multiplying the above two matrices, we get the encoded data as follows:

$$E = \begin{vmatrix} -5 & 30 & 0 & 41 & 35 \\ 2 & -49 & -14 & -35 & -9 \\ 6 & 23 & 15 & 7 & -6 \end{vmatrix}$$

This encoded message is transmitted through the channel in a serial form as follows:

$$-5, \quad 2, \quad 6, \quad 30, \quad -49, \quad 23, \quad 0, \quad -14, \quad 15, \quad 41, \quad -35, \quad 7, \quad 35, \quad -9, \quad -6$$

This encoded version is then received at the receiver side where the decryption process is performed. Here, the decoding matrix, A^{-1}, is multiplied by the received encrypted data where it is formed as a sequence of 3 by 1 vectors.

$$D = A^{-1} \times E = \begin{vmatrix} 1 & 1 & 2 \\ 1 & 1 & 1 \\ 2 & 3 & 4 \end{vmatrix} \begin{vmatrix} -5 & 30 & 0 & 41 & 35 \\ 2 & -49 & -14 & -35 & -9 \\ 6 & 23 & 15 & 7 & -6 \end{vmatrix}$$

$$= \begin{vmatrix} 9 & 27 & 16 & 20 & 14 \\ 3 & 4 & 1 & 13 & 20 \\ 20 & 5 & 18 & 5 & 19 \end{vmatrix}$$

Writing the matrix in a linear form, we get the following:

9	3	20	27	4	5	16	1	18	20	13	5	14	20	19
I	C	T	*	D	E	P	A	R	T	M	E	N	T	S

10 Linear Congruence

A linear congruence (LC) [13] is a congruence of the following form to find $x \in \mathbb{Z}$

$$ax \equiv b \ (\bmod \ c)$$

where a, b, c are the set of integers, i.e., $a, b, c \in \mathbb{Z}$.

LC has the following properties.

1. If $\gcd(a, c) = 1$, LC has a unique solution.
2. If $\gcd(a, c) \nmid b$, LC has no solution.
3. If $\gcd(a, c) \mid b$, LC has b solutions.

Example 1 Find all the solutions for $17x \equiv 3 \ (mod \ 29)$.

We will calculate the gcd(17, 29) at first, using Euclid's algorithm.

$$29 = 1 \times 17 + 12$$
$$17 = 1 \times 12 + 5$$
$$12 = 2 \times 5 + 2$$
$$5 = 2 \times 2 + 1$$
$$2 = 2 \times 1 + 0$$

Therefore, $\gcd(17, 29) = \gcd(1, 0) = 1$. So, it has one solution. We will now find the multiplicative inverse of 17 *mod* 29 as follows:

$$1 = 5 - 2 \times 2$$
$$= 5 - 2 \times (12 - 2 \times 5)$$
$$= 5 \times 5 - 2 \times 12$$
$$= 5 \times (17 - 1 \times 12) - 2 \times 12$$
$$= 5 \times 17 - 7 \times 12$$
$$= 5 \times 17 - 7 \times (29 - 1 \times 17)$$
$$= 12 \times 17 - 7 \times 29$$

So, the multiplicative inverse of 17 *mod* 29 is 12. Multiplying both sides of the linear congruence $17x \equiv 3 \ (mod\ 29)$ by 12, we get

$$12 \times 17x \equiv 12 \times 3 \ (mod\ 29)$$
$$x \equiv 36 \ (mod\ 29)$$
$$\equiv 7 \ (mod\ 29) \ since\ 36 - 29 = 7$$

We can check the value of x whether it is correct or wrong as follows:

$$17 \times 7 \equiv 119 = 3 \ (mod\ 29) \ since\ 119 = 4 \times 29 + 3$$

Notice that $12 \times 17x = x$. Say $m = 17$ and multiplicative inverse of 17 *mod* $29 = m^{-1} = 12$. We know that $m \times m^{-1} = 1$. Therefore, $m \times m^{-1} = 17 \times 12 = 1$. That is why, $12 \times 17x = x$.

Example 2 Find all the solutions to the linear congruence $22x \equiv 7(mod\ 143)$.

We will calculate the gcd(22, 143) at first, using Euclid's algorithm.

$$143 = 6 \times 22 + 11$$
$$22 = 2 \times 11 + 0$$

Hence, $\gcd(22, 143) = \gcd(11, 0) = 11$ and $11 \nmid 7$. So, the linear congruence $22x \equiv 7(mod\ 143)$ does not have any solution.

11   Prime and Relative Prime Numbers

When two or more numbers are multiplied to get another number, then those numbers are called factors. For example, the factors of 12 are 1, 2, 3, 4, 6, and 12. We can find all the factors of 12 as follows:

$$1 \times 12 = 12$$
$$2 \times 6 = 12$$
$$3 \times 4 = 12$$

An integer number is called prime number if it has two factors, 1 and the number itself. In other words, an integer number $n > 1$ is called a prime if it has two factors ± 1, and $\pm n$. For example, 5 is a prime number as it has only two factors, 1 and 5; number 6 is not a prime number as it has more than two factors, 1, 2, 3, and 6. The first 25 prime numbers are as follows:

2	3	5	7	11	13	17	19	23	29	31	37	41	43	47	53

59	61	67	71	73	79	83	89	97

Numbers X and Y are relatively primes if they have no common factor other than 1. Relatively prime numbers are also called mutually prime or coprime [13]. For example, 15 and 17 are relatively prime numbers since they have only one common factor.

$$1 \times 3 \times 5 = 15$$
$$1 \times 17 = 17$$

Factors of 15 are $\{1, 3, 5\}$ and of 17 are $\{1, 17\}$. Hence, the common factor is 1.

Numbers 16 and 20 are not relatively prime numbers since they have more than 1 common factor. Factors of 16 are $\{1, 2, 4, 8, 16\}$ and of 21 are $\{1, 2, 4, 5, 10, 20\}$. Hence, the common factors are $\{1, 2, 4\}$.

$$1 \times 2 \times 2 \times 2 \times 2 = 16$$
$$1 \times 2 \times 2 \times 5 = 20$$

The product of primes is called as the prime factorization, e.g., $15 = 1 \times 3 \times 5$

$$16 = 1 \times 2^4$$
$$20 = 1 \times 2^2 \times 5$$

12   Greatest Common Divisor (Euclid's Algorithm, Bezout's Algorithm, Extended Euclid's Algorithm)

12.1   Greatest Common Divisor (GCD)

GCD of two or more non-zero numbers is the largest positive number that divides each of the numbers with zero remainders [13]. There are two ways to find GCD. At first, it can be calculated by finding the greatest factor from the common divisors/factors of the numbers.

$$\text{All factors of } 12 = 1, 2, 3, 4, 6, 12$$

$$\text{All factors of } 30 = 1, 2, 3, 5, 6, 10, 15, 30$$

Common divisors of 12 and 30 are 1, 2, 3, and 6. The greatest of these divisors is 6. So, GCD (12, 30) = 6.

Secondly, we can calculate GCD by multiplying all the common prime factors.

$$12 = 2 \times 2 \times 3$$
$$30 = 2 \times 3 \times 5$$

The common prime factors of 12 and 39 are 2 and 3. GCD $(12, 30) = 2 \times 3 = 6$.

12.2   Euclidean/ Euclid's Algorithm

An efficient method of quickly finding the GCD between two numbers is Euclidean algorithm [13].

Consider two numbers x and y where $x > y$. Then, GCD can be calculated using Euclidean algorithm as follows:

$$\gcd(x, y) = \begin{cases} x, & if\ y = 0 \\ \gcd(y, x \bmod y), & otherwise \end{cases}$$

Here, if $y \neq 0$, the larger number x is divided each time by the smaller number y and then x is replaced by y, and y will be replaced by the remainder. If $y = 0$, x will be the GCD between x and y. Otherwise, we will continue this process until $y = 0$.

Example 3 Let us find the GCD between 12 and 30 using Euclidean algorithm.

Consider $x = 30$, and $y = 12$. Here, $y \neq 0$, then applying the above algorithm we get,

$$30 \div 12 = 2 \text{ remainder } 6 \text{ or } 30 = 2 \times 12 + 6$$

$$12 \div 6 = 2 \text{ remiander } 0 \text{ or } 12 = 2 \times 6 + 0$$

Hence, the remainder is zero, so $GCD(12, 30) = GCD(6, 0) = 6$.

We will now explain it more elaborately. We know that $x \ mod \ y$ works like $x \ \% \ y$ which will return the remainder. In the above example, $gcd \ (x, y) = gcd \ (30, 12)$ where $x = 30$, $y = 12 \neq 0$ so, $x \ mod \ y = 30 \ mod \ 12 = 6$, then $gcd \ (y, x \ mod \ y) = gcd \ (12, \ 6)$ where $x = 12$, $y = 6 \neq 0$, so again $x \ mod \ y = 12 \ mod \ 6 = 0$. Now, $gcd \ (y, \ x \ mod \ y) = gcd \ (6, 0)$, hence $x = 6$ and $y = 0$. From the Euclidean algorithm, $y = 0$ indicates that $x = 6$ is the GCD between 12 and 30.

Example 4 Find GCD of 123 and 36 using Euclid's algorithm.

$$123 = 3 \times 36 + 15$$
$$36 = 2 \times 15 + 6$$
$$15 = 2 \times 6 + 3$$
$$6 = 2 \times 3 + 0$$

Therefore, $GCD \ (123, 36) = GCD(3, 0) = 3$.

12.3 Bezout's Algorithm

For two non-zero integers x and y, GCD can be represented as a linear combination $px + qx = gcd \ (x, y)$ where integers p and q are the Bezout's coefficients [13].

If x, y are coprime, then $gcd \ (x, y) = px + qx = 1$. We can calculate the value of x and y using Euclid's algorithm and then applying backward and substitution process called extended Euclid's approach.

Example 5 Consider $x = 123$ and $y = 36$ to calculate $gcd(123, 36)$. We apply Euclid's algorithm at first as follows:

$$123 = 3 \times 36 + 15$$
$$36 = 2 \times 15 + 6$$
$$15 = 2 \times 6 + 3$$
$$6 = 2 \times 3 + 0$$

So, $gcd(123, 36) = gcd \ (3, 0) = 3$

Then, we apply backward and substitution process as follows:

$$3 = 15 - 2 \times 6$$
$$= 15 - 2 \times (36 - 2 \times 15)$$
$$= 5 \times 15 - 2 \times 36$$
$$= 5 \times (123 - 3 \times 36) - 2 \times 36 \quad \text{(iii)}$$
$$= 5 \times 123 - 15 \times 36 - 2 \times 36$$
$$= 5 \times 123 - 17 \times 36$$

Hence, $gcd(123, 36) = 3 = 123p + 36q$ (iv)

Comparing (iii) and (iv), we get $p = 5$ and $q = -17$.

12.4 Extended Euclid's Algorithm

This algorithm is an extension of Euclid's method. It is widely used in cryptography to compute the multiplicative inverse of the polynomial expressions. The RSA encryption approach uses the multiplicative inverse. Extended Euclidean approach follows the steps of Euclid's algorithm reciprocally/reversely. It is mostly used to compute the Bezout's coefficients p and q in the following equation [13].

$$px + qx = \gcd(x, y)$$

Example 3 shows the explanation of finding p and q.

If x and y are relative primes, then $gcd(x, y) = 1$. Hence, $gcd(x, y) = px + qy = 1$. Then, $1 \equiv qy \bmod p$ so that y is the multiplicative inverse of $q \bmod p$.

Now, we will see the implementation of computing multiplicative inverse using extended Euclid's approach in the following example.

.

Example 6 Find the multiplicative inverse of 8 *mod* 11.

We will apply the Euclid' algorithm to find the $gcd(8, 11)$ to check whether it is equal to 1. If $gcd(8, 11) = 1$, then it has the multiplicative inverse, otherwise not.

$$11 = 1 \times 8 + 3$$
$$8 = 2 \times 3 + 2$$
$$3 = 1 \times 2 + 1$$
$$2 = 2 \times 1 + 0$$

Therefore, $gcd(8, 11) = gcd(1, 0) = 1$. So, 8 *mod* 11 has the inverse. Now, we proceed in the reverse way as follows:

$$1 = 3 - 1 \times 2$$
$$= 3 - 1 \times (8 - 2 \times 3)$$
$$= 3 \times 3 - 1 \times 8$$
$$= 3 \times (11 - 1 \times 8) - 1 \times 8$$
$$= 3 \times 11 - 4 \times 8$$
$$= 3 \times 11 + 7 \times 8$$

(since the additive inverse of 4 is 7 mod 11)

We ignore multiples of 11. The multiplicative inverse of 8 *mod* 11 is 7. Notice that $8(7) = 56 = 1 + 5(11) \equiv 1 \ mod \ (11)$.

13 Conclusion and Future Work

Functional encryption (FE) is an efficient encryption system where what things a user can learn from the ciphertext is calculated by a function of the encrypted data and its secret key. FE is used in different applications, e.g., cloud storage system, spam filtering, to ensure the data security. In near future, we would like to propose and implement a secure FE system so that the challenges of FE can be overcome. We also would like to implement the face recognition approach in FE system to ensure the security of the data.

Acknowledgments I would like to thank my dear wife Farhana Rea Deeba for her support and my beloved parents for their dedication.

References

1. Wee, H. (2014). Functional encryption and its impact on cryptography. In M. Abdalla & R. De Prisco (Eds.), *Security and cryptography for networks. SCN 2014. Lecture notes in computer science* (Vol. 8642). Cham: Springer. https://doi.org/10.1007/978-3-319-10879-7_18.
2. Hossen, S. M., & Hossen, S. M. (2020). Implementation of encryption techniques in secure communication model. In *5th international conference on advanced computing and intelligent engineering, Université des Mascareignes (UdM)* (p. 2020). Mauritius: Springer.
3. Kiennert, C., Bouzefrane, S., & Thoniel, P. (2015). Authentication systems. In *Digital identity management* (pp. 95–135). Elsevier. https://doi.org/10.1016/b978-1-78548-004-1.50003-1.
4. Hossen, M. S. (2017). A Java based GUI application for substitution encryption techniques. In *International Conference on Computer, Communication, Chemical, Material and Electronic Engineering*. Bangladesh: University of Rajshahi.
5. Boneh, D., Sahai, A., & Waters, B. (2011). Functional encryption: Definitions and challenges. In Y. Ishai (Ed.), *Theory of cryptography. TCC 2011. Lecture notes in computer science* (Vol. 6597). Berlin, Heidelberg: Springer. https://doi.org/10.1007/978-3-642-19571-6_16.
6. Bonch, D., Sahai, A., & Waters, B. (2012). Functional encryption: A new vision for public-key cryptography. *Communications of the ACM, 55*(11), 56–64. https://doi.org/10.1145/2366316.2366333.

7. Zhao, Q., Zeng, Q., & Liu, X. (2018). Improved construction for inner product functional encryption. *Security and Communication Networks, 2018*, 1–12. https://doi.org/10.1155/2018/6561418.
8. Gentry, C. (2010). Computing arbitrary functions of encrypted data. *Commun, ACM 53*(3), 97–105. https://doi.org/10.1145/1666420.1666444.
9. Alwen, J., et al. (2013). On the relationship between functional encryption, obfuscation, and fully homomorphic encryption. In M. Stam (Ed.), *Cryptography and coding. IMACC 2013. Lecture notes in computer science* (Vol. 8308). Berlin, Heidelberg: Springer.
10. Katz, J., Sahai, A., & Waters, B. (2008). Predicate encryption supporting disjunctions, polynomial equations, and inner products. In *EUROCRYPT* (pp. 146–162).
11. Waters, B. (2013). Functional encryption: Origins and recent developments. In K. Kurosawa & G. Hanaoka (Eds.), *Public-key cryptography – PKC 2013. PKC 2013. Lecture notes in computer science* (Vol. 7778). Berlin, Heidelberg: Springer.
12. Garg, S., Gentry, C., Halevi, S., Raykova, M., Sahai, A., & Waters, B. (2013). Candidate Indistinguishability Obfuscation and Functional Encryption for all Circuits. In *2013 IEEE 54th Annual Symposium on Foundations of Computer Science*. https://doi.org/10.1109/focs.2013.13.
13. Stallings, W. (2013). *Cryptography and network security: Principles and practice* (6th. ed.). USA: Prentice Hall Press.

Chapter 2
Impact of Group Theory in Cryptosystem

Priyanka Singh, Manju Khari, and Nikhil S. Kaundanya

Abstract This paper presents a new approach of group theory toward cryptosystem with its application of mathematical methods and functions for cryptographic regions. This new group theory in cryptographic paradigm is brought together with relation to existing functional groups, fields, rings, mathematical theorems, mathematical tools, etc. this study concentrated on semantic security solutions evaluation for cryptographic systems. Such cryptographic systems are mostly designed to use the group's applications such as fields, ECC, functionality tests, cyclic groups, and Schnorr groups. In this paper, we discussed some possible applications of groups and group theory by introducing different forms of groups, fields, rings, theorems, etc. A brief description of some mathematical tools which include functionalities tests, methods, discrete logarithmic problems, ECC, factoring polynomials, and baby giant algorithm is also presented in this paper. Solutions for cryptographic problems inspired by group theory and its applications in this field are also described in the study.

Keywords Group · Fields · Schnorr group · Lagrange's theorem · Bilinear mapping · Functionalities test · Discrete logarithm problem in subgroup of Z_p^* Discrete logarithmic problem · Sieve method

1 Introduction

Group Theory, in science and conceptual variable-based math, contemplates the existence of the arithmetical constructions that are denoted as Groups. The idea of a gathering is fundamental to digest polynomial math: other notable logarithmic constructions. Examples of such arithmetic constructs include the existence of rings, fields, and vector spaces. These constructs can theoretically all be viewed as forms of Groups. However, their nature differs at the point that they are invested with

P. Singh (✉) · M. Khari · N. S. Kaundanya
Department of Computer Science, AIACTR, Delhi, India

© Springer Nature Switzerland AG 2021
K. A. B. Ahmad et al. (eds.), *Functional Encryption*, EAI/Springer Innovations in
Communication and Computing, https://doi.org/10.1007/978-3-030-60890-3_2

extra activities and aphorisms. Groups repeat all through science, and the strategies for Group Theory have impacted numerous pieces of variable-based math. Straight mathematical Groups and Lie bunches are two parts of gathering hypothesis that have encountered propel and have become branches of knowledge in their own right.

If need be, we can take examples of different physical frameworks that exist in the natural physical world. Examples of such structures would include precious stones, the hydrogen particles, and such novelties. They might be displayed coincidentally by balance Groups. Hence as a result, Group Theory can be said to have enormous impacts on many paths, and the firmly related portrayal hypothesis has been proven to have numerous significant impacts in generating advancements for many areas. Such areas include material science, theoretical physics, and many such areas of science. The gathering hypothesis is likewise key to open key cryptography.

Group-based cryptography can be introduced as a method for the utilization of groups to construct various cryptographic sequences and primitives. A group in cryptographic context is defined as a very general algebraic structure. It is also seen that most cryptographic algorithms and plans make use of groups in some way or the other. If we take a particular cryptographic method as example, Diffie–Hellman key exchange makes use of finite cyclic groups in its key exchange method and proves as an excellent example for the benefit groups bring to cryptographic concepts. In reality, the term group-based cryptography has found its existence mostly in reference to cryptographic protocols that make use of infinite non-abelian groups in their working methodology such as a braid group.

In the course of this chapter, Section 2 talks about the Group Theory, its properties, and examples followed with Sect. 2.1 which presents Abelian group, Sect. 2.2 briefs the Lagrange's Theorem, Sect. 2.3 shows the Schnorr group with its properties in cryptosystem, Sect. 2.4 briefs the finite field, and Sect. 2.5 states bilinear mapping; Section 3 discusses Functionalities (Predicate Encryption, Equality test, Inequality Test, Inner Product Evaluation) and other methods; Section 4 states the roles of group theory in cryptosystem; and finally the whole chapter is concluded in Sect. 5 with the future scope.

2 Group Theory

The term group was first coined by Galois during 1830s to define sets function on finite sets that could be grouped together to form a closed set [1]. According to the modern mathematics groups, the non-empty set lets "G" conjointly with some operation holds two elements of G (say a and b) and when integrate them together they generate element, i.e., third element f such that $f = a*b$, where $*: G \times G \to G$ is a binary operation define on G as $(a, b) \mapsto a * b$, $\forall$ a, b $\in$ G [2].

2.1 Group Axiom

1. *Identity of elements*: There exists an element "e" in G where e ∈ G holds a * e = e * a = a ∀ a ∈ G, where "e" is considered as an identity element. In arithmetic, an identity component, or neutral component, is an uncommon kind of component of a set that is mainly impactful for a binary procedure on that set, which leaves any component of the set unaltered when joined with it.
2. *Associativity of elements*: for any element e, f, g ∈ G, (e * f) * g = e * (f * g). The associative property expresses that you can perform addition or multiplication on elements of a set, paying little mind to how the numbers are assembled or grouped. By "assembled," we signify "how you use the parenthesis" that signifies the step-by-step progress of the operation. At the end of the day, on the off-chance that you are adding or multiplying, it does not make a difference where you put the bracket.
3. *Inverse of element*: for any e ∈ G, there exists an element f ∈ G, thus e*f = f*e = a, where "a" is considered as an identity element. Element "a" also belongs to G. In the event that we consider a as any possible integer, at that point 0 + a = a + 0 = a. Zero is known as the identity element or neutral element of addition operation in light of the fact that adding it to any integer returns back the same integer. For each a, it should definitely contain an integer b with the end aim that b + a = a + b = 0. The integer b is known as the inverse element of the integer a and is signified with −a.
4. *Closure of elements:* for any element such as e, f ∈ G, the resultant of e*f also belongs to G. The Closure Property expresses that when you play out an activity (for example, addition, multiplication, and so on.) on any two elements in a set, the consequent result of the calculation is another element in a similar set. The fourth axiom, i.e., "closure," is put at the fourth place for prominence. If group constructions contain I as a subset of G group, with same operation like G, these types of cases must include the closure check: for certain element x, y in I, x*y is a component of group G but may or may not belongs to I.

2.2 Illustrations

The groups can be specified by defining the "set" along with the operations of group. The groups are illustrated by following examples:

1. "Z" (Integers set) contains addition as group operation
2. $\mathbb{R}^*$ (A set of all real numbers where $R\,\hat{}$ *is not − zero*), with multiplication as group operation
3. Z_n (A set of Integers: 1, 2......, n-1), through addition modulo *n* as group operation.
4. $\mathbb{Z}_n^*$ (set of integers where {1 ≤ a ≤ n − 1 : GCD(a, n) = 1}, multiplication modulo *n* as group operation

5. S_n(set of bijective functions $[n] \to [n]$ where $[n] = \{1, 2, \ldots \ldots, n\}$) with the function composition as group operation.

2.3 Properties of Group

Groups contain some basic properties and few terms which illustrate groups and related elements. Associativity condition of elements shows the dropping of the parentheses completely and making the product of "n" elements of group G such as $x_1 *x_2 * x_3 \ldots x_n$. Since the arrangement of parenthesis does not matter. If the operation is clear, the product can be written without operator $*$, i.e., $x_1 \, x_2 \, x_3 \ldots x_n$. Although what matters most is the "order" of elements, whereas normally it is not that ef = fe for all e, f $\in$ G. for any e, f in G where ef = fe is termed as abelian group. The abelian groups are the commutative groups in a group theory.

2.3.1 Abelian Group

Groups can be categorized as commutative and non-commutative, and the commutative groups are also known as abelian groups. In abelian groups, the law of composition is commutative such as for all e, f $\in$ G satisfies the property, i.e., $e*f = f*e$. In the formal definition of abelian group set G and operation o are combined together, i.e., G x G $\to$ G. Two elements of set G are taken by operation and returns element of G which satisfies the certain properties or axioms as group [3]. The commutative axiom of Abelian group is stated as follow:

Commutativity Axiom: $\forall$ e, f $\in$ G holds relation e $\circ$ f = f $\circ$ e. In math, the associative and commutative properties are designed to be laws that are applied to addition and multiplication operations that consistently exist. The associative property expresses that in practice you can re-bunch numbers and perform the operation to find that you will find a similar solution. The commutative property expresses similar view point that you can move numbers around to change the order of the operations taking place on the numbers and still show up at a similar answer.

Examples

- The most common example of abelian groups is *cyclic* group. These groups are created by single element, and a finite cyclic group is isomorphic to Z_n, for some positive integer n. The generators use the successive implementation of law of groups.
- The generator creates a cycle among the various elements of the group. For example, the generator g with powers of set of integers Z6 are $\{g0, g1, g2, g3, g4\} = \{g0, g1, g2, g3\}$. So it makes element $\{g0, g1, g2, g3, g4\}$ since $g^x \, g^y = g^y$

$g^x = g^{x+y}$ are abelian groups. All cyclic groups are abelian, but not all abelian groups are cyclic [4].

- Another example of abelian groups is *Rings*, according to their additive operations. An abelian group can be formed by units of rings with respect to their multiplicative operation. For eg: a set of integers modulo n is Zn = {0, 1. .. n − 1} and forms an Abelian group.

Remarks: In groups if two objects contain the same structure, they are termed as isomorphic. An isomorphism ϕ: G → H amid dual sets GG and HH by group operation mapping ∗H and ∗G. This mapping must also satisfy the two conditions such as Φ is a bijection and ∀ e, f ∈ G, holds $\phi(x*Gy) = \phi(x)*H\phi(y)$. Two groups such as group G and H are isomorphic if they contain isomorphism between them, i.e., G ≅ H are isomorphic. Each group which is cyclic of order m can be isomorphic to (Zn, +). If |G| is considered prime, then group G must be cyclic and ∀ g ∈ G\{1} remain generators.

2.3.2   Lagrange's Theorem

The Lagrange's theorem is an extension of Euler's theorem which can be dealt with the power of integers, i.e., modulo positive integers. The Euler's concept can be used as an application in RSA cryptosystem for the generation of random number generators. Euler's theorem defines that for positive integer n, let x be some integer which is relatively prime to n, then $x^{\varphi(n)} = 1 \bmod n$ [5].

- ∀ G group, G itself and a is an identity element of G. Let some element g ∈ G and also take all powers of element g. This is defined as a subgroup generated by g and represented by: <g > = {... g-3, g-2, g-1, a, g1, g2, g3 ...}.
- The order of element g can be defined for the smallest positive n for which $g^n = a$, o(g) = n. If G contains subgroup H and g ∈ G, then the set gH = {gh: h ∈ H} is referred as left Coset of H in G. The range of function ϕ_g is $H \rightarrow G$, $\phi_g(H) = g$ of coset gH.

According to Lagrange's theorem when H is a subgroup of G and group G is a finite group, then |H| divides |G|. When G is a finite group and g ∈ G, then o(g) ∣ G∣ and $g^{|G|} = a$ ∀ g ∈ G [6].

2.3.3   Schnorr Group

Claus P. Schnorr initially proposed the idea of the group that was later known as Schnorr group. Its composition was designed as such that a big prime order Z_p^* subgroup of Z_p^* was depicted as the Schnorr Group. The group was essentially the multiplicative group that consists of integers modulo p that acts for a random prime element p. If a group of such nature has to be generated, generate the elements p, q, r in such a manner that they form the relation p = qr. + 1.

Inputting the elements p, q as prime elements is the next step. The next step is to choose any element h that originates in the range of $1 < h < p$ until that we find a suitable element that satisfies the relationship given below:

$$h^r \not\equiv 1 \, (mod\, p)$$

This relationship value is reached subsequently

$$g = h^r \bmod p$$

It is a representation of generator of a subgroup of Z_p^* of order q.

In many depictions of such applications such as Schnorr signatures and DSA, typically element p is taken as the chosen value. The value of the element p has to be big enough that it is able to resist index calculus. It is also ensured that it combats the various related methods that are available for solving the discrete log problem. These discrete log problems all lie in the range of 1024–3072 bits. On the other hand, the q is also large enough that it is able to resist the birthday attack. This attack was made to work on discrete log problems. This attack can work in any group whose range might be between 160 and 256 bits.

The Schnorr group in their essence was devised to be a group of prime order. It contains none of the non-trivial proper subgroups which help in thwarting the confinement attacks. These confinement attacks may take place due to the presence of small subgroups. If there are to be implementations of the various protocols that may employ the use of Schnorr groups, it is necessary that it must also state and verify clearly where it is deemed appropriate that integer's elements provided by other groups are in truth the elements that consist the Schnorr group. It is said that x can be said to be the member of the group, in the circumstances that it satisfies the relation $0 < x < p$ and $x^q \equiv 1$ (mod p). The other elements that pop up as members of the group can be determined to be the generator of the group, except if the element 1 is chosen [7].

2.3.4 Finite Field

When seen from the perspective in arithmetic, a finite field or Galois field is one of the types of a field. This variation of field is for the most part included a set number of segments. All things considered with some other type of a field, the exercises of increase; expansion, deduction, and partition are portrayed by their capacities and fulfill particular fundamental rules on finite field. The utmost notable occurrences of finite fields remain specified through the entire numbers mod p, while p stands as prime number. Limited fields are fundamental in different zones of Mathematics and programming building, including number theory, Algebraic geometry, Galois speculation, finite geometry, cryptography, and coding theory [8].

2.4   Properties of Finite Field

A finite field is a limited set that acts as a field; this recommends duplication, advancement, end, and division (aside from division by zero) are portrayed and fulfill the measures of number rearranging known as the field articulations. The 1 measure of parts of a determinate field is termed as one order q, a product of the time and its size. A Finite field of order q occurs if the order q stays a prime power p^k (somewhere p remains a prime number besides a positive whole number k). Now within a field of requesting pk, including p duplicates of any part dependably acknowledges zero, i.e., the idea is p field.

On the off chance that each field of sales q is isomorphic. Furthermore, a field cannot contain two grouped Finite subfields with practically identical sales. One may thusly see each Finite field with an equivalent sale, and they are unambiguously suggested, Fq or GF(q), where the letters GF connote the "Galois field." In sales q finite field, the polynomial $X^q - X$ takes altogether q fragments as roots of the Finite field. The non-zero portions of a Finite field structure a multiplicative social event. This get-together remains cyclic, thus every non-zero sections could be passed on as forces of a solitary portion called an unpleasant fragment of the field. (With everything considered, there are several harsh fragments designed for a certain field.)

The least inconvenient instances of Finite fields are the fields of prime demanding: for each prime number p, the prime field of sales p, showed GF(p), Z/pZ, or Fp, might be made as the whole numbers modulo p. The fragments of the prime field of sales p might be tended to be numbers in the range 0, ..., p − 1. The total, the separation, and the thing are the rest of the division by p of the deferred outcome of the relating whole number activity. The multiplicative in opposite of a section might be dealt with by utilizing the exhaustive Euclidean calculation (see Extended Euclidean figuring § Modular whole numbers).

Dismissal F is a Finite field. For any part x in F and any number n, infers by n x the whole of n duplicates of x. The least positive n with a definitive target that n 1 = 0 is the trademark p of the field. This awards portraying the development of a section k of GF(p) by a fragment x of F by picking a whole number expert for k. This extension makes F into a GF(p)-vector space. It follows that the measure of sections of F is pn for some number n. The character [27]

$$(x + y)^p = x^p + y^p$$

(now and then called the first-year recruit's fantasy) is valid in a field of trademark p. This follows from the binomial hypothesis, as every binomial coefficient of the development of $(x + y)^p$, except for the first and the last, is numerous of p. By Fermat's little hypothesis, if p is a prime number and x is in the field GF(p) at that point $x^p = x$. This suggests the equity [27]

$$X^p - X = \prod_{a \in GF(p)} (X - a)$$

for polynomials over GF(p). All the more, by and large, every component in GF(p^n) fulfills the polynomial condition $x^{p^n} - x = 0$. Some Finite enlargement of a Finite field remains separable and fundamental. That is, if E remains as Finite field and F remains a subfield of E, by then E is obtained from F by adjoining a singular part whose irrelevant polynomial is distinct. To use a language, Finite fields are extraordinary.

An undeniably expansive arithmetical arrangement that satiates the different maxims of a field, anyway whose increase is not requisite to be commutative, is known as a division ring (or on occasion incline field). By Wedderburn's little theory, any Finite division ring is commutative and from now on is a Finite field.

2.5 Bilinear Mapping

In arithmetic, a bilinear Map is a capacity consolidating modules of binary vector places to produce a component of a third vector place and is straight in its each of one contention. Grid augmentation is a model. Its main purpose is the combination of elements of two vector spaces. This combination is done to subsequently obtain an element of the third vector space. The entire function strives to be linear in all of its arguments [9]. One of the main examples of this function is the Matrix Multiplication Method. A mapping technique considers as a bilinear map that deals with two factors. If you have a straight Map L which is used to map a X vector space along with one more space Y at that point you can compose, [28].

$$L(a\vec{u} + b\vec{v}) = aL(\vec{u}) + bL(\vec{v})$$

Using a bilinear Map, one's mapping of a vector space's cartesian results to certain further vector space. Thus let B: X × Y → Z stand as bilinear map. Then one can write, $B(a\vec{u} + b\vec{v}, c\vec{s} + d\vec{t}) = acB(\vec{u}, \vec{s}) + adB(\vec{u}, \vec{s}) + bcB(\vec{v}, \vec{s}) + bdB(\vec{v}, \vec{t})$, [28].

A basic case of this type of a Map is the mapping $f : R × R → R : (x, y) → x \cdot y$. Where we believe that genuine numbers to be a vector space over the genuine e field. Another case of such a mapping is the internal item (when it is confined to just have genuine qualities). If freely observed, a bilinear Map fulfills [28]:

$B(x + y, z) = B(x, z) + B(y, z)$ (added substance in the primary "organize"),

$B(x, y + z) = B(x, y) + B(x, z)$ (added substance in the second "organize"),

$B(cx, y) = cB(x, y) = B(x, cy)$ (jelly scaling in each "arrange").

Consider B increase of genuine numbers for instance : $B(a, b) = a \cdot b$

$B(x + y, z) = (x + y) \cdot z = x \cdot z + y \cdot z = B(x, z) + B(y, z)$

$B(x, y + z) = x \cdot (y + z) = x \cdot y + x \cdot z = B(x, y) + B(x, z)$.

$B(cx, z) = (cx) \cdot z = c \cdot (xz) = x \cdot (cz) = B(x, cz)$

3   Mathematical Tools

3.1   Functionalities (Equality Test, Predicate Encryption, Inequality Test, Inner Product Evaluation) and Other Methods

Functional Encryption techniques are designed to confirm whether the created schemes are real in existent applications. Functional Encryptions focus on practical functionalities. Practical Functionality Testing Methods are often used for Product Evaluation. Techniques available for these evaluations include Predicate Encryption, Inequality Test, Equality test, and Inner Product Evaluation.

One of the new paradigms for public key encryption is Predicate Encryption. This method involves a process that generalizes identity-based encryption. In this process, the secret keys show a direct correspondence to predicates. On the other hand, cipher texts display their association with attributes. This can be better understood using a certain example:

Secret Key (SK_f) directly corresponds to predicate f. It is used to decrypt a ciphertext associated with attribute I. This will be successful only if $f(I) = 1$.

To guarantee the built plans are down to earth in genuine applications, we center around useful encryption for common sense functionalities, for example, correspondence tests, inequality tests, and inward item assessment, which are the significant functionalities that can be applied in security safeguarded information search and protection saved information sharing. Decisively, equality tests' practical encryption and tests of inequality can be functional in accessible encryption, while 296 the utilitarian encryption intended for inward item assessment can be applied toward 1 accomplishes leveled information sharing. These utilizations of practical encryption are extremely valuable and basic in distributed computing security [10–12].

3.1.1   Equality Test

Functional Encryption used for Equality Tests (FE-ET) exists as a subcategory of Functional Encryption using the usefulness displayed as follows [29]:

$$f\ (k \in K; x \in X) = 1\ if\ k = x;\ Otherwise\ f\ (k \in K; x \in X) = 0$$

Here we assume K to be a function key space, X represents the message space, k represents function key, and x represents a message.

3.1.2 Inequality Test

When in doubt, the inequality test alludes to an association that expresses that two parts are not comparable. Conclusively, it is progressively noteworthy if the segments are in an orchestrated define with the objective that they can be taken a gander at. Disregard S an organized set with a hard and fast solicitation $\leq$ and a; b 2 S. What is more, we in like manner describe a > b if b < an, and a $\geq$ b if b $\leq$ a.

3.1.3 Inner Product Evaluation

Utilitarian Encryption used for Interior Products are clarified through the accompanying model. Let K = X = V be a similar vector space. Utilitarian Encryption intended for Inner Products (FE-IP) stands as a subclass of FE by the usefulness [29].

$$f\left(\vec{k} \in V, \vec{x} \in V\right) = \left\langle \vec{k}, \vec{x} \right\rangle = \vec{k} \cdot \vec{x}.$$

3.2 Wiener Theorem

Paley–Wiener hypothesis is any hypothesis that exists in mathematics. The main focus of the hypothesis is that it identifies with rot properties of capacity or dissemination at vastness with the analyticity of its Fourier change. The old-style Paley–Wiener hypotheses utilize the holomorphic Fourier change on classes of square-integral capacities bolstered on the genuine line. Officially, the thought is to take the fundamental characterizing the (converse) Fourier change [13].

3.3 Primality Test | Set 1 (Introduction and School Method)

Check if the number is prime or not when provided a positive integer. A natural number that is greater than 1 and one which has no positive divisors other than 1 and itself is classified as a prime number [14].

Earliest examples of prime number series are {2, 3, 5 . . . }.

3.3.1   School Method

A straightforward arrangement is to emphasize through all numbers from 2 to n-1 and for each number check on the off chance that it isolates *n*. If we locate any number that partitions, we return bogus.

3.3.2   Fermat's Strategy

Fermat's Little Theorem is the basis of this strategy. This strategy is a probabilistic technique. The Fermat primality test is classified as a probabilistic test which decides if a number is likely to be a prime number.

3.3.3   Fermat's Little Theorem

In the event that a and n is co-prime number, at that point for each a, 1 < a < n-1.
   While assuming a and n are co-prime, the following formula is generated [30]:

$$a \equiv 1 \,(\mathrm{mod}\, n) \ \mathrm{OR} \ a\%n = 1$$

Fermat's little hypothesis expresses that if p is prime and 'a*n' is not separable by p, at that point [30].

$$a^{p-1} \equiv 1 \,(\mathrm{mod}\, p)$$

3.3.4   Miller-Rabin Method

The Miller–Rabin primality test or Rabin–Miller primality test is a primality test. It is a way of figuring that chooses if a given number is prime or not. In its essence, it is similar to the Fermat primality test and the Solovay–Strassen primality test.
   This procedure is a probabilistic method (Like Fermat); anyway, it generally preferred over Fermat's methodology. It returns false in the case n is composite and returns legitimate if n It is likely to prime. k is an information parameter that chooses the exactness level. A higher estimation of k shows more precision.

3.3.5   The Solovay–Strassen Primality Test

Robert M. Solovay and Volker Strassen created the Solovay–Strassen primality test. It is well known as a probabilistic test that decides whether a number is a prime or most likely composite. Baillie-PSW primality test and the Miller–Rabin primality test have taken over its popularity nowadays, but it is still valued as an important mathematical tool. However, it still has incredible verifiable significance in indicating the down to earth attainability of the RSA cryptosystem. The Solovay–Strassen test was found to be a subset of the Euler–Jacobi pseudoprime test. When considering any prime number p and a whole number a, Euler demonstrated that [31],

$$a^{(p-1)/2} \equiv \left(\frac{a}{p}\right) (\bmod\ p)$$

3.4 Discrete Logarithm Problem in Subgroup of Z_p^*

Discrete logarithmic (DL) tends to be connected through the multiplicative cyclic set. On the off chance that "g" says some generator multiplicative set G, around then every component "e" in set G composed such as g^x used for certain x. Discrete logarithmic problem (DLP) contains three most basic parts, for example, a gathering or set G, a component "e" and a generator "g" to decide the base "g" discrete logarithm of a component "e" in the G set. For example, on the off chance that a set Z_6^* with generator 3, then the discrete logarithm of 1 is 3. $6^3 \equiv 0$ mod 6 scheming of the discrete logarithm is not always a monotonous job [15]. It hinges on the group arrangement. Z_6^* contributes an improved combination of gatherings planned for discrete logarithm where p is considered as a prime number. The computation of the backward of logarithmic worth is exceptionally extreme. A framework takes a stab at every type until the condition matches. An issue in finding discrete logarithms be dependent upon the hardness of gatherings, e.g., a general selection of sets intended for discrete logarithm grounded crypto-frameworks stands Z_p^* in which "p" remains as a prime number. Though, if $p - 1$ can be considered as a result of little prime numbers, at that point the Pohlig–Hellman calculation is used to determine the discrete logarithm issue capably in this gathering. Because of this explanation, p should be a protected prime while utilizing Z_p^* such as the base of discrete logarithm manufactured crypto-frameworks. A safe prime number is a number that matches $2q + 1$ wherever q remains as a big prime number. It infers that p should be enormous (as a base 1024-piece) to make the safe crypto-frameworks [16].

3.5 Baby-Step Giant-Step Algorithm

Baby-Step Giant-Step Algorithm is an important tool in group theory. In the baby-step giant, the progression is a trade-off count for enlisting the discrete logarithm or solicitation of a part in a restricted abelian gathering. In many systems, the most 1 usually used cryptography systems rely on the assumption made using the basis of a discrete log. The assumption is that the discrete log is exceptionally difficult to process; the more inconvenient it is, the more noteworthy security it gives a data move. One way to deal with growing the difficulty of the discrete log issue is to assemble the cryptosystem for a greater social occasion [17].

The computation relies upon a space-time exchange off. It is a really direct difference in primer increase, an honest method for finding discrete logarithms. Given a cyclic social event G of solicitation n, a generator α of the get-together, and a get-together segment β, the issue is to find a number x with the ultimate objective that [32].

$$\alpha^x = \beta.$$

The baby-step giant-step algorithm is based on rewriting x :

$$x = im + j$$
$$m = \left[\sqrt{n}\right]$$
$$0 \le i < m$$
$$0 \le j < m$$

Therefore, we have:

$$\alpha^x = \beta$$
$$\alpha^{im+j} = \beta$$
$$\alpha^j = \beta\left(\alpha^{-m}\right)^i$$

The calculation pre-computes a^j for a few estimations of j. At that point, it fixes an m and attempts an estimation of I in the right-hand side of the coinciding above, in the way of preliminary duplication. It tests to check whether the compatibility is fulfilled for any estimation of j utilizing the pre-computed estimations of a^j.

3.6   Functional Field Sieve

In science, the functional field sieve was presented in 1994 as an effective method for separating discrete logarithms over limited fields of the little trademark. Sieving for focuses at which a polynomial-esteemed capacity is distinguishable by a given polynomial is not significantly more troublesome than sieving over the whole numbers—the fundamental structure is genuinely comparative, and Gray code gives a helpful method to step through products of a given polynomial proficiently [18].

3.7   Elliptic Curve Factorization

Elliptic Curve Factorization technique stands as the fastest factorization technique. It is also termed as Lenstra elliptic curve process among various professionals. This method is used for the factorization of integers, which employs the elliptic curve. ECM designed for the general-purpose factoring, in which it ranks the third fastest factoring method. ECM is found to be the best appropriate method for calculating small factors. Recently, it is considered as the best algorithm of divisors of decimal digits ranging from 10 to 40 digits, as per its smallest factor p size is dominating on the running time rather than the number n size which can be factored. This method is capable in deleting the minor factors commencing a group of integers having multiple factors.

This process is on its essential enhancement of the traditional p-1 algorithm. This algorithm determines p such as p-1 stands b-power smooth in lieu of smaller values of b. For every "e," several of p-1 besides some a comparatively prime to "p," through Fermat's little theorem one needs $a^e \equiv 1(modp)$. Then $gcd\ (a^e - 1, n)$ exists as possible factors of n. The p-1 algorithm has failed if p-1 contains large factors, like the case of numbers having strong primes, e.g., ECM becomes about this problem through seeing the group of the arbitrary elliptic curve above the Z_p finite field somewhat than allowing for the Z_p multiplicative group of order p-1 each time [19].

3.8 Random Square Factoring

Dixon's factorization strategy (additionally Dixon's random squares factoring or Dixon's calculation) is a universally useful whole number factorization calculation in the number theory. It is the prototypical factor base strategy. Different from other factor base strategies, its run-opportunity bound accompanies a thorough verification that does not depend on guesses about the perfection properties of the qualities taken by polynomial [20].

Dixon's technique depends on finding a coinciding match of squares modulo to the whole number N which is planned to factorize. Fermat's factorization strategy finds such a compatibility by choosing arbitrary or pseudo-irregular x esteems and trusting that the number x^2 mod N is an ideal square (in the whole numbers).

3.9 Quadratic Sieve Method

The quadratic sieve algorithm (QS) is whole number factorization estimation and, in a little while, another speediest procedure identified (later the all-purpose number field sifter). It is as of recently the quickest for whole numbers below 100 decimal numbers or else close, and is generally less complex than field sifter. This is a broadly accommodating factorization tally, recommending that this one's successive time relies absolutely over the magnitude of the whole number to be figured, also not on excellent configuration or possessions. The calculation tries to established a concordance of squares modulo n (whole number remains factorized), which routinely prompts a factorization of n. The figuring works in two stages: the information gathering stage, where it assembles data that might incite a closeness of squares; and the information preparing to arrange, where it puts all the information it has amassed into a structure and handles it to acquire a consistency of squares [21].

3.10 Factoring Polynomials over a Finite Field

In mathematics and computer algebra, the factorization of a polynomial over a finite field comprises of breaking down it into a result of final elements. This disintegration is hypothetically conceivable and is novel for polynomials with coefficients in any field, but instead solid limitations on the field of the coefficients are expected to permit the calculation of the factorization by methods for a calculation. By and by, calculations have been planned distinctly for polynomials with coefficients in a limited field, in the field of reasons or in a limitedly created field expansion of one of them. All factorization calculations, including the instance of multivariate polynomials over the level-headed numbers, diminish the issue to this case. It is additionally utilized for different uses of limited fields, for example, coding hypothesis (cyclic excess codes and BCH codes), cryptography (open key cryptography by the methods for elliptic bends), and computational number hypothesis. The decrease in the factorization of multivariate polynomials to that of univariate polynomials does not have any particularity [22].

3.11 Irreducible Polynomials over a Finite Field Zp

Let Z be considered as finite field. As per general fields, a non-constant "f" polynomial in finite field Z[x] is supposed to be irreducible above finite field Z if this one is not the artifact of two polynomials of some positive degree. This is followed in finite field computation of non-prime degree, so it required producing an irreducible polynomial. Irreducible polynomials over $Z_p[x]$ used to bring the mathematics in the field extension of Z_p. Complexity, and coding of cryptography theories use the computation of such extensions. To determine the irreducible polynomials of any degree above Z_p, there exists a random polynomial algorithm. Non-prime order built finite fields is constructed by using irreducible polynomial [23].

4 Role of Group Theory in Cryptosystem

The heart of crypto encryption system RSA (Rivest, Shamir, Adleman), Diffie–Hellman key exchange, and other schemes is algebraic structures. The algebraic structures and statical properties of group theory are used in cryptography. The cryptography is a concept of transmitting secret data securely, and for this purpose security experts use algebraic structures. These algebraic structures help to make the cryptosystem unbreakable by using structures and properties of groups [24]. Some of the cryptographic examples of group theory are mentioned in the following.

4.1 Use of Group Theory in RSA-Based Cryptosystem

(a) The encryption system RSA uses the unbreakable secret key for encryption and decryption. For the generation of key two properties are used such as (a) take two discrete prime sums q and p and (b) some integer value e which is rather prime toward (q-1) and (p-1). respectively [25]. The p and q prime numbers through some "e" integer which satisfies $\gcd(e, (q - 1)(p - 1)) = 1$. Then calculate some d integer which satisfies $e \times d = 1 \mod (p - 1)(q - 1)$. For encryption we calculate: $N = M^e \mod n$ and to decrypt N calculate: $M = N^d \mod n$.

(b) RSA uses the base of Lagrange's theorem, i.e., the Euler's theorem of group theory for the theoretical implementation.

(c) The Z^*_n group contain elements $\varphi(n)$, and when $n = pq$ which is considered as a product of prime number p and q, then Z^*_{pq} contain $(p - 1)(q - 1)$ elements. If group $G = Z^*_n$, $n = pq$, then we can hold, $|G| = \varphi(n) = (q-1)(p-1)$ and having e integer which satisfies $\gcd(e, (q-1)(p-1)) = 1$, then some d integer satisfies $de = 1 \mod \varphi(n)$. So we can say that in RSA, for some message M, we hold $(M^e)^d \mod n = 1$. We can also claim that message M cannot be divisible through p or q since we get $M^{\varphi(n)} = 1$ from the corollary of Lagrange's theorem. The original message during decryption is recovered by $(M^e)^d = M$.

4.2 Schnorr Group in Log-Based Cryptosystem

Schnorr groups of group theory depict their usefulness in various types of discrete log-based cryptosystems. These cryptosystems include many different kinds of derivatives such as Schnorr signatures and DSA [26].

4.3 Elliptic Curve and Bilinear Mapping-Based Cryptosystem

(a) The basis of DHP, RSA, elliptic curve, and bilinear curves is finite fields, and baby-step giant algorithm is used to solve the problems such as discrete logarithmic problem of elliptic curves and MIM attack in RSA.

(b) Bilinear mapping can be used in pairing-based cryptography scheme which is specifically designed to solve the discrete logarithmic problem and MOV attack in elliptic curve cryptography.

(c) The bilinear-based cryptosystem is used in many situations:

- IBE-based encryption (identity-based encryption)—hierarchal IBE, ID-based ring signatures, ID-based blind signatures, ID-based signcryption, ID-based hashes, etc.

- Signature scheme—short signatures (half of original signatures), multi-signatures, aggregate-signatures, blind signatures, ring signatures, and authentication tree-based signatures, etc.

(d) The Diffie–Hellman key agreement variants can be constructed by using group theory. This scheme used cyclic subgroup of finite groups and also possesses cyclic subgroups with discrete logarithmic problem which is so hard [26].

(e) The key exchange agreements and public key cryptosystems are the main fields where group theory can be used. Some algebraic and statistical properties of group can also be satisfied by block ciphers used in AES and DES.

5 Conclusions

Group theory has long helped us with the activities that involve the usage of polynomial math and other notable logarithmic constructions. The group theory and its various properties have always helped in promoting the efficiency of solving various variables-based mathematics. The variants formed out of group theory such as Abelian group, Schnorr group, and Lagrange's theorem all have made their own impacts and contributions in various fields. Currently group theory shows enormous promises in the field of cryptosystems. Its contributions have been proven through various practical applications such as the RSA, Diffie–Hellman Key Exchange, IBE schemes, and signature schemes. Based on further advancements, it is shown that there are a lot of possibilities for application of group theory in cryptosystems in future.

References

1. Blackburn, S. R., Cid, C., & Mullan, C. (2011). Group theory in cryptography. In *Proceedings of Group St Andrews 2009 in Bath* (pp. 133–149).
2. Vasco, M. I. G., & Steinwandt, R. (2015). *Group theoretic cryptography (Vol. 9)*. CRC Press.
3. https://brilliant.org/wiki/abelian-group/#properties-of-abelian-groups
4. Bannai, E., & Bannai, E. (1994). Spin models on finite cyclic groups. *Journal of Algebraic Combinatorics, 3*(3), 243–259.
5. Sun, Z. W. (2016). A result similar to Lagrange's theorem. *Journal of Number Theory, 162*, 190–211.
6. Conrad, K., 2018. Cosets and Lagrange's theorem. http://www.math.uconn.edu/~kconrad/blurbs/grouptheory/coset.
7. Maxwell, G., Poelstra, A., Seurin, Y., & Wuille, P. (2019). Simple schnorr multi-signatures with applications to bitcoin. *Designs, Codes and Cryptography, 87*(9), 2139–2164.
8. Savas, E., & Koç, Ç. K. (2010). Finite field arithmetic for cryptography. *IEEE Circuits and Systems Magazine, 10*(2), 40–56.
9. Mahalanobis, A., & Shinde, P. (2017). Bilinear cryptography using groups of Nilpotency class 2. In *IMA International Conference on Cryptography and Coding (pp. 127–134). Springer, Cham. Computing Technologies* (pp. 543–554). Singapore: Springer.

10. Sun, J., Bao, Y., Nie, X., & Xiong, H. (2018). Attribute-hiding predicate encryption with equality test in cloud computing. *IEEE Access, 6*, 31621–31629.
11. Fu, X., Nie, X., & Li, F. (2017). Outsource the Ciphertext decryption of inner product predicate encryption scheme based on prime order bilinear map. *IJ Network Security, 19*(2), 313–322.
12. Baltico, C. E. Z., Catalano, D., Fiore, D., & Gay, R. (2017). Practical functional encryption for quadratic functions with applications to predicate encryption. In *Annual International Cryptology Conference* (pp. 67–98). Cham: Springer.
13. Kumar, V., Kumar, R., & Pandey, S. K. (2017). An enhanced and secured RSA public key cryptosystem algorithm using Chinese remainder theorem. In *International Conference on Next Generation*.
14. Rajput, J. and Bajpai, A., 2019. Study on deterministic and probabilistic computation of primality Test. Available at SSRN 3358737.
15. Teng, L., & Li, H. (2018). A high-efficiency discrete logarithm-based multi-proxy blind signature scheme via elliptic curve and bilinear mapping. *IJ Network Security, 20*(6), 1200–1205.
16. Ahmad, K., Kamal, A., & Ahmad, K. A. B. (2019). Discrete logarithm problem. *Emerging Security Algorithms and Techniques, 1*, 1.
17. Kushwaha, P. and Mahalanobis, A., 2017. A probabilistic baby-step giant-step algorithm. arXiv preprint arXiv:1701.07172.
18. Dhumal, G., & Padmavathy, R. (2017). Sparse linear algebra in function field sieve. In *2017 8th International Conference on Computing, Communication and Networking Technologies (ICCCNT)* (pp. 1–6). IEEE.
19. Shirase, M. (2017). Condition on composite numbers easily factored with elliptic curve method. *IACR Cryptology ePrint Archive, 2017*, 403.
20. Brown, D. R. (2016). Breaking RSA may be as difficult as factoring. *Journal of Cryptology, 29*(1), 220–241.
21. Kleinjung, T. (2016). Quadratic sieving. *Mathematics of Computation, 85*(300), 1861–1873.
22. Poonen, B., 2017. Using zeta functions to factor polynomials over finite fields.
23. Ferraguti, A., Micheli, G., & Schnyder, R. (2018). Irreducible compositions of degree two polynomials over finite fields have regular structure. *The Quarterly Journal of Mathematics, 69*(3), 1089–1099.
24. Patel, K. H., & Patel, S. S. (2016). Implementing digital signature with RSA encryption algorithm to enhance the data security of cloud in cloud. *Computing, 4*.
25. Tomescu, A., Chen, R., Zheng, Y., Abraham, I., Pinkas, B., Gueta, G. G., & Devadas, S. (2020). Towards Scalable Threshold Cryptosystems. In *2020 IEEE Symposium on Security and Privacy (SP)*.
26. Cremers, C., & Jackson, D. (2019). Prime, order please! Revisiting small subgroup and invalid curve attacks on protocols using Diffie-Hellman. In *2019 IEEE 32nd Computer Security Foundations Symposium (CSF)* (pp. 78–7815). IEEE.
27. R Lidl, H Niederreiter - 1997 Finite Fields. books.google.com
28. Yamakawa, T., Yamada, S., Hanaoka, G., & Kunihiro, N. (2017). *Self-bilinear map on unknown order groups from indistinguishability obfuscation and its applications Algorithmica*. Springer.
29. Zhang, S. (2017). *Practical Functional Encryption Techniques and Their Applications*. The University of Wollongong, School of School of Computing and Information Technology.
30. EW Weisstein – Fermat's Little Theorem 2004. mathworld.wolfram.com
31. Burt Rosenberg The Solovay-Strassen Primality Test, 1993. www.cs.miami.edu
32. Matsuo, K., Chao, J., & Tsujii, S. (2002). An improved baby step giant step algorithm for point counting of hyperelliptic curves over finite fields. In *The 2002 Symposium on Cryptography and Information Security Shirahama, Japan*. The Institute of Electronics, Information and Communication Engineers.

Chapter 3
XTR Algorithm: Efficient and Compact Subgroup Trace Representation

Pinkimani Goswami, Madan Mohan Singh, and Dimpi Biswas

Abstract In Crypto 2000, Lenstra et al. introduced the concept of XTR. The term XTR is used for efficient and compact subgroup trace representation. The security of the XTR cryptosystem is based on the discrete logarithm problem (DLP) and its variants, defined over a subgroup of multiplicative group of a finite field. It is considered as favorable alternative to RSA and elliptic curve cryptosystem. In this book chapter, we will thoroughly discuss about the mathematics behind XTR and its applications in cryptography.

Keywords Cryptography · XTR · ElGamal scheme · Signature scheme · DSA · NR scheme · Finite field · Cyclotomic polynomial · Conjugate element · Trace

1 Introduction

Cryptography is a method of storing and transmitting data in such a way that only those it is intended for can read messages and processes. Symmetric and asymmetric key cryptography are two main branches of cryptography that provide confidentiality. One of the main disadvantages of symmetric key cryptosystem is that before communicating with each other user has to agree on a shared key and hence requires a prior communication between two parties. This is called the key distribution problem. In 1976 [1], Diffie et al. proposed the idea of a key exchange protocol called Diffie-Hellman (DH) key exchange protocol and hence introduced a new branch of cryptography called asymmetric key cryptography.

The DH key exchange protocol is the first practical solution to the key distribution problem. In the DH scheme, a large prime number p and a generator g of the

P. Goswami (✉) · D. Biswas
Department of Mathematics, University of Science & Technology, Meghalaya, India

M. M. Singh
Department of Basic Sciences and Social Sciences, North-Eastern Hill University, Shillong, Meghalaya, India

© Springer Nature Switzerland AG 2021
K. A. B. Ahmad et al. (eds.), *Functional Encryption*, EAI/Springer Innovations in Communication and Computing, https://doi.org/10.1007/978-3-030-60890-3_3

multiplicative group $GF(p)^*$ of a finite field $GF(p)$ are fixed as a system parameter. If two parties want to share a common secret key, then each of them will generate a random key x and y, respectively, such that $0 \leq x, y < p - 1$. Each will send $g^x \bmod p$ and $g^y \bmod p$ to the other party. In this way, they will generate the common secret key $g^{xy} \bmod p$. The scheme will be secured if the size of p is at least 1024 bits such that $p - 1$ contains a prime factor of 160 bits. Hence, each party sends 1024 bits to the other party.

In 1985 [2], ElGamal suggested that a finite extension field $GF(p^r)$, $r > 0$ can be used instead of prime field $GF(p)$. In 1991 [3], Schnorr proposed a variant of the DH scheme where a subgroup G of the multiplicative group $GF(p)^*$ is considered. The order of G is a prime q, which is considered as a very small compared to p. This reduces the computational cost of the DH scheme, but the number of bits to be exchanged between each party remains same. In 1997 [4], Lenstra suggests that one can generalize the Schnorr scheme by considering any multiplicative subgroup of prime order of the finite extension field $GF(p^r)$, $r > 0$. In 1999 [5], Brouwer et al. proposed a variant DH scheme in which the number of bits exchanged is reduced to one-third of the number of bits required by the DH scheme. However, the security of both the schemes are the same. It was shown that elements of a subgroup of order q of $GF(p^6)$ can be represented using $2\log_2(p)$ if $q \mid (p^2 - p + 1)$. In the same paper, they generalized the scheme for $GF(p^r)$. In 2000 [6], Lenstra et al. improved the method proposed by Brouwer et al. [5]. This method is called the XTR.

The term XTR represents for Efficient and Compact Subgroup Trace Representation (ECSTR) [6]. In XTR, a subgroup of prime order q(say G) of $GF(p^6)^*$ is considered, where $q \mid (p^2 - p + 1)$. This subgroup is called the XTR group (or XTR subgroup). In this method, an element α of $G \subseteq GF(p^6)^*$ is represented by their trace $Tr(\alpha)$ over $GF(p^2)$. As the arithmetics are computed over $GF(p^2)$, so to represent an element of G needed $2\log_2 p$ bits. That is, XTR provided a more compact representation of the element of G. This method achieves the same communication advantage compared to the method proposed by Brouwer et al. [5], but with a less computational cost. Note that XTR is not the only one approach where the elements are represented by their trace; LUC [7] also used the trace representation of the elements, where the underlying field is $GF(p^2)$.

It is proved that the discrete logarithm (DL), Diffie-Hellman (DH), and decisional Diffie-Hellman (DDH) problem in the XTR group are as secured as the problems are defined over $GF(p^6)$. Therefore, the XTR group can be used in any cryptographic protocol which is based on DLP and or its variants. It achieves the same security with traditional DLP-based cryptographic protocol. In the same paper, Lenstra et al. showed that XTR can be used as an alternative to both RSA [8] and ECC ([9, 10]). It is shown that the security of 170-bit XTR is equivalent to the security of 1024-bit RSA. Also, the key and parameter selection of a XTR cryptosystem is more efficient than the ECC. Therefore, one can conclude that XTR needed fewer data storage, computation and communication overhead compared to other cryptosystems with equivalent security.

The purpose of this chapter is to discuss thoroughly about the mathematics behind XTR and its applications. After the preliminaries section of the chapter, the next section is devoted to describe the mathematics behind XTR. In the same section, we discuss the algorithms for parameter selection and finding the generator of a XTR group. The applications of XTR in cryptography are discussed in Sect. 4. Section 5 summarizes the recent development of XTR. The chapter concludes in Sect. 6.

2 Some Definitions and Results from Finite Field

In this chapter, for a prime p, we denote a finite field of characteristic p by $GF(p^r)$, where $r > 0$. The cardinality of $GF(p^r)$ is p^r. Also, the multiplicative group of $GF(p^r)$ is denoted by $GF(p^r)^*$, whose cardinality is $p^r - 1$. For $t < r$, $GF(p^t) \subseteq GF(p^r)$ is a finite extension. A polynomial $f(X) = a_n X^n + a_{n-1} X^{n-1} + \cdots + a_1 X + a_0 \in GF(p^t)[X]$ is called an irreducible polynomial if it has no roots in $GF(p^t)$. $f(X)$ is called monic polynomial if $a_n = 1$.

Suppose $f(X)$ is a monic irreducible polynomial of degree $s = \frac{r}{t} \in \mathbb{Z}$ over $GF(p^t)$ and $\xi \in GF(p^r)$ is a root of $f(X)$, then

$$GF\left(p^r\right) \cong GF\left(p^t\right)[\xi] \cong \frac{GF\left(p^t\right)[X]}{f(X)}$$

i.e., $GF(p^r) \cong \{g(X) \bmod f(X) : g(X) \in GF(p^t) \ \& \ \deg(g(X)) < s\}$.

Therefore, to represent the element of $GF(p^r)$, it is enough to find the representation of the element of $GF(p^t)$. Again, an element $GF(p^r)$ can be expressed as a vector of length s over $GF(p^t)$, where the entries are the coefficient of $g(X)$. Therefore, to represent an element of $GF(p^r)$ needs $r \log p$ bits.

Again, if $f(X)$ is a monic irreducible polynomial over $GF(p^t)$ and $\xi \in GF(p^r)$ is a root of $f(X)$, then $\xi^{p^i} \in GF\left(p^r\right)$ is a root of $f(X)$, where $i = 0, 1, 2, \cdots$. Therefore, the minimal polynomial of $\xi \in GF(p^r)$ over $GF(p^t)$ is

$$m(X, \xi) = (X - \xi)\left(X - \xi^p\right)\left(X - \xi^{p^2}\right) \cdots \left(X - \xi^{p^{d-1}}\right) \in GF\left(p^t\right)[X]$$

where d is the smallest positive integer such that $\xi^{p^d} = \xi$. Clearly, $\deg m(X, \xi) \leq \frac{r}{t}$. The elements ξ^{p^i}, for $i = 1, 2, \cdots, d - 1$, are called the conjugate of $\xi \in GF(p^r)$. The sum of the conjugate elements of ξ, i.e. $\sum_{i=0}^{d-1} \xi^{p^i}$ is called trace of ξ over $GF(p^t)$, and it is denoted by $Tr(\xi)$.

Let $m(X, \xi) = X^d + a_{d-1} X^{d-1} + \cdots + a_0$. Since $m(X, \xi) \in GF(p^t)[X]$, so $a_i \in GF(p^t)$, $i = 0, 1, \cdots, d-1$. In particular, $a_{d-1} = -\sum_{i=0}^{d-1} \xi^{p^i} \in GF\left(p^t\right)$ and hence $Tr(\xi) \in GF(p^t)$.

Note that one can represent an element $\alpha \in GF(p^r)$ by its minimal polynomial $m(X, \alpha) \in GF(p^t)$ too. Here, $\deg m(X, \alpha)$ must be $\frac{r}{t}$, otherwise $\alpha \in GF(p^k)$, where $GF(p^t) \subseteq GF(p^k) \subseteq GF(p^r)$. Hence, this approach also requires $r \log p$ bits to represent an element of $GF(p^r)$. However, in some cases there exist a relationship between the coefficients of the minimal polynomial, which allows to reduce the number of coefficients and hence provides a compact representation [11]. This concept is used in XTR where the elements are represented by its trace.

For the completeness of this section, let us define the following definitions:

Normal Basis [12]: An element $\xi \in GF(p^r)$ is called a normal element if the conjugate $\xi, \xi^p, \cdots, \xi^{p^{d-1}}$ is linearly independent over $GF(p^t)$. For a normal element $\xi \in GF(p^r)$, $\xi, \xi^p, \cdots, \xi^{p^{d-1}}$ from a basis of $GF(p^r)$ over $GF(p^t)$ and is called normal basis.

Cyclotomic Polynomial [12]: For any positive integer n, the nth cyclotomic polynomial, $\Phi_n(X) = (X - \omega_1)(X - \omega_2)\cdots(X - \omega_t)$, where $\omega_1, \omega_2, \cdots, \omega_t$ are the primitive nth root of unity, i.e., $\Phi_n(X) = \prod_{\substack{1 \le k \le n \\ \gcd(k,n) = 1}} \left(X - e^{\frac{2\pi i}{n}k} \right)$. It can also be expressed as $\Phi_n(X) = \prod_{d|n}(X^d - 1)^{\mu\left(\frac{n}{d}\right)}$, where $\mu\left(\frac{n}{d}\right)$ is the Mobius function of $\frac{n}{d}$.

Cyclotomic Subgroup [11]: A subgroup of order q of a multiplicative group $GF(p^r)^*$ over a finite field $GF(p^r)$ is called a cyclotomic subgroup if $q \mid \Phi_r(p)$ and $q \nmid r$. It is denoted by $G_{q,\,p,\,r}$.

3 Fundamental of XTR

In this section, we will discuss about the mathematics behind the XTR in detail. We will also discuss the parameter selection algorithms and the algorithm to generate a generator of the XTR group. This chapter is based on [4, 6, 13].

3.1 XTR Group

In Crypto 2000 [6], Lenstra et al. introduced the idea of XTR. In XTR, the finite field $GF(p^6)$ and a cyclotomic subgroup $G_{q,\,p,\,6}$ are considered, where $p \equiv 2 \bmod 3$. The subgroup $G_{q,\,p,\,6}$ is called the XTR group (or XTR subgroup). By definition of cyclotomic subgroup, $q \nmid 6$ is a prime number and $q \mid \Phi_6(p)$, i.e., $q \mid p^2 - p + 1$ and $q > 3$. In the remaining part of the chapter, we denote $G_{q,\,p,\,6}$ as G. Clearly, G is a cyclic group and let $G = (\mu)$, $\text{ord}(\mu) = q$, where $\mu \in GF(p^6)$.

As for any element $\eta \in G$, the conjugate elements of η are η, η^{p^2}, and η^{p^4} and hence $Tr(\eta) = \eta + \eta^{p^2} + \eta^{p^4} \in GF(p^2)$. Moreover, for $\eta_1, \eta_2 \in GF(p^6)$, and c_1, $c_2 \in GF(p^2)$

$$Tr(c_1\eta_1 + c_2\eta_2) = c_1 Tr(\eta_1) + c_2 Tr(\eta_2)$$

In particular if $\eta = \mu^n$, then

$$Tr\left(\mu^n\right) = \mu^n + \mu^{np^2} + \mu^{np^4}$$

As $p^2 \equiv p - 1 \mod (p^2 - p + 1)$ and $p^4 \equiv -p \mod (p^2 - p + 1)$, so

$$Tr\left(\mu^n\right) = \mu^n + \mu^{n(p-1)} + \mu^{-np}$$

Also, it is easy to verify that $\mu^n \mu^{n(p-1)} + \mu^n \mu^{-np} + \mu^{n(p-1)} \mu^{-np} = Tr(\mu^n)^p$.

Again the minimal polynomial of $\mu^n \in GF(p^2)$ is $(X - \mu^n)(X - \mu^{n(p-1)})$ $(X - \mu^{-np})$, which is equal to $X^3 - Tr(\mu^n)X^2 + Tr(\mu^n)^p X - 1 \in GF(p^2)[X]$. Therefore, the minimal polynomial of $\mu^n \in G$ can be determined uniquely by $Tr(\mu^n) \in GF(p^2)$.

Now, let us define a function $G \to GF(p^2)$ as $\mu^n \mapsto Tr(\mu^n)$, $n \in \mathbb{Z}$. Then by the above discussion, the function is well-defined and hence to represent an element of G needs only $2\log_2 p$ bits. Clearly the function is not one–one function as $Tr(\mu^n) = Tr(\mu^{n(p-1)}) = Tr(\mu^{-np})$. Also, the function is not group homomorphism. In order to complete the description of the implementation of the arithmetic of G, the following two issues need to be discussed.

1. How one can implement arithmetic operation in $GF(p^2)$?
2. How one can translate the arithmetic operation in G to $GF(p^2)$ w.r.t. the function defined above?

3.2 Arithmetic Operation in $GF(p^2)$

For $p \equiv 2 \mod 3$, $f(X) = X^2 + X + 1$ is an irreducible polynomial over $GF(p)$. If $\beta \in GF(p^2)$ is a root of $f(X)$, then

$$GF\left(p^2\right) \cong GF(p)[\beta] \cong \frac{GF(p)[X]}{\langle f(X) \rangle}$$

So, $GF(p^2) \cong \{a_1 + a_2\beta : a_1, a_2 \in GF(p) \,\&\, \beta^2 + \beta + 1 = 0\}$. Since $\beta^2 + \beta + 1 = 0$, so $a_1 + a_2\beta = a_1(-\beta^2 - \beta) + a_2\beta = (a_2 - a_1)\beta + (-a_1)\beta^2$. Therefore,

$$GF\left(p^2\right) \cong \left\{a_1\beta + a_2\beta^2 : a_1, a_2 \in GF(p) \,\&\, \beta^2 + \beta + 1 = 0\right\}$$

Note that as $p \equiv 2 \mod 3$ and $\beta^3 = 1$, so the basis $\{\beta, \beta^2\}$ is same with the normal basis $\{\beta, \beta^p\}$. Therefore, for $a_1, a_2 \in GF(p)$, $a_1\beta + a_2\beta^p = a_1\beta + a_2\beta^2$. An element $t \in GF(p)$ is represented as $-t\beta - t\beta^2$.

For $p \equiv 2 \bmod 3$ and $a, b, c \in GF(p)$, where $a = a_1\beta + a_2\beta^2$, $b = b_1\beta + b_2\beta^2$ and $c = c_1\beta + c_2\beta^2$, the following Table 3.1 considered from [6] summarized the number of multiplications in $GF(p)$ needed for the operation a^p, a^2, ab, and $ac - bc^p$.

3.3   Translation of Arithmetic Operation of G to $GF(p^2)$

In this section, we will discuss how the arithmetic operation of G is translated to the arithmetic operation of $GF(p^2)$ w.r.t. the trace representation $Tr(\mu^n) \in GF(p^2)$ for $\mu^n \in G$. For this, we first define a polynomial $F(\lambda, X) = X^3 - \lambda X^2 + \lambda^p X - 1$ in $GF(p^2)$ for $\lambda \in GF(p^2)$ and discuss its properties. For the roots $\gamma_1, \gamma_2, \gamma_3 \in GF(p^6)$ of $F(\lambda, X)$ and $n \in \mathbb{Z}$, let us define $\lambda_n = \gamma_1^n + \gamma_2^n + \gamma_3^n$.

Some of the properties of $F(\lambda, X)$ and its roots are discussed below. These results are considered from [6].

Lemma 3.3.1 [6] For $n, m \in \mathbb{Z}$

(i) $\lambda_0 = 3$
(ii) $\lambda = \lambda_1$
(iii) $\gamma_1\gamma_2\gamma_3 = 1$
(iv) $\gamma_1^n\gamma_2^n + \gamma_1^n\gamma_3^n + \gamma_2^n\gamma_3^n = \lambda_{-n}$
(v) $F\left(\lambda, \gamma_i^{-p}\right) = 0$, $i = 1, 2, 3$.
(vi) $\lambda_{-n} = \lambda_{np} = \lambda_n^p$
(vii) Either all $\gamma_i \in GF(p^2)$ or $\operatorname{ord}(\gamma_i) \mid (p^2 - p + 1)$ and $\operatorname{ord}(\gamma_i) > 3$, $\forall i$.
(viii) $\lambda_n \in GF\left(p^2\right)$.

Lemma 3.3.2 [6] For $n, m \in \mathbb{Z}$ and $i = 1, 2, 3$

(i) $\lambda_{n+m} = \lambda_m\lambda_n - \lambda^p\lambda_{m-n} + \lambda_{m-2n}$
(ii) $F\left(\lambda_n, \gamma_i^n\right) = 0, \forall i$.
(iii) $F(\lambda, X)$ is reducible over $GF(p^2)$ if $\lambda_{p+1} \in GF(p)$.

Remark 3.3.3 By using Lemmas 3.3.1 and 3.3.2 and Table 3.1, we can conclude the following table (see corollary 2.35 of [6]) forgiven λ, λ_n, λ_{n-1}, and λ_{n+1} in $GF(p^2)$.

Note that from Lemmas 3.3.1 and 3.3.2, λ_n is the nth term of a second-order linear recurrence relation [14], where $\lambda_0 = 3$, $\lambda_1 = \lambda$, $\lambda_{-1} = \lambda^p$ and $\lambda_{n+m} = \lambda_m\lambda_n - \lambda^p\lambda_{m-n} + \lambda_{m-2n}$.

Now, we will examine whether one can compute $\lambda_n \in GF(p^2)$ for given $\lambda \in GF(p^2)$ or not, and if it is, then what will be the cost to compute it. Detailed analysis on it is given in [6]. In [6], the authors defined the following notation:

Definition 3.3.4 [6] Let $S_n(\lambda) = (\lambda_{n-1}, \lambda_n, \lambda_{n+1}) \in GF(p^2)^3$.

Table 3.1 Cost analysis of a^p, a^2, ab & $ac - bc^p$ in $GF(p)$

Operation	Number of multiplications in $GF(p)$
$a^p = a_2\beta + a_1\beta^2$	0
$a^2 = a_2(a_2 - 2a_1)\beta + a_1(a_1 - 2a_2)\beta^2$	2
$ab = (a_2b_2 - a_1b_2 - a_2b_1)\beta + (a_1b_1 - a_1b_2 - a_2b_1)\beta^2$	3
$ac - bc^p = (c_1(b_1 - a_2 - b_2) + c_2(a_2 - a_1 + b_2))\beta$ $+\left(c_1(a_1 - a_2 + b_1)\right.$ $\left. +c_2(b_2 - a_1 - b_1)\beta^2\right)$	4

For $n \in \mathbb{Z}$ and $\lambda \in GF(p^2)$, one can compute $S_n(\lambda)$ by using the following algorithm:

Algorithm 3.3.5 [6] For given $\lambda \in GF(p)$ and $n \in \mathbb{Z}$, the algorithm computes $S_n(\lambda) = (\lambda_{n-1}, \lambda_n, \lambda_{n+1})$ as follows:

1. If $n = 0$, then by (i), (ii), and (v) of Lemma 3.3.1, $S_0(\lambda) = (\lambda_{-1}, \lambda_0, \lambda_1) = (\lambda^p, 3, \lambda)$.
2. If $n = 1$, then by (i), (ii), and (viii) of Lemma 3.3.1, $S_1(\lambda) = (\lambda_0, \lambda_1, \lambda_2) = (3, \lambda, \lambda^2 - 2\lambda^p)$.
3. If $n = 2$, then $S_2(\lambda) = (\lambda_1, \lambda_2, \lambda_3)$, which can be computed by using (vii) of Lemma 3.3.1 and $S_1(\lambda)$.
4. For $n \geq 3$, one has to compute the following steps:

(a) Compute $m = \begin{cases} \frac{n-2}{2} & \text{if } n \text{ is even} \\ \frac{n-1}{2} & \text{if } n \text{ is odd} \end{cases}$

(b) Convert $m = (1\,m_{t-1}m_{t-2}\cdots m_1m_0)_2$

(c) Set $j = 1$

(d) Compute $S_{2j+1}(\lambda) = (\lambda_{2j}, \lambda_{2j+1}, \lambda_{2j+2})$

(e) For $i = t - 1$,

Compute $S_{4j+1}(\lambda) = (\lambda_{4j}, \lambda_{4j+1}, \lambda_{4j+2})$, if $m_i = 0$
or $S_{4j+3}(\lambda) = (\lambda_{4j+2}, \lambda_{4j+3}, \lambda_{4j+4})$, if $m_i = 1$
Compute $j = 2j + m_i$ and go to step (iv).

(f) Repeat step (e) for $i = t - 2, \cdots, 1, 0$.

(g) If $j = m$, compute $S_{2j+1}(\lambda) = \begin{cases} S_{n-1}(\lambda) & \text{if } n \text{ is even} \\ S_n(\lambda) & \text{if } n \text{ is odd} \end{cases}$.

5. For $n < 0$, $-n \geq 1$. So, apply the algorithm for $-n$ and use (vi) of Lemma 3.3.1.

Now, one can conclude the following statement by using the Algorithm 3.3.5 and Table 3.2

Theorem 3.3.6 [6] To compute λ_n for given λ require $8\log_2(n)$ multiplication in $GF(p)$, where λ and λ_n are defined as above.

By (vii) of Lemma 3.3.1, it is clear that if $F(\lambda, X)$ is irreducible over $GF(p^2)$ then the roots of $F(\lambda, X)$ are of the form γ, γ^{p^2} and γ^{p^4}, where $\gamma \in GF(p^6)$ and

Table 3.2 Cost analysis of λ_{2n}, λ_{n+2}, λ_{2n-1}, λ_{2n+1}

Operation	Number of multiplications in $GF(p)$
$\lambda_{2n} = \lambda_n^2 - 2\lambda_n^p$	2
$\lambda_{n+2} = \lambda\lambda_{n+1} - \lambda^p\lambda_n + \lambda_{n-1}$	4
$\lambda_{2n-1} = \lambda_{n-1}\lambda_n - \lambda^p\lambda_n^p + \lambda_{n+1}^p$	4
$\lambda_{2n+1} = \lambda_{n+1}\lambda_n - \lambda\lambda_n^p + \lambda_{n-1}^p$	4

$\mathrm{ord}(\gamma) \mid (p^2 - p + 1)$ with $\mathrm{ord}(\gamma) > 3$. This implies $\lambda = \gamma + \gamma^{p^2} + \gamma^{p^4} = Tr(\gamma)$. As $\mathrm{ord}(\mu) = q > 3$ and $q \mid (p^2 - p + 1)$, so $\lambda = Tr(\mu) \in GF(p^2)$ and for $n \in \mathbb{Z}$, $\lambda_n = \mu^n + \mu^{np^2} + \mu^{np^4} = Tr(\mu^n) \in GF(p^2)$. Also, from (ii) of Lemma 3.3.2, $F(Tr(\mu^n), \mu^n) = 0$, for $n \in \mathbb{Z}$. Moreover,

$$S_n(Tr(\mu)) = \left(Tr\left(\mu^{n-1}\right), Tr\left(\mu^n\right), Tr\left(\mu^{n+1}\right)\right).$$

Note that, for given $Tr(\mu)$ one can use Algorithm 3.3.5 to compute $S_n(Tr(\mu))$. Therefore, if $p \equiv 2 \bmod 3$, then by the Theorem 3.3.6 one can conclude that for given $Tr(\mu)$, computing $Tr(\mu^n) \in GF(p^2)$ needs $8\log_2(n)$ multiplication in $GF(p)$. This is three times faster than the traditional exponentiation method for computing μ^n from given μ [6]. Thus, in XTR, $\mu^n \in G$ is replaced by $Tr(\mu^n) \in GF(p^2)$ for $n \in \mathbb{Z}$ and for given $Tr(\mu)$, one can efficiently compute $Tr(\mu^n)$. Hence one can be able to use this representation efficiently for cryptographic protocols.

In some cryptographic protocol (see Sects. 4.4 & 4.5), it is required to efficiently compute $Tr(\mu^a \mu^{bd})$ for given $Tr(\mu) \in GF(p^2)$, $S_d(Tr(\mu)) \in GF(p^2)^3$ and unknown d, where $a, b \in \mathbb{Z}$. Next, we will discuss an algorithm for computing $Tr(\mu^a \mu^{bd})$ efficiently. We considered the following definition and lemmas from [6, 13].

Definition 3.3.7 [6] For $n \in \mathbb{Z}$, let us define two 3×3 matrix $A(\lambda)$ and $M_n(\lambda)$ over $GF(p^2)$ as follows:

$$A(\lambda) = \begin{pmatrix} 0 & 0 & 1 \\ 1 & 0 & -\lambda^p \\ 0 & 1 & \lambda \end{pmatrix} \text{ and } M_n(\lambda) = \begin{pmatrix} \lambda_{n-2} & \lambda_{n-1} & \lambda_n \\ \lambda_{n-1} & \lambda_n & \lambda_{n+1} \\ \lambda_n & \lambda_{n+1} & \lambda_{n+2} \end{pmatrix}$$

Lemma 3.3.8 [6] For $n, m \in \mathbb{Z}$,

(i) $S_n(\lambda) = S_m(\lambda) A(\lambda)^{n-m}$
(ii) $M_n(\lambda) = M_m(\lambda) A(\lambda)^{n-m}$

Proof: (By induction on $n - m$)

(i) For $n - m = 1$, $S_{n-1}(\lambda)A(\lambda) = (\lambda_{n-1}, \lambda_n, \lambda_{n+1}) = S_n(\lambda)$. If the result is true for $n - m = k$, then for $n - m = k + 1$

$$S_{n-(k+1)}(\lambda) A(\lambda)^{k+1} = (\lambda_{n-k-1}, \lambda_{n-k}, \lambda_{n-k+1}) A(\lambda)^k = S_n(\lambda).$$

Hence, by induction on $n - m$ gives the required result.

(ii) For $n - m = 1$, $M_{n-1}(\lambda) A(\lambda) = \begin{pmatrix} \lambda_{n-2} & \lambda_{n-1} & \lambda_n \\ \lambda_{n-1} & \lambda_n & \lambda_{n+1} \\ \lambda_n & \lambda_{n+1} & \lambda_{n+2} \end{pmatrix} = M_n(\lambda)$. If the result

is true for $n - m = k$, then for $n - m = k + 1$

$$M_m(\lambda) A(\lambda)^{k+1} = M_{n-(k+1)}(\lambda) A(\lambda)^{k+1}$$

$$= \begin{pmatrix} \lambda_{n-k-2} & \lambda_{n-k-1} & \lambda_{n-k} \\ \lambda_{n-k-1} & \lambda_{n-k} & \lambda_{n-k+1} \\ \lambda_{n-k} & \lambda_{n-k+1} & \lambda_{n-k+2} \end{pmatrix} A(\lambda)^k$$

$$= M_{n-k}(\lambda) A(\lambda)^k = M_n(\lambda).$$

Hence, by induction on $n - m$ gives the required result.

Corollary 3.3.9 [6] Suppose $C(A)$ denote the second column of a 3×3 matrix A. Then

$$\lambda_n = S_m(\lambda) C\left(A(\lambda)^{n-m}\right).$$

Proof: Suppose, $I_{3 \times 3}$ be the 3×3 identity matrix.

$$S_m(\lambda) C\left(A(\lambda)^{n-m}\right) = S_m(\lambda) A(\lambda)^{n-m} C(I_{3\times3})$$

$$= S_n(\lambda) \begin{pmatrix} 0 \\ 1 \\ 0 \end{pmatrix}$$

$$= \lambda_n$$

Lemma 3.3.10 [6] Suppose $\triangle_n$ (or $\triangle_n(\lambda)$) denote the determinant of $M_n(\lambda)$. Then

(i) $\triangle_0 = \lambda^{2p+2} + 18\,\lambda^{p+1} - 4\left(\lambda^{3p} + \lambda^3\right) - 27 \in GF(p)$

(ii) If $\triangle_0 \neq 0$ then $M_0(\lambda)^{-1}$ is equal to

$$\frac{1}{\triangle_0} \begin{bmatrix} 2\lambda^2 - 6\lambda^p & 2\lambda^{2p} + 3\lambda - \lambda^{p+2} & \lambda^{p+1} - 9 \\ 2\lambda^{2p} + 3\lambda - \lambda^{p+2} & \left(\lambda^2 - 2\lambda^p\right)^{p+1} - 9 & \left(2\lambda^{2p} + 3\lambda - \lambda^{p+2}\right)^p \\ \lambda^{p+1} - 9 & \left(2\lambda^{2p} + 3\lambda - \lambda^{p+2}\right)^p & \left(2\lambda^2 - 6\lambda^p\right)^p \end{bmatrix}$$

(iii) $\triangle_0\left(Tr(\mu)\right) = \left(Tr\left(\mu^{p+1}\right)^p - Tr\left(\mu^{p+1}\right)\right)^2 \neq 0$

Proof:

(i) Using (i), (ii), (vi) of Lemma 3.3.1 and (i) of Lemma 3.3.2, we get $\lambda_{-2} = \lambda^{2p} - 2\lambda$, $\lambda_{-1} = \lambda^p$, $\lambda_2 = \lambda^2 - 2\lambda^p$, which implies

$$
\triangle_0 = \begin{vmatrix} \lambda^{2p} - 2\lambda & \lambda^p & 3 \\ \lambda^p & 3 & \lambda \\ 3 & \lambda & \lambda^2 - 2\lambda^p \end{vmatrix}
$$
$$
= \lambda^{2p+2} + 18\,\lambda^{p+1} - 4\left(\lambda^{3p} + \lambda^3\right) - 27
$$

(ii) Suppose $\triangle_0 \neq 0$, then $M_0(\lambda)^{-1}$ exists, and some simple computation gives the required result.

(iii) Clearly, $M_0\left(Tr\left(\mu\right)\right) = \begin{pmatrix} Tr\left(\mu^{-2}\right) & Tr\left(\mu^{-1}\right) & Tr\left(\mu^0\right) \\ Tr\left(\mu^{-1}\right) & Tr\left(\mu^0\right) & Tr\left(\mu\right) \\ Tr\left(\mu\right) & Tr\left(\mu\right) & Tr\left(\mu^2\right) \end{pmatrix}$

$$
= \begin{pmatrix} \mu^{-1} & \mu^{-p^2} & \mu^{-p^4} \\ 1 & 1 & 1 \\ \mu & \mu^{p^2} & \mu^{p^4} \end{pmatrix} \begin{pmatrix} \mu^{-1} & \mu^{-p^2} & \mu^{-p^4} \\ 1 & 1 & 1 \\ \mu & \mu^{p^2} & \mu^{p^4} \end{pmatrix}^T
$$

where T denotes transpose. As the determinant of $\begin{pmatrix} \mu^{-1} & \mu^{-p^2} & \mu^{-p^4} \\ 1 & 1 & 1 \\ \mu & \mu^{p^2} & \mu^{p^4} \end{pmatrix}$ is

$Tr(\mu^{p+1})^p - Tr(\mu^{p+1})$, so $\triangle_0(Tr(\mu)) = (Tr(\mu^{p+1})^p - Tr(\mu^{p+1}))^2$.

Remark 3.3.11 As $\triangle_0(Tr(\mu)) \neq 0$, so $M_0(Tr(\mu))^{-1}$ exist. Therefore, one can compute $A(Tr(\mu))^n = M_0(Tr(\mu))^{-1}M_n(Tr(\mu))$ by using (ii) of Lemma 3.3.8 for $\lambda = Tr(\mu)$ and $m = 0$. The following lemma state that computing $A(Tr(\mu))^n$ in $GF(p^2)$ requires a small constant number of operations.

Lemma 3.3.12 [6]: For given $Tr(\mu)$ and $S_n(Tr(\mu))$, computing $A(Tr(\mu))^n$ needs a small constant number of operations in $GF(p^2)$.

Corollary 3.3.13 [6]

$$
C\left(A(Tr\left(\mu\right))^n\right) = M_0(Tr\left(\mu\right))^{-1}(S_n\left(Tr\left(\mu\right)\right))^T.
$$

Proof: By (ii) of Lemma 3.3.8,

$$
\begin{aligned}
C\left(A(Tr\left(\mu\right))^n\right) &= C\left(M_0(Tr\left(\mu\right))^{-1}M_n\left(Tr\left(\mu\right)\right)\right) \\
&= M_0(Tr\left(\mu\right))^{-1} C\left(M_n\left(Tr\left(\mu\right)\right)\right) \\
&= M_0(Tr\left(\mu\right))^{-1}\left(Tr\left(\mu^{n-1}\right), Tr(\mu)^n, Tr(\mu)^{n-1}\right)^T \\
&= M_0(Tr\left(\mu\right))^{-1}\left(S_n(Tr\left(\mu\right))\right)^T.
\end{aligned}
$$

Now, for given $Tr(\mu)$, $S_d(Tr(\mu))$ and $a, b \in \mathbb{Z}$ with $0 < d < q$, where d is unknown, one can compute $Tr(\mu^a \mu^{bd})$ as follows:

Algorithm 3.3.14

1. Compute $e = ab^{-1} \mod q$
2. Compute $S_e(Tr(\mu))$
3. Compute $C(A(Tr(\mu))^e)$ based on $Tr(\mu)$ and $S_e(Tr(\mu))$.
4. Compute $Tr(\mu^{e+d}) = S_d(Tr(\mu))C(A(Tr(\mu))^e)$
5. Compute $S_b(Tr(\mu^{e+d}))$ and return $Tr(\mu^{(e+d)b}) = Tr(\mu^a \mu^{bd})$.

Using Theorem 3.3.6, we conclude that.

Theorem 3.3.15 [6] For Given $M_0(Tr(\mu))^{-1}$, $Tr(\mu)$ and $S_k(Tr(\mu))$, one can compute $Tr(\mu^a \mu^{bd})$ at a cost of $8\log_2(ab^{-1} \mod q) + 8\log_2 b + 34$ multiplications in $GF(p)$.

In [6], Lenstra et al. mentioned that $Tr(\mu^a \mu^{bd})$ can be computed at a cost of $16\log_2 q$ multiplication in $GF(p)$ if $M_0(Tr(\mu))^{-1}$ is computed once and for all. For details, see [6].

Now, there are two issues to be discussed:

1. Method to generate primes p and q such that $q \mid (p^2 - p + 1)$ and $p \equiv 2 \mod 3$, where DLP is hard to solve.
2. Method to generate a subgroup $G \in GF(p^6)$ of order $q > 3$ such that $q \mid (p^2 - p + 1)$.

In the following two sections, we will consider these two issues.

3.4 Parameter Selection

In order to achieve the advantage of the representation of elements with their trace and for the fast arithmetic over $GF(p^2)$, one needs to select the primes p and $q > 3$ in such a way that $q \mid p^2 - p + 1$ and $p \equiv 2 \mod 3$. Also, p and q must be chosen in such a way that DLP over $GF(p^6)^*$ and its subgroup of order q is difficult to solve by a known algorithm. Hence, it is suggested to consider $6\ell_p \approx 1024$ and $\ell_q \approx 160$, where ℓ_p, ℓ_q denotes the size of p and q, respectively [6]. In [13], Lenstra et al. described four algorithms to choose proper primes p and q. We have considered the following algorithm from [13]. For details, see [6, 13].

Algorithm 3.4.1

1. Choose a random number $\xi \in \mathbb{Z}$ such that $q = \xi^2 - \xi + 1$ is a prime of size ℓq.
2. Choose a random number $\eta \in \mathbb{Z}$ such that $p = \xi + q\eta$ is a prime of size ℓ_p and $p \equiv 2 \mod 3$.

The above algorithm is very fast. It can be used to find p such that it satisfies a polynomial of degree 2 with small coefficient [6]. Particularly, if $\eta = 1$, then one have to find ξ such that $\xi^2 - \xi + 1$ and $\xi^2 + 1$ are prime. Also, $\xi^2 + 1 \equiv 2 \bmod 3$, which implies ξ is even and $p \equiv 1 \bmod 4$. But such p may not be suitable to consider as it may allow DLP variant of Number Field Sieve [6]. This disadvantage does not work in the following algorithm.

Algorithm 3.4.2 [13]

1. Choose a prime $q \equiv 7 \bmod 12$ of size ℓ_q.
2. Find the roots ξ_1, ξ_2 of $X^2 - X + 1 \bmod q$.
3. Compute $\eta \in \mathbb{Z}$ such that $p = \xi_i + q\eta$ for $i = 1$ or 2 is a prime of size ℓ_p and $p \equiv 2 \bmod 3$.

Note that, if ξ is a root of $X^2 - X + 1 \bmod q$, then $\xi^2 \equiv \xi - 1 \bmod q$, which implies the roots of $X^2 - X + 1 \bmod q$ are the quadratic residue. As $q \equiv 7 \bmod 12 \equiv 1 \bmod 3$, this guarantees the existence of quadratic residues and hence the existence of the root ξ_1 and ξ_2 (see 5. (b) of [15] page 174). Again $q \equiv 3 \bmod 4$, so roots can be determined by using 5. (c) of page 174 of [15].

To resist the subgroup attacks, Lenstra et al. in [13] described the following two algorithms for the selection of p and q. For details, see [13].

Algorithm 3.4.3 [13]:

1. Choose a prime p of size ℓ_p such that $p^2 - p + 1 = qt$, where q is a prime and $t \in \mathbb{Z}$ is small.

Algorithm 3.4.5 [13]:

1. Choose a prime q of size ℓ_q and $q \equiv 7 \bmod 12$.
2. Compute the roots ξ_1, ξ_2 of $X^2 - X + 1 \bmod q$.
3. Choose $\eta \in \mathbb{Z}$ such that $p = \xi_i + q\eta$ for $i = 1$ or 2 is a prime of size ℓ_p with $p \equiv 2 \bmod 3$ and $\frac{p^2 - p + 1}{q} = rt$, where r is a prime of size at least ℓ_q and $t \in \mathbb{Z}$ of small size.

3.5 Subgroup Selection

In this section, we will discuss the method to find a subgroup $G = \langle \mu \rangle$ of $GF(p^6)$ such that for $\mu \in GF(p^6), \mathrm{ord}(\mu) = q > 3$, and $q \mid p^2 - p + 1$. To find such a subgroup, it is enough to find $Tr(\mu) \in GF(p^2)$ as for given $Tr(\mu)$, a generator μ can be determined from any roots of $F(Tr(\mu), X)$. Therefore by (vii) of Lemma 3.3.1, to find $Tr(\mu)$, one has to find an irreducible polynomial $F(\lambda, X) \in GF(p^2)[X]$. In that case, $\lambda = Tr(x)$ for some $x \in GF(p^6)$ with $\mathrm{ord}(x) \mid (p^2 - p + 1)$ and $\mathrm{ord}(x) > 3$. Moreover if $\lambda_{p^2-p+1/q} \neq 3$, then $Tr(\mu) = \lambda_{p^2-p+1/q}$ [6]. The resulting $Tr(\mu)$ is the trace of an element $\mu \in GF(p^6)$ of order q. Therefore, to find $Tr(\mu)$ one has to find $\lambda \in GF(p^2)$ such that $F(\lambda, X)$ is irreducible over $GF(p^2)$. In [6], Lenstra et al. proved that for a randomly chosen $\lambda \in GF(p^2)$, the probability that the polynomial $F(\lambda, X)$ is an irreducible polynomial over $GF(p^2)$ is about 1/3. The following algorithm from [6] describes a method to compute $Tr(\lambda)$ for a randomly chosen $\lambda \in GF(p^2) \backslash GF(p)$.

Algorithm 3.5.1 [6]:

1. Choose a random number $\lambda \in GF(p^2) \backslash GF(p)$.
2. Compute λ_{p+1} (using Algorithm 3.3.5).
3. If $\lambda_{p+1} \in GF(p)$, then go to step (1).
4. Compute $\lambda_{(p^2-p+1/q)}$ (using Algorithm 3.3.5).
5. Set $Tr(\mu) = \lambda_{(p^2-p+1/q)}$, if $\lambda_{(p^2-p+1/q)} \neq 3$. Otherwise, go to step (1).

This algorithm needs almost $\frac{3q}{q-1}$ applications of Algorithm 3.3.5 with $n = p+1$ and $\frac{q}{q-1}$ applications with $n = (p^2 - p + 1)/q$ (see Theorem 3.23 of [6]).

In order to compute $Tr(\mu)$, some other faster methods have been discussed in [16–19], where $p \equiv 2 \bmod 3$. These methods are based on the method to test $F(\lambda, X)$ is irreducible. In [19], Lenstra et al. showed that for $p \equiv 3 \bmod 4$, $Tr(\mu)$ can be computed as effectively as when for $p \equiv 2 \bmod 3$. Another approach to find $Tr(\mu)$ and hence generator of an XTR group is found in [20], where $p \equiv 1 \bmod 3$.

4   Cryptographic Applications of XTR

In this section, we will discuss some encryption schemes and signature schemes, which we have considered from [6, 13]. Note that, any scheme based on Discrete Logarithm (DL) Problem can use XTR [6]. In the next section, we will introduce the XTR version of DH problem and its variants. The XTR-DH key exchange scheme, XTR-ElGamal scheme, XTR-NR signature scheme, and XTR-DSA scheme are discussed in the remaining sections.

4.1   XTR Version of DH Problem and its Variants

From [4, 5], it is clear that the DL problem in a multiplicative subgroup (say (h)) of a field (say $GF(p^r)$) is as difficult as the DLP in $GF(p^r)^*$ if (i) ord(h) is a sufficiently large prime and (ii) the minimal surrounding subfield of (h) is $GF(p^t)$ for a large prime p. Therefore, whenever $GF(p^6)$ is the minimal surrounding subfield of the XTR group G and q is sufficiently large, then the DLP in $G = (\mu)$ is as difficult as the DLP in $GF(p^6)$ and hence the DH and DDH problems too.

Now, we consider the XTR version of DH, DDH, and DL problems [6]. It is called the XTR version as each element g of G is represented by their trace, $Tr(g)$.

XTR-DH Problem: For given $Tr(\mu^a)$ and $Tr(\mu^b)$, the XTR-DH problem asked to compute $Tr(\mu^{ab})$.

XTR-DDH Problem: For given $Tr(\mu^a)$, $Tr(\mu^b)$, and $Tr(\mu^c)$, XTR-DDH problem is the problem to determine whether $Tr(\mu^{ab}) = Tr(\mu^c)$.

XTR-DL Problem: For given $Tr(\mu)$ and $Tr(\mu^a)$, the XTR-DL problem is the problem of finding a such that $0 \leq a < q$.

In [6], Lenstra et al. proved that the XTR-DH, XTR-DDH, and XTR-DL problem is equivalent to the DH, DDH, and DL problem, respectively, in XTR group G (see Theorem 5.21 of [6]).

Note that there exists an efficiently computable injective homomorphism from CTP curves over $GF(p^2)$ [21] onto the XTR group. At Crypto 2000 rump session [22], Menezes et al. suggested that there may exist a PPT algorithm to compute the inverse of such homomorphism (called X2C hypothesis [23]) and in that situation, XTR group is similar to the group of a supersingular elliptic curve. Hence, the security XTR cryptosystem is not better than the security of the Elliptic curve cryptosystem. In [23], Verheul showed that if X2C hypothesis hold, then one can solve several other problems, which are believed to be hard. Hence the XTR group provides better security than the isomorphic group on a supersingular elliptic curve. For details see [23].

4.2 XTR-Diffie-Hellman (XTR-DH) Key Exchange Scheme

The XTR-DH protocol is described in [6]. It is based on the XTR-DH problem. Suppose, two parties (say Alice and Bob) want to agree on a common secret key K. Then for given XTR public key data $(p, q, Tr(\mu))$, both Alice and Bob have to execute the following steps:

1. Alice chooses a random integer a such that $1 < a < q - 2$ and compute

$$S_a\left(Tr\left(\mu\right)\right) = \left(Tr\left(\mu^{a-1}\right), Tr\left(\mu^a\right), Tr\left(\mu^{a+1}\right)\right)$$

2. Send $Tr(\mu^a)$ to Bob.
3. Bob receives $Tr(\mu^a)$.
4. Bob chooses a random integer b such that $1 < b < q - 2$ and compute

$$S_b\left(Tr\left(\mu\right)\right) = \left(Tr\left(\mu^{b-1}\right), Tr\left(\mu^b\right), Tr\left(\mu^{b+1}\right)\right)$$

5. Bob sends $Tr(\mu^b)$ to Alice.
6. Bob computes $S_b(Tr(\mu^a)) = (Tr(\mu^{a(b-1)}), Tr(\mu^{ab}), Tr(\mu^{a(b-1)}))$ and finds a secret key K based on $Tr(\mu^{ab})$.
7. Alice receives $Tr(\mu^b)$ from Bob.
8. Alice computes $S_a(Tr(\mu^b)) = (Tr(\mu^{b(a-1)}), Tr(\mu^{ba}), Tr(\mu^{b(a+1)})$.
9. Alice finds the secret key K based on $Tr(\mu^{ba})$.

Efficiency: Both the communication and computational overhead of the XTR-DH key exchange protocol are about 1/3 of the DH key exchange protocol based on a multiplicative subgroup of a finite field [6].

4.3 XTR-ElGamal Encryption Scheme

The XTR version of ElGamal scheme is found in [6]. The scheme is based on the XTR-DL problem. As a symmetric key is also used in this scheme, so the scheme is considered as a hybrid version of the ElGamal scheme in a XTR group. The key generation, encryption, and decryption algorithm are defined as follows:

Key Generation:

1. Choose two primes p and $q > 3$ with $q \mid (p^2 - p + 1)$ and $p \equiv 2 \bmod 3$.
2. Compute $Tr(\mu)$ (see Sect. 3.5).
3. Choose a such that $2 \leq a \leq q - 3$.
4. Compute $Tr(\mu^a)$.

The secret key of the receiver is a and $(p, \ q, \ Tr(\mu), Tr(\mu^a))$ is the corresponding public key.

Encryption: Suppose, for the given public key $(p, \ q, \ Tr(\mu), Tr(\mu^a))$, the sender wants to encrypt a message M. The sender has to execute the following steps:

1. Choose a random number k such that $1 < k < q - 2$.
2. .Compute $S_k(Tr(\mu)) = (Tr(\mu^{k-1}), Tr(\mu^k), Tr(\mu^{k+1}))$
3. Compute $S_k(Tr(\mu^a)) = (Tr(\mu^{(k-1)a}), Tr(\mu^{ka}), Tr(\mu^{(k+1)a}))$
4. Choose a symmetric encryption key K_s based on $Tr(\mu^{ka}) \in GF(p^2)$.
5. Choose a symmetric encryption technique E with key K_s to encrypt M (say $E(M)$).

The pair $(Tr(\mu^k), E(M))$ is the ciphertext of M.

Decryption: For a given ciphertext $(Tr(\mu^b), E(M))$ and the secret key a, the receiver execute the following steps to recover M:

1. Compute $S_a(Tr(\mu^k)) = (Tr(\mu^{(a-1)k}), Tr(\mu^{ak}), Tr(\mu^{(a+1)k}))$.
2. Determines symmetric encryption key K_s based on $Tr(\mu^{ak}) \in GF(p^2)$.
3. Choose the symmetric decryption technique D with key K_s to decrypt $E(M)$, which results the message M.

Efficiency: In the XTR-ElGamal scheme, both the communication and computational overhead of encryption and decryption method are about 1/3 of the ElGamal encryption and decryption algorithms that are based on a multiplicative subgroup of a finite field [6]. The scheme achieves the same security level as that of the ElGamal scheme. For details see ([6, 13]).

4.4 XTR-Nyberg-Rueppel (NR) Signature Scheme

The XTR version of NR message recovery signature scheme is discussed in [6]. The key generation algorithm is same with the XTR-ElGamal scheme. The public key of the signer is $(p, \ q, \ Tr(\mu), \ Tr(\mu^a))$, and a is the corresponding secret key. Also, it is assumed that $Tr(\mu^{a-1})$, $Tr(\mu^{a+1})$ and hence $S_a(Tr(\mu))$ is known to the verifier. A hash function H is considered which is public for all.

Signature Generation: For a message M and secret key a, the signer executes the following steps:

1. Choose a random integer k such that $2 \le k \le q-3$.
2. Compute $S_k(Tr(\mu)) = (Tr(\mu^{k-1}), \ Tr(\mu^k), \ Tr(\mu^{k+1}))$.
3. Choose a symmetric encryption key K_s based on $Tr(\mu^k)$.
4. Choose a symmetric encryption technique with key K_s and encrypt the message M (say E).
5. Compute $s = (a \, H(E) + k) \bmod q$.

(E, s) is the signature on the message M.

Signature Verification: For a given signature (E, s) and public key of the signer, the verifier executes the following steps:

1. Check that $0 < r, s < q$, if no then reject the signature. Otherwise proceed step (2).
2. Compute $H(E)$ and replace by $-H(E) \in \{0, 1, \cdots, q-1\}$.
3. Compute $Tr(\mu^s \mu^{aH(E)})$ (by using Algorithm 3.3.14).
4. Determine a symmetric key K_s based on $Tr(\mu^s \mu^{aH(E)}) \in GF(p^2)$.
5. Choose a symmetric encryption technique with key K_s and decrypt the ciphertext E, results M.
6. If M contains the agreed upon redundancy, then (E, s) is a valid signature.

Efficiency: The signature generation and verification algorithms of the XTR-NR signature scheme is faster than the NR signature Scheme [24] based on the multiplicative group of a finite field with the same security level. Note that the XTR-NR signature generation and verification algorithm, respectively, are about three times and about 1.75 faster than the signature generation and verification algorithm of the NR signature Scheme [13].

4.5 XTR-DSA Signature Scheme

The XTR-DSA signature scheme is defined in (Lenstra & Verheul, 2001). The public key of the signer is $(p, \ q, \ Tr(\mu), \ S_a(Tr(\mu)))$, and the corresponding secret key is a. Like the DSA signature Scheme [25], the size of the prime q is considered as 160-bits, and the hash function SHA-1 [26] is used.

Signature Generation: For a message M and a secret key a, the signer executes the following steps:

1. Choose a random number k such that $2 \leq k \leq q - 3$.
2. Compute $Tr(\mu^k) \in GF(p^2)$
3. Express $Tr(\mu^k) = x_1\beta + x_2\beta^2$
4. Compute $r = (x_1 + px_2) \bmod q$
5. Compute $s = (H(M) + ar)k^{-1} \bmod q$, if $r \neq 0$. Otherwise, repeat the steps for a new choice of k.

(r, s) is the resulting signature on M.

Signature Verification: For a signature (r, s), a message M and the signer public key $(p, q, Tr(\mu), \; S_a(Tr(\mu)))$, the verifier executes the following steps:

1. Check that $0 < r, \; s < q$, if no then reject the signature. Otherwise proceed to step (2).
2. Compute $u = s^{-1} \bmod q$.

$$t_1 = uH(M) \bmod q$$
$$t_2 = ur \bmod q$$

3. Compute $t_3 = Tr\left(\mu^{t_1}\mu^{t_2a}\right) \in GF\left(p^2\right)$ by using Algorithm 3.3.14.
4. Express $t_3 = y_1\beta + y_2\beta^2$.
5. Compute $t = (y_1 + py_2) \bmod q$
6. If $t = r$, then the signature is a valid signature.

5 Recent Development of XTR

Numerous works had been done since the XTR was introduced. In 2000 [16], Lenstra et al. improved the XTR key generation and parameter generation method for the signature scheme. They provide a faster subgroup selection method based on an irreducible test. They showed that by this improved method, an irreducible polynomial can be found on average $7.2\log_2 p$ multiplications over $GF(p)$. The irreducibility test proposed in [16] further improved in [17]. By this new approach, an irreducible polynomial can be determined on an average of $2.7\log_2 p$ multiplications in $GF(p)$. The methods to speed up XTR implementation are also discussed in [18, 19]. Another improvement of the irreducibility test can be found in [27]. In XTR, basically $p \equiv 2 \bmod 3$ is used to improve the time complexity of irreducibility testing. In 2004 [20], Kwon et al. presented an algorithm for fast irreducibility testing, which can be used for $p \equiv 1 \bmod 3$. They also provide a method to compute a generator of the XTR group without having any irreducible test by using a Gaussian normal basis in $GF(p^2)$. A new key generation algorithm is also found in [28].

Various cryptographic protocols based on XTR has been proposed. Some of the public key cryptosystem based on XTR are found in [29, 30], etc., and some signature schemes are proposed in [31–35], etc. Some other cryptographic protocols are proposed in [36–38] and [39].

The security of XTR is also considered in several papers such as [30, 40–44], whereas, some applications of XTR cryptographic protocol are discussed in [45–47], etc.

In 2000 [6], Lenstra et al. also states that the idea of the XTR group over $GF(p^6)$ can be extended to any field $GF(p^{6e})$, for some $e \in \mathbb{Z}$. In [48], it is showed that to get the same communication and computational advantage of the generalized XTR over $GF(p^{6e})$, either both e and $2e + 1$ are primes or $2e + 1$ is a Fermat prime. In [11], Bosma et al. proved that if BPH conjecture [5] holds, then a more compact representation of the element over the field of degree 30 is possible. In 2006 [49], Wang et al. pointed out that XTR has parameter corresponding problem. To solve this problem, Wang et al. proposed the concept of XTR^+. It is based on the finite field $GF(p^8)$, where the arithmetic is over $GF(p^4)$. They proved that XTR^+ has achieved the same security level as XTR, and it needed less amount of data storage, computation and communication overhead. Some other works related to XTR are found in [50–52], etc.

6 Conclusion

In XTR, elements of a subgroup of order $q > 3$ of the multiplicative group $GF(p^6)^*$ are represented by their trace over $GF(p^2)$, where $q \mid (p^2 - p + 1)$. It is proved that XTR cryptosystem is more efficient than those cryptosystems which are based on DLP on a multiplicative group over a finite field. In this chapter, we have studied the methods and techniques for XTR. We have discussed the algorithm for parameter selection and to generate XTR group. We also discussed the application of XTR in cryptography. Basically, we discussed the XTR-DH key exchange protocol, XTR-ElGamal scheme. We also discussed the XTR-NR message recovery scheme and XTR-DSA. A short overview on the recent development of XTR is also presented.

References

1. Diffie, W., & Hellman, a M. E. (1976). New directions in cryptography. *IEEE Transaction on Information Theory, 22*(6), 644–654.
2. ElGamal, T. (1985). A public key cryptosystem and a signature scheme based on discrete logarithms. *IEEE Transactions on Information Theory, 31*(4), 469–472.
3. Schnorr, C. P. (1991). Efficient signature generation by smart cards. *Journal of Cryptology, 4,* 161–174.
4. Lenstra, A. K. (1997). Using cyclotomic polynomials to constract efficient discrete logarithm cryptosystems over finite fields. In V. Varadharajan, J. Pieprzyk, & Y. Mu (Eds.), *Information Security and Privacy - ACISP 1997.LNCS 1270, (pp. 127–138).* Berline, Heidelberg: Springer.
5. Brouwer, A. E., Pellikaan, R., & Verheul, E. R. (1999). Doing more with fewer bits. In K. Y. Lam, E. Okamoto, & C. Xing (Eds.), *ASIACRYPT 1999.LNCS 1716* (pp. 321–332). Berlin Heidelberg: Springer.

6. Lenstra, A. K., & Verheul, E. R. (2000). The XTR public key system. In M. Bellare (Ed.), *Advances in cryptology - CRYPTO 2000.LNCS 1880* (pp. 1–19). Berlin, Heidelberg: Springer.
7. Smith, P., & Skinner, C. (1995). A public key cryptosystem and a digital signature system based on the Lucus function analogue to discrete logarithms. In J. Pieprzyk & R. Safavi-Naini (Eds.), *Advances in cryptology - ASIACRYPT 1994.LNCS 917* (pp. 355–364). Berlin, Heidelberg: Springer.
8. Rivest, R. L., Shamir, A., & Adleman, L. (1978). A method for obtaining digital signatures and public key cryptosystem. *Communications of ACM, 21*(2), 120–126.
9. Koblitz, N. (1987). Elliptic curve cryptosystems. *Mathematics of Computation, 48*, 203–209.
10. Miller, V. (1986). Use of elliptic curves in cryptography. In W. H. C (Ed.), *Advances in Cryptology - CRYPTO 1985.LNCS 218* (pp. 417–426). Berlin, Heidelberg: Springer.
11. Bosma, W., Hutton, J., & Verheul, E. R. (2002). Looking byond XTR. In Y. Zheng (Ed.), *Advances in Cryptology - ASIACRYPT 2002.LNCS 2501, (pp. 46–63)*. Heidelberg: Springer.
12. Das, A., & Veni Madhavan, C. E. (2009). *Public-key Cryptography : Theory and Practice*. Delhi: Pearson Education.
13. Lenstra, A. K., & Verheul, E. R. (2001). An overview of the XTR public key system. In *Public key cryptography and computational number theory (Warsaw, 2000)* (pp. 151–180). Berlin: de Gruyter.
14. Stam, M. (2003). *Speeding up subgroup cryptosystems. Ph. D Thesis*. Eindhoven: Technische Universiteit.
15. Burton, D. M. (2012). *Elementary number theory*. TATA McGraw Hill.
16. A. K. Lenstra and E. R. Verheul. Key improvements to XTR. ASIACRYPT 2000.LNCS 1976, (pp. 220–233), Springer, Heidelberg 2000.
17. Lenstra, A. K., & Verheul, E. R. (2001). Fast irreducibility and subgroup membership testing in XTR. In K. K (Ed.), *Public Key Cryptography - PKC 2001.LNCS 1992* (pp. 73–86). Berlin, Heidelberg: Springer.
18. Stam, M., & Lenstra, A. K. (2001). Speeding up XTR. In C. Boyd (Ed.), *Advances in cryptology - ASIACRYPT 2001.LNCS 2248* (pp. 125–143). Heidelberg: Springer.
19. Stam, M., & Lenstra, A. K. (2003). Efficient subgroup exponentiation in quadratic and sixth degree extensions. In B. S. Kaliski, K. Koc, & C. Paar (Eds.), *Cryptographic Hardware and Embedded Systems - CHES 2002.LNCS 2523, (pp. 318–332)*. Berlin: Springer.
20. Kwon, S., Kim, C. H., & Hong, C. P. (2005). Fast irreducibility testing for XTR using Gaussian normal basis of low complexity. In H. H & A. Hasan (Eds.), *Selected Areas in Cryptology - SAC 2004.LNCS 3357* (pp. 144–158). Berlin, Heidelberg: Springer.
21. Menezes, A. (1993). *Elliptic curve public key cryptosystems*. Boston: Kluwer Academic Publishers.
22. Menezes, A., & Vanstone, S. (2000). ECSTR (XTR): Elliptic curve singular trace representation. In *Rump session of Crypto*.
23. Verheul, E. R. (2004). Evidence that XTR is more secure than supersingular elliptic curve cryptosystems. *Journal of Cryptology, 17*, 277–296.
24. Nyberg, K., & Rueppel, R. (1995). Message recovery for signature schemes based on the discrete logarithm problem. In A. De Santis (Ed.), *Advances in cryptology - EUROCRYPT 1994.LNCS 950* (pp. 182–193). Berlin, Heidelberg: Springer.
25. Digital Signature Standard. Federal Information Processing Standards Publications 186, NIST. 1994.
26. Secure Hash Function (SHA-1). Fderal information processing standards publications 180-1, NIST (1995).
27. Kim, J. M., Yie, I., Oh, S. I., & Ryu, J. (2001). First generation of cubic irreducible plynomials for XTR. In RCP & C. Ding (Eds.), *Progress in Cryptology - INDOCRYPT 2001.LNCS 2247* (pp. 73–78). Berlin, Heidelberg: Springer.
28. Grzeskowiak, M. (2006). New key generation algorithms for the XTR cryptosystem. In R. Meersman, Z. Tari, & P. Herrero (Eds.), *On the move to meaningful internet systems 2006 : OTM 2006 workshop. OTM 2006.LNCS 4277* (pp. 439–449). Berlin, Heidelberg: Springer.

29. Akleylek, S., & Kirlar, B. B. (2015). New methods for public key cryptosystems based on XTR. *Security and Communication Networks, 8*, 3682–3689.
30. Chen, X., Feng, F., & Wang, Y. (2003). New key improvements and its application to XTR system. In *The 17th international conference on Advanced Information Networking & Applications - AINA 2003. IEEE*.
31. Chen, X. F., Wang, J. L., & Wang, Y. M. (2004). Signature scheme based on extended XTR system. *Journal of Electronics & Information Technology, 26*(4), 563–567.
32. Tang, Q., & Shen, F. (2013). Identity-based XTR blind signature scheme. *Intelligent Automation and Soft Computing, 19*(2), 143–149.
33. Wang, J. L., Wu, Q. H., & Gao, H. M. (2004). Schnorr and ring signatures based on XTR. *Journal of XiDian University, 31*(3), 454–458.
34. Yan, Y. J. (2007). Verifiable encryption of XTR signatures. *Journal of Xi'An University of Post and Telecommunication, 12*(5), 109–111.
35. Yan, Y. J., Ma, W. P., & Wang, X. M. (2005). Blind group signature based on XTR. *Application Research of Computers, 5*, 108–109.
36. Hoeper, K., & Gong, G. (2006). Integrated DH-like key exchange protocols from LUC, GH and XTR. In *International Symposium on Information Theory* (pp. 922–926). Seattle, WA: IEEE.
37. Jung, K. S., & Rao, I. A. (2006). Design of user authentication protocol based onXTR-ElGamal. In *International Conference on Hybrid Information Technology* (pp. 677–683). Cheju Island: IEEE.
38. Li, B. (2009). A forward-secrecy WTLS handshake protocol based on XTR. In J. H. Park, H. H. Chen, M. Atiquzzaman, C. Lee, T. Kim, & S. S. Yeo (Eds.), *Advances in Information Security and Assurance - ISA 2009.LNCS 5576, (pp. 635–643)*. Berlin: Springer.
39. S. Zhang,J. Chen, J. Xia, andX. Ai,"An XTR based constant round key agreement scheme",Mathematical Problems in Engineering, pp. 1–9, 2013.
40. Bevan, R. (2005). Improved zero value attack on XTR. In BC & J. M. G. Nieto (Eds.), *Information Security and Privacy.LNCS 3574* (pp. 207–217). Berlin, Heidelberg: Springer.
41. Chung, J., & Hasan, A. Security analysis of XTR exponentiation algorithms against simple power analysis attack. University of Waterloo. In *Preprint of CACR, CACR 2004–05* (p. 2004).
42. Ciet, M., & Giraud, C. (2004). Transient fault induction attacks on XTR. In J. Lopez, S. Qing, & E. Okamoto (Eds.), *Information and Communications Security - ICICS 2004.LNCS 3269, (pp. 440–451)*. Heidelberg: Springer.
43. Han, D. G., Takagi, T., & Lim, J. (2006). Further security analysis of XTR. In K. Chen, R. Deng, X. Lai, & J. Zhou (Eds.), *Information Security Practice and Experience - ISPEC 2006.LNCS 3903, (pp. 33–44)*. Berlin, Heidelberg: Springer.
44. Han, D. G., Takagi, T., Kim, T. H., Kim, H. W., & Chung, K. I. (2005). Collision attack on XTR and a countermeasure with a fixed pattern. In *Embedded and Ubiquitous Computing-EUC 2005*. Berlin, Heidelberg: Springer.
45. Cai, X. F., & Zhang, Y. S. (2006). The advantages and applications of XTR public key cryptosystem in secure e-commerce. *Network & Computer Security, 1*, 25–28.
46. Nanda, A. K., & Awasthi, L. K. (2012). XTR cryptosystem for SMS security. *IACSIT International Journal of Engineering and Technology, 4*(6), 836–839.
47. Peeters, E., Neve, M., & Ciet, M. (2004). XTR implementation on reconfigurable hardware. In M. Joye & J. J. Quisquater (Eds.), *Cryptographic Hardware and Embedded Systems - CHES 2004.LNCS 3156, (pp. 386–399)*. Berlin: Springer.
48. Lim, S., Kim, S., & Yie, I. (2001). XTR extended to GF(P6m). In S. Y. Vaudenay (Ed.), *Selected Areaa in Cryptography - SAC 2001.LNCS 2259, (pp. 301–312)*. Berlin: Springer.
49. Wang, Z., & Zhang, Z. (2007). XTR+: A provably security public key cryptosystem. In Y. Wang, Y. Cheung, & H. Liu (Eds.), *CIS 2006.LNAI 4456* (pp. 534–544). Berlin: Springer.
50. Giuliane, K. J. (2003). *Analogues to gong-horn and XTR cryptosystems*. University of Waterloo: Combinatorics and Optimizing Research Report CORR 2003–34.
51. Shirase, M., Han, D. G., Hibino, Y., Kim, H. W., & Takagi, a T. (2008). A more compact representation of XTR cryptosystem. *IEICE Trans. Fundamentals, E91-A*, 2843–2850.

52. Shirase, M., Han, D. G., Hibino, Y., Kim, H. W., & Takagi, T. (2007). Compressed XTR. In J. Katz & M. Yung (Eds.), *Applied Cryptography and Network Security - ACNS 2007.LNCS 4521, (pp. 420–431)*. Berlin: Springer.
53. Han, D., & Lim, J. S. (2004). On the security of XTR public key cryptosystem against side channel attacks. In H. Wang, J. Pieprzyk, & V. Varadharajan (Eds.), *Information Security and Privacy ACISP 2004.LNCS 3108, (pp. 454–465)*. Berlin, Heidelberg: Springer.

Chapter 4
HECC (Hyperelliptic Curve Cryptography)

Taspia Salam and Md. Sharif Hossen

Abstract Cybersecurity plays a very important role in our daily lives to protect our data against unauthorized access. Cryptography as the core functionality can provide key intelligence services such as protection, authentication, key setup, and integrity of data. We seek higher safety, efficiency, faster implementation, and low power consumption in the field of cryptography technology. There are already many cryptographic algorithms available that can meet these standards. Nevertheless, the new technological systems are smaller in size and have very limited capacity for processing and storage. The key size of RSA is somewhat large, and it takes much longer to handle and consumes a lot of power to implement RSA in small devices. Therefore, in practice, cryptosystems are favored which use smaller key sizes, including many of which depend on a discrete logarithm issue throughout a cumulative elliptical curve unit indicated in finite fields. For this, the concept of cryptography with the elliptic curve came. It is centered over the abelian group of curve points. The benefit of using the cryptography process of elliptic curve is that its key size is smaller than RSA. On the other hand, the complexity of the elliptic curve in mathematics is more involved than that of systems based at RSA. Then, the idea of a discrete logarithm problem with hyperelliptical curves and a cryptosystem was constructed over the Jacobian. Smaller base field size enables hyper-elliptical curves a better choice for lightweight cryptosystems. In this chapter, we have demonstrated the details of the cryptography of the hyper-elliptic curve and its level of security to protect data.

Keywords Cryptography · Elliptic curve cryptography · Jacobian · Hyperelliptic curve cryptography · Mumford representation

---

T. Salam · Md. S. Hossen (✉)
Dept. of Information and Communication Technology, Comilla University, Comilla, Bangladesh

© Springer Nature Switzerland AG 2021
K. A. B. Ahmad et al. (eds.), *Functional Encryption*, EAI/Springer Innovations in Communication and Computing, https://doi.org/10.1007/978-3-030-60890-3_4

1   Introduction

While historians split human history into periods before and after Jesus, a similar distinction can be made for cryptographic history, i.e., before and after the key exchange of the Diffie-Hellman scheme. Humanity has long been grappling with the central issue of distribution. Cryptography is the intellectual pursuit of creating a crypto-system for cybersecurity. It deals with the individual protection of digital data. This refers to the development of computer algorithm-based systems that provide basic safety facilities for information. The symmetric (or secret key) cryptography dominated over the people at that time, so before any verification, face-to-face gathering or use of a trusted courier was required. Both were also impractical and dangerous at all times. In 1976, a new way of transmitting cryptographic keys was implemented, and the era of asymmetric cryptography (or public-key cryptography) was shared. Their mechanism is focused on one-way trapdoor functions and allows the user to share a private key over an unsafe network. Throughout the dual-key cryptography, the individual requires just a few keys, one is termed the public key, and the other is called the private key. Hence, the private key is always numerically connected to the public. Then the public key must be issued, whereas the private key appears hidden, suggesting a trapdoor. The most common cryptosystem for the public key is RSA. The RSA cryptosystem is still widely used although many years have passed since its first implementation. Undoubtedly, in our modern information environment, the most significant instrument is public-key cryptography. Outdated cryptosystems such as the listed RSA depends on issues such as factoring or discrete logarithm based on hard number theory. But due to the possible quantum computer attacks, there is a need for additional well-organized cryptosystems for minor strategies; creating new public-key cryptosystems, particularly those that could withstand upcoming outbreaks using quantum computers, is a great challenge. Therefore, for the past three decades, the question of developing a new public-key cryptosystem has dominated the cryptographic research fields [1–3].

Elliptic curve cryptography (ECC) is a technique to encode data files so that they can be decoded by specific individuals, and it is formulated by Neal Koblitz and Victor Miller in 1985. ECC is dependent on elliptic curve mathematics and utilizes point position to protect data on the elliptic curve. It uses fairly short keys for encryption and decryption. This key is quicker, requiring less energy in computation. But its major drawback is that it raises the size of encrypted messages over RSA encryption. Besides, its algorithm is far more complicated and much more difficult to manage than RSA that arises implementation errors. Thus, it reduces the security of algorithms [3–6]. Neal Koblitz and Victor Miller offered the usage of elliptical structures that construct public-key cryptographic frameworks. After that time, a wealth of literature has been conducted mostly along with the safety and active employment of elliptic curve cryptography. But against digital threats, we need to incorporate more protection. The concept of the hyper-elliptic curve has been developed to provide further security. This is used to fix passwords, factorization of the integer, and encryption of the public key. Along these lines, the foremost objectives of this chapter are to clarify the fundamental factors of the divisor group, hyperelliptic encryption and decryption algorithm, signature generation algorithm, and HECC security [7, 8].

1.1 Basic Mathematical Terminologies

1.1.1 Arithmetical Closure

A field B is assumed to be arithmetically closed if B has a root in B for each non-constant polynomial with coefficients. The arithmetical closure is an arithmetic expansion of a field B signified as $\overline{B}$ which is arithmetically closed.

1.1.2 The Cryptography of Hyperelliptic Curve

Suppose, i is a region, $\overline{i}$ is the arithmetical closure of i, and A is a hyperelliptic curve. This hyperelliptic curve is on genus g, and it is over i ($g \geq 1$) with a set of all points (X, Y). This curve is denoted by a formula of the subsequent procedure:

$$A : Y^2 + H(X)Y = f(X) \text{ where } i\,[X, Y] \tag{4.1}$$

Here, $H\,(X) \in i\,[X]$ is also a polynomial of degree g. Here, a polynomial of degree $2\,g + 1$ is $f\,(X)$ which concurrently gratify $Y^2 + H\,(X)\,Y = f\,(X)$. The fractional derivative of the equation is $2\,Y + H\,(X) = 0$ and $H'\,(X)\,f'\,(X) = 0$. A single point over A is an explanation $(X, Y) \in i^2$ which concurrently gratifies all these circumstances. Therefore, there are no singular points in the hyperelliptic curve.

Hyperelliptic curvature is used in many important research fields such as generators of pseudo-random numbers, coding theory, number theory algorithms, and cryptography. We can use hyperelliptic curvatures rather than elliptic bends for modeling cryptographic strategies. Expanded types of elliptic curves are known as hyperelliptic curves.

1.1.3 Finite Field

Suppose, F_q is a finite field with q elements. Here, q is an integer. It satisfies $q = p^n$, where p and n are the prime number (characteristic of the field) and positive integer (dimension or the extension degree), respectively. The prime number is well-defined as the smallest probable integer such that $\underbrace{1 + \cdots + 1}_{p} = 0$. Here, 1 can be in any field, and the summation is a closed procedure. We can say that $1, 1 + 1, 1 + 1 + 1$, $1 + 1 + 1 + 1, 1 + 1 + 1 + 1 + 1$, and so on are components of the field. There are two prospects for this categorization of components, such as either some addition of 1's will ascend to 0. Here the series cycles via a few finite set of values. In different circumstances, no components of the series are similar, and we can have an infinite quantity of components in the field. The characteristic of the field is the smallest positive number of 1's with sum $= 0$. Without this sum, we can say that the field has characteristic zero.

1.1.4 Interpretation

Let us look at the curve,

$A = Y^2 + (X^2 + X) Y - X^5 - X^4 - X^2 - 1 = 0$ throughout the finite field F_2^5.

In this case, $g = 2$, $H(x) = X^2 + X$ and $f(x) = X^5 + X^4 + X^2 + 1$.

So, the partial derivatives of the curve A are as follows:

$$\frac{\partial A}{\partial x} = (2X + 1) Y - 5X^4 - 4X^3 - 2X = 0 \text{ and } \frac{\partial A}{\partial y} = 2Y + X^2 + X = 0$$

As there is no point B in the field that fulfills the expression of the curve A and therefore $\frac{\partial A}{\partial x}$ and $\frac{\partial A}{\partial y}$ are not equivalent to zero in any B, then A is simply a hyperelliptic curve [9, 10].

1.2 Divisors

Suppose, F_q is a list of points of the curve A. Thus, the divisor of A will be used as an alternative. Aimed at real-world operation, it is also significant to present an appropriate symbol of the group fundamentals.

1.2.1 Explanation 1 (Divisor, Degree, Order at a Point)

According to the above explanation, a divisor DIV can be determined by.

$$DIV = \sum_{B \in A} m_B B, \text{ here } m_B \in Z \tag{4.2}$$

In this case, m_B contains non–zero value. The deg DIV is signified as the degree of DIV. The order of DIV can be determined by $ord_B(DIV) = m_B$.

1.2.2   Explanation 2 (GCD of Divisors)

Suppose, the two divisors are $DIV_1 = \sum_{B \in A} m_B B$ and $DIV_2 = \sum_{B \in A} n_B B$. Here, a finite number of m_B and n_B are non-zero. There is a point at infinity. That is, GCD is a divisor of degree 0. According to the explanation, the greatest common divisor (GCD) of these two divisors is given through.

$$GCD\,(DIV_1, DIV_2) = \sum_{B \in A} \min\,(m_B, n_B)\,B - \sum_{B \in A} \min\,(m_B, n_B)\,\infty$$

(4.3)

It is informal to understand that $GCD(DIV_1, DIV_2) \in DIV^0$. DIV^0 is the collection of divisors that has degree 0.

1.2.3   Explanation 3 (Rational Function Divisor)

Let B is a field, R is a rational function, A is a hyperelliptic curve, and $B\,(A)$ is the function field of the curve is represented by the field of the segment $\overline{B}(A)$. Likewise, the function field $\overline{B}(A)$ over $\overline{B}$ is specified by the field of segments of $\overline{B}(A)$. The components of $\overline{B}(A)$ are termed as rational functions on A.

Let, $R \in \overline{B}\,(A)^*$. Here, R is the rational function, B is a field, and $\overline{B}$ is an algebraic closure. So, the divisor of the rational function R can be determined by

$$DIV\,(R) = \sum_{B \in A} (ord_B\,R)\,B$$

(4.4)

1.2.4   Explanation 4 (Primary Divisor)

Divisor $DIV \in DIV^0$ is referred to as primary divisor if $DIV = DIV\,(R)$ for any logical function $R \in \overline{B}\,(A)^*$. A subgroup of DIV^0 is the set of all primary divisors, signified as B.

1.2.5   Explanation 5 (Divisor of Jacobian)

The curve A for Jacobian (Jac) is specified as the quotient unit [11, 12].

$$Jac = \frac{DIV^0}{B}$$

(4.5)

When $DIV_1, DIV_2 \in DIV^0$, then we can say $DIV_1 \sim DIV_2$ if $DIV_1 - DIV_2 \in B$. So, we can say that Divisors DIV_1 and DIV_2 are equivalent.

1.3 The Jacobian of Hyperelliptic Curve

Under this group, a divisor DIV can be determined by the following accumulation rule:

$$\sum_{B \in A} m_B \, B + \sum_{B \in A} n_B \, B = \sum_{B \in A} (m_B + n_B) \, B \qquad (4.6)$$

Here, DIV^0 is a list of all degree 0 divisors. It becomes a subset of DIV. Likewise, the addition of key divisors looks like multiplication of their corresponding rational functions in $A \, (\overline{Fq})$. Here, A is the hyperelliptic curve and $\overline{Fq}$ provides us with the closure property we need to conclude the set of principal divisors, denoted B, is indeed a subgroup of DIV^0.

Each period we have a subcategory of materials similar to principal divisors rooted inside a more robust group. It is an exposed appeal to take the quotient. At this point, there is no exception. We can identify the hyperelliptic curve of Jacobian as $Jac = DIV^0/B$. And it is a category of degree 0 divisors with B as identity [13–14].

1.3.1 A Jacobian Instruction

The Jacobian's cardinality is the key factor that establishes the safety of a hyperelliptic crypto algorithm. Therefore, to choose a good curve and a good underlying finite field for cryptographic reasons, it is necessary to know the orders of the Jacobian.

1.3.2 Theory of Hasse-Weil

Suppose, Jac is the Jacobian of hyperelliptic curve A of genus g described above F_b. In this case, the order of Jacobian $Jac \, (F_b^n)$, denoted as $\#Jac \, (F_b^n)$, is confined through

$$\left(\sqrt{b^n} - 1\right)^{2g} \leq \#\text{Jac} \, (F_b^n) \leq \left(\sqrt{b^n} + 1\right)^{2g} \qquad (4.7)$$

Henceforth, $\#\text{Jac} \, (F_b^n) \approx b^{ng}$.

1.3.3 Specification of Zeta Function

Suppose, a hyperelliptic curve described over F_b is A, and $m_n = \#A \, (F_b^n)$, where $n \geq 1$. It is the rational number of points F_b^n on A. The power sequence of the zeta function is as follows:

$$zeta_A(T) = EXP\left(\sum_{n\geq 1} m_n \frac{T^n}{n}\right) \tag{4.8}$$

The preceding procedure outlines the methodology used to calculate the Jacobian ordering in the context of the hyperelliptic curve of genus 2. This methodology is focused upon the zeta function principles.

1.3.4 Methodology for Ordering Genus 2 of the Jacobian Hyperelliptic Curve

Input: Rational points m_1 and m_2, a field F_b^n.
  Output: Jacobian order is Jac_n

1. Place $c_1 \leftarrow m_1 - 1 - b$; $c_2 \leftarrow \left(m_2 - 1 - b^2 + c_1^2\right)/2$.
2. Discover η_1, η_2 by approaching the quadratic formula $x^2 + c_1(x) + (c_2 - 2b) = 0$.
3. Figure out $x^2 - \eta_1(x) + b = 0$ to get a result β_1.
4. Figure out $x^2 - \eta_2(x) + b = 0$ to get a result β_2.
5. Calculate $Jac_n = \left|1 - \beta_1^n\right|^2 \left|1 - \beta_2^n\right|^2$.
6. Go to $Jac_n.$.

1.4 Semi-Reduced and Reduced Divisors

We can identify a semi-reduced divisor by the following formula:

$$DIV = \sum m_i B_i - \left(\sum m_i\right)\infty \tag{4.9}$$

In this case, every $m_i \geq 0$ and B_i are fixed points of the curve A. It is in the sense that, if $B_i \in DIV$ then and there $\tilde{B}_i \notin DIV$, except $B_i = \tilde{B}_i$, in the instance of $m_i = 1$. Representation of fundamentals of the Jacobian by semi-reduced divisors, i.e., by couples of polynomials, undergoes from a big difficulty—the depiction is not sole. This trouble is detached if we contemplate semi-reduced divisors of this superior kind. We can signify the semi-reduced divisor as reduced divisor by the following formula [15]:

$$\sum m_i \leq g, \text{ for genus } g \tag{4.10}$$

1.4.1   The Accession of Reduced Divisors

The inclusion of reduced divisors describes a collective principle of hyperelliptic curves on Jacobians. Thereby Jacobian and the inclusion activity fulfills the concepts of the unit, namely associativity, identity, and inverse.

So, we slightly specify the generalized Euclidean method through polynomials because it is necessary for the move of composition in the Jacobian divisor inclusion algorithm.

Subsequently, $GCD(c_1, c_2, \ldots, c_n) = GCD(c_1, GCD(c_2, \ldots, c_n))$; however, we limit our vision to the issue where there are double polynomials $c, e \in I[x]$, and $DEG_x c \geq DEG_x e$ are provided. We would like to deal with polynomials m, n, $o \in I[x]$ where $m = GCD(c, e) = nc + oe$.

1.4.2   Methodology for Extension Euclidean Polynomial Methods

Input: $c(x)$, $e(x)$ are two non-zero polynomials where $DEG_x c \geq DEG_x e$.

Output: $m(x)$, $n(x)$, $o(x)$ are the polynomials where $M = GCD(c, e) = nc + oe$.

1. Place $n_2 \leftarrow 1$; $n_1 \leftarrow 1$; $o_2 \leftarrow 0$; $o_1 \leftarrow 1$.
2. Whereas $e \neq 0$ do.
3. Using rest section to determine $c = qe + r$.
4. $n \leftarrow n_2 - qn_1$; $o \leftarrow o_2 - qo_1$
5. $c \leftarrow e$; $e \leftarrow r$; $n_2 \leftarrow n_1$; $n_1 \leftarrow n$; $o_2 \leftarrow o_1$; $o_1 \leftarrow o$.
6. End while.
7. $M \leftarrow c$; $n \leftarrow n_2$; $o \leftarrow o_2$.
8. Return $m(x)$, $n(x)$, $o(x)$.

Suppose, $Rd_1 = DIV(c_1, e_1)$ and $Rd_2 = DIV(c_2, e_2)$ are two reduced divisors. A reduced divisor calculation $Rd_1 + Rd_2$ continues in dual stages. Initial stage (termed stage of formation) is defined as the algorithm of inclusion during which a semi-reduced divisor $Sd(c, e)$ in the source of $Rd_1 + Rd_2$ is calculated. The reduction methodology is being used in the second (mitigation) stage, to convert acquired semi-reduced divisor $Sd(c, e)$ into such an appropriate reduced divisor.

1.4.3   Methodology for Accession

Input: $Rd_1 = DIV(c_1, e_1)$ and $Rd_2 = DIV(c_2, e_2)$ are the reduced divisors; A is the hyperelliptic curve where $A = Y^2 + H(X)Y - f(X) = 0$.

Output: $Sd = (c, e)$ is the semi-reduced divisor where $Sd \sim Rd_1 + Rd_2$.

1. Calculate the greatest common divisor G_d,

$$G_d = GCD(c_1, c_2, e_1 + e_2 + h) = n_1 c_1 + n_2 c_2 + n_3(e_1 + e_2 + h).$$

2. Place $c \leftarrow \frac{c_1 c_2}{G_d^2}$ and $e \leftarrow \dfrac{n_1 c_1 e_2 + n_2 c_2 e_1 + n_3 \left(e_1 e_2 + f \right)}{G_d} \mod c$.

3. Return (c, e).

1.5 Semi-Reduced Sums Via Mumford Arithmetic

Suppose, a semi-reduced divisor is $DIV = \sum m_i B_i$ where $B_i = (x_i, y_i)$. The Mumford identification of the divisor is a couple of polynomials of $(s\ (x), q\ (x))$ that entirely decides DIV. In this case, $s(x)$ captures all the coordinates of x with multiplicities, and $y = q(x)$ is the outburst function through all the B_i.

Legitimately,

$$s(x) = \prod_{i=1}^{r} (x - x_i)^{m_i} \tag{4.11}$$

$$\left(\frac{d}{dx} \right)^j \left[q(x)^2 + q(x)\, h(x) - f(x) \right]_{x=x_i} = 0; \ (0 \le j \le m_i - 1) \tag{4.12}$$

Highlights of the above conditions

1. $s(x_i) = 0$, $q(x_i) = y_i$ through multiplicity m_i where $1 \le i \le r$.
2. $s(x)$ is monic and differences $q(x)^2 + h\ (x)\ q\ (x) - f\ (x)$.
3. DIV completely controls $s(x)$ and $q(x)\ mod\ s(x)$.
4. $s(x)$, $q(x) \in \bar{I}[x]$ with $s(x)$ monic. It splits $q(x)^2 + h\ (x)\ q(x) - f\ (x)$ and decides the semi-reduced divisor.

Again, suppose, $DIV_1 = (s_1, q_1)$, $DIV_2 = (s_2, q_2)$.

Targeted at any $[B]$ happening in DIV_1, $[\overline{B}]$ does not occur in DIV_2 and vice versa. Formerly $DIV_1 + DIV_2 = (s, q)$ is semi-reduced and.

$$s = s_1 s_2, q = \begin{cases} q_1 \ (\text{mod } s_1) \\ q_2 \ (\text{mod } s_2) \end{cases} \tag{4.13}$$

Undertake $B = (x_0, y_0)$ occurs in DIV_1 and $\overline{B}$ occurs in DIV_2. Then, $s_1(x_0) = s_2(x_0) = 0$ and $q_1(x_0) = y_0 = -q_2(x_0) - h(x_0)$, so $x-x_0$ divides $s_1(x), s_2(x), q_1(x) + q_2(x) + h\ (x)$.

$$DIV = GCD\ (s_1, s_2, q_1 + q_2 + h) = p_1 s_1 + p_2 s_2 + p_3\ (q_1 + q_2 + h) \tag{4.14}$$

$$s = s_1 s_2 / DIV^2 \tag{4.15}$$

$$q \equiv 1/DIV \ (p_1 s_1 q_2 + p_2 s_2 q_1 + p_3 \ (q_1 q_2 + f)) \ (\text{mod } s) \tag{4.16}$$

1.6   Reduction Via Mumford Arithmetic

Suppose, a semi-reduced divisor is $DIV = (s, q)$ on $A : y^2 + h(x)y = f(x)$.
The pseudocode of this calculation is as follows [16, 17]:
While $DEG \ (s) > g$
Then, do the followings:

1. Shift the x coordinates of the points in DIV by those of the order intersections points of A with q:

$$s \leftarrow \frac{f - qh - q^2}{s}.$$

2. Shift the new points by their reverses:

$$q \leftarrow (-q - h) \, (\text{mod } s)$$

2   Hyperelliptic Curves over Finite Fields

Explanation (a): Suppose, a field is I, the arithmetic closure is $\overline{I}$ and a hyperelliptic curve is A. It is on genus $g \geq 1$ with a set of all points (X, Y). This curvature can be identified by equation (4.1) which is as follows:

$$A : Y^2 + H \ (X) \ Y = f(X) \text{ where } I \ [X, Y]$$

Here, $H(X) \in I[X]$, and it is of degree $2g + 1$. There is no $(X, Y) \in \overline{I} \times \overline{I}$.
Sample 1: Suppose, A is a hyperelliptic curve over I.

1. For $H \ (X) = 0$, $char(I) \neq 2$.
2. If $char(I) \neq 2$, then an alteration of parameters $X \rightarrow X, Y \rightarrow (Y - H(X)/2)$ alters A to $Y^2 = f(X)$ where $DEG_X f(X) = 2g + 1$.
3. Suppose, a hyperelliptic curve of $H(X) = 0$ and $char(I) \neq 2$. The curve A will be hyperelliptic if and only if $f(X)$ does not have frequent roots in $\overline{I}$.

2.1   Proof of Hyperelliptic Curve

1. Suppose, $H(X) = 0$ and $char(I) = 2$, $f'(X) = 0$ is the partial derivative of the equation. Here, $DEG_X f'(X) = 2 \, g$, the root of this equation is $m \in I$, and the root

of $Y^2 = f(X)$ is $n \in \overline{I}$. After that we can see that the singular point on the curve is (m, n).

2. If the variables on the equation of a hyperelliptic curve are changed, then

$$(Y - H(X)/2)^2 + H(X)(Y-H(X)/2) = f(X),$$

It truncates that $Y^2 = f(X) + (H(X))^2/4$; so $DEG_X \left(f + \frac{H^{2T}}{4} \right) = 2g + 1$.

3. We have seen that (m, n) is a singular point on the curve. It should gratify $n^2 = f(m), 2n = 0$, and $f'(m) = 0$.

Henceforth, the term $n = 0$ and the repetitive root of $f(X)$ is m.

Explanation (b): Suppose, an augmentation of field I is L, which is the assortment of L rational points on curve A. That means $A(L)$ is the assortment of all points $B = (x, y) \in L \times L$. It fulfills equation (4.1) alongside exceptional point at∞. In this way, the assortment of points $A\left(L\right)$ can be distinguished by A and these points other than ∞ are known as finite points.

Sample 2: Two cases of hyperelliptic curves over the real number field are shown in the following figures. The genus of these curves is $g = 2$ and $H(x) = 0$.

1. The graph on the actual plane is $A1 : y^2 = x^5 + x^4 + 4x^3 + 4x^2 + 3x + 3 = (x + 1)$ $(x^2 + 1)(x^2 + 3)$ which is shown in Fig. 4.1.
2. The graph on the actual plane is $A2 : y^2 = x^5 - 5x^3 + 4x = x(x - 1)(x + 1)(x - 2)$ $(x + 2)$ which is shown in Fig. 4.2.

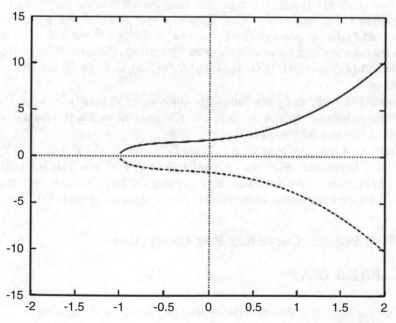

Fig. 4.1 Graph of A1 : $y^2 = x^5 + x^4 + 4x^3 + 4x^2 + 3x + 3 = (x + 1)(x^2 + 1)(x^2 + 3)$

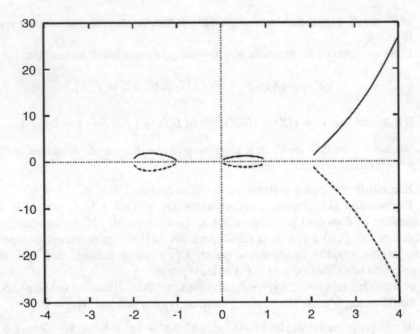

Fig. 4.2 Graph of $A2 : y^2 = x^5 - 5x^3 + 4x = x(x - 1)(x + 1)(x - 2)(x + 2)$

Explanation (c): Suppose, $B = (x, y)$ is a finite point on the curve A. The reverse of this point is $\tilde{B} = (x, -y-h(x))$. $\tilde{B}$ is on curve A. $\tilde{\infty} = \infty$ is the reverse of ∞. If $B = \tilde{\infty}\tilde{\infty}$, then this point is exceptional; otherwise, it is an ordinary point.

Sample 3: Suppose, a curve $A : y^2 + xy = x^5 + 5x^4 + 6x^2 + x + 3$ is over the finite field F_7. At this point, $H(x) = x, f(x) = x^5 + 5x^4 + 6x^2 + x + 3$, $g = 2$, and A does not have any singular point without ∞. Henceforth, the curve is hyperelliptic.

Here, $A (F_7) = \{\infty, (1, 1), (1, 5), (2, 2), (2, 3), (5, 3), (5, 6), (6, 4)\}$ are the rational points on A.

Sample 4. Suppose, F_2^5 is a finite field. Here, $F_2^5 = F_2[u] / (u^5 + u^2 + 1)$. The primitive polynomial is $u^5 + u^2 + 1$, and β is the root of this polynomial. Here, Table 4.1 contains the powers of β.

Suppose, A is a curve of genus $g = 2$ over the region F_2^5. $A : y^2 + (x^2 + x)y = x^5 + x^3 + 1$. In this case, $H(x) = x^2 + x$ and $f(x) = x^5 + x^3 + 1$. This is established that other than ∞, curve A does not have any singular point, and it is hyperelliptic. $A (F_2^5)$ are the finite points, so the rational points of the curve are [16–18]:

3 Hyperelliptic Curve Key Pair Generation

3.1 Divisor Order

The divisor order D is distinct to be the least probable integer r such that

Table 4.1 The powers of β for the above field $F_2^5 = F_2[u]/(u^5 + u^2 + 1)$

N	β^N	N	β^N	N	β^N
0	1	11	$\beta^2 + \beta + 1$	22	$\beta^4 + \beta^2 + 1$
1	β	12	$\beta^3 + \beta^2 + \beta$	23	$\beta^3 + \beta^2 + \beta + 1$
2	β^2	13	$\beta^4 + \beta^3 + \beta^2$	24	$\beta^4 + \beta^3 + \beta^2 + \beta$
3	β^3	14	$\beta^4 + \beta^3 + \beta^2 + 1$	25	$\beta^4 + \beta^3 + 1$
4	β^4	15	$\beta^4 + \beta^3 + \beta^2 + \beta + 1$	26	$\beta^4 + \beta^2 + \beta + 1$
5	$\beta^2 + 1$	16	$\beta^4 + \beta^3 + \beta + 1$	27	$\beta^3 + \beta + 1$
6	$\beta^3 + \beta$	17	$\beta^4 + \beta + 1$	28	$\beta^4 + \beta^2 + \beta$
7	$\beta^4 + \beta^2$	18	$\beta + 1$	29	$\beta^3 + 1$
8	$\beta^3 + \beta^2 + 1$	19	$\beta^2 + \beta$	30	$\beta^4 + \beta$
9	$\beta^4 + \beta^3 + \beta$	20	$\beta^3 + \beta^2$	31	1
10	$\beta^4 + 1$	21	$\beta^4 + \beta^3$		

$(0,1)$	$(1,1)$	(β^5, β^{15})	(β^5, β^{27})	(β^7, β^4)	(β^7, β^{25})
(β^9, β^{27})	(β^9, β^{30})	(β^{10}, β^{23})	(β^{10}, β^{30})	(β^{14}, β^8)	(β^{14}, β^{19})
$(\beta^{15}, 0)$	(β^{15}, β^8)	(β^{18}, β^{23})	(β^{18}, β^{29})	(β^{19}, β^2)	(β^{19}, β^{28})
(β^{20}, β^{15})	(β^{20}, β^{29})	$(\beta^{23}, 0)$	(β^{23}, β^4)	(β^{25}, β)	(β^{25}, β^{14})
$(\beta^{27}, 0)$	(β^{27}, β^2)	(β^{28}, β^7)	(β^{28}, β^{16})	$(\beta^{29}, 0)$	(β^{29}, β)
$(\beta^{30}, 0)$	(β^{30}, β^{16})				

$$\underbrace{D + \cdots + D}_{r} = \text{div}\,(1, 0) \tag{4.17}$$

Suppose, A is a hyperelliptic curve on $\mathbb{F}p$, the Jacobian of A is J well-defined on $\mathbb{F}p^n$, and D_1 is a reduced divisor of order r. The public domain parameters are prime p, field extension n, the equation of the hyperelliptic curve, and the reduced divisor D_1. k is a private key of the interval $[1, r-1]$, and it is an integer number. The public key is, $D_2 = kD_1$.

Procedure: The key pair generation of a hyperelliptic curve.

Input: The public domain parameters.

Output: The two keys, especially private (k) with the public (D_2) key.

1. Define an integer which is $k \in [1, r-1]$
2. Estimate $D_2 = kD_1$
3. Arrival at the key couple (D_2, k).

3.2 Principle of Hyperelliptic Curve Cryptographic Arrangement

In this case, we consider a divisor as M which is considered as a clear text m. As a clear text, it is encrypted by D_2, and here is a randomly selected integer such as

$a \in [1, r]$. Then the sender conveys the divisors $C_1 = aD_1$ and $C_2 = M + aD_2$ to the receiver and figures out the private key k [19].

$$kC_1 = k\,(aD_1) = a\,(kD_1) = kD_2 \tag{4.18}$$

Then, we recover

$$M = C_2 - aD_2 \tag{4.19}$$

4 Hyperelliptic Curve Encryption and Decryption

The encryption and decryption procedures of hyperelliptic curve are briefly discussed as follows [19]:

4.1 Procedure 1. Hyperelliptic Curve Encryption

Input: Domain parameters, public key D_2, m as a message.
  Output: The ciphertexts are C_1 and C_2

1. Symbolize m like a reduced divisor of the hyperelliptic curve's Jacobian J.
2. Define a as a random integer which is $a \in [1, r-1]$.
3. Estimate $C_1 = aD_1$ and $C_2 = M + aD_2$.
4. Arrival at the ciphertext(C_1, C_2).

4.2 Procedure 2. Hyperelliptic Curve Decryption

Input: Domain parameters, private key k, ciphertext(C_1, C_2).
  Output: Message m

1. Provide a private key to figure $M = C_2 - kC_1$.
2. Handover M into m.
3. Arrival at the clear text m.

Sample 5. Let us assume an example. Suppose, A is a hyperelliptic curve, where

$$A : Y^2 + \left(X^2 + X\right) Y - X^5 - X^4 - X^2 - 1 = 0$$

over a finite field F_2^5. Here, genus $g = 2$, $H(X) = X^2 + X$, and $F(X) = X^5 + X^4 + X^2 + 1$. Reduced divisor $D_1 = DIV\,(X - 21, 5)$, order $R = 482$, and another reduced divisor is a message m, where $m = DIV(X - 7, 3)$.

The way out is

1. From the start, we need to generate key pair:
   We have taken K as a private key where $K = 2$. It is selected uniformly as $K \in [1, R - 1]$.
   According to the algorithm, the public key D_2 can be determined by

$$D_2 = kD_1 = DIV\left(X^2 + 28, 2X + 10\right)$$

2. Next, we need to start the encryption process:
   A randomly selected integer, $A = 1, A \in [1, R - 1]$.

$$\text{Cipher text, } C_1 = A\ D_1 = D_1 = DIV\ (X{-}21, 5).$$

$$\text{Cipher text, } C_2 = M + A\ D_2 = DIV\left(X^2 + 13\ X + 4, 22\ X + 6\right).$$

3. At last, we need to perform decryption as follows:
   Message, $m = C_2 - KC_1$

$$\begin{aligned}
m &= C_2 - 2\ C_1 \\
&= DIV\left(X^2 + 13X + 4, 22X + 6\right) + 2\left(-DIV\left(X^2 + 28, 2X + 10\right)\right) \\
&= DIV\left(X^2 + 13X + 4, 22X + 6\right) + 2\left(DIV\left(X^2 + 28, 2X + 10\right)\right) \\
&= DIV\left(X + 7, 3\right)
\end{aligned}$$

5 HECC Signature Algorithm

5.1 Elgamal Signature Method

Elgamal is precisely aimed at the digital signature. This method is used to compute a discrete logarithm problem that helps in both encryption and digital signature. It is also a renowned safety scheme afterward RSA, and it is really good enough than RSA. This is a public key encryption method and has certain investigations with applications in digital signatures, electronic certification, and safety rules nowadays. But, the Elgamal signature process is combined with other encryption approaches which makes it structural. So, we can call it a hybrid encryption system that contains an encryption algorithm for encrypting messages, and for constructing signatures we need the Elgamal signature process. For the design issues, one can use the elliptic curves for encrypting messages, after that produce the signature by using the Elgamal signature process. At that moment we need to execute the digital signature on plain text m. At first, we can choose p as a large prime number and create a finite field for p, namely $GF(p)$. Through it, we can also achieve a primitive root as

$g\ (mod\ p)$. Now, in this field, we can randomly choose a private key x as an integer and figure out the public key, which is $y = g^x(\ mod\ p\ )$ [20].

The signature procedure is given as follows:

1. Arbitrarily we choose an integer k that is a prime by p in the field.
2. Evaluate $a = g^k(mod\ p)$.
3. Discover s and meet the equation $s = k^{-1}(m - xa)\ mod\ p$.

Now, we have the signature (a, s) from the sender. Next, we need to send the public parameters (g, y, p) and the signature to the receiver. After receiving the signature, it will be verified by the receiver by the following procedure:

1. Estimate $m = (xa + ks)\ mod\ p$.
2. Resolve $g^m = (\ y^a a^s\ )\ mod\ p$.

By getting the equation $g^m = (\ y^a a^s\ )\ mod\ p$, one can recognize that this signature (a, s) of the message is legal, so the signature can be acknowledged, or it is forbidden.

The Elgamal signature system is safe because we know about the public parameters(g, y, p) and $y = g^x(mod p)$. But this is difficult to explain x in the mode of the opposite method over the procedure $y = g^x(\ mod\ p\ )$.

5.2   HECC Signature Generation

To use the Elgamal method into an authentic device, the hybrid security mechanism is generally associated with other encryption methods. Throughout this method how HECC and Elgamal systems are combined and being used to create the HECC-Elgamal signature model: first, we use HECC to encrypt messages, and then produce a secure digital signature with Elgamal algorithm for ciphertext [20].

If we want to combine spontaneously HECC and Elgamal method, we would define a one-to-one mapping between the Jacobian HECC quotient group and finite field which are $J(Fq)$ and $GF(p)$, respectively. That is, $J(Fq) \overset{\theta(Z)}{\to} GF\ (p)$. It suggests that it can map divisor D of the Jacobian quotient group to only integer polynomial mixture of the finite field, and $\theta(Z)$ is the mapping function.

As a matter of first importance, embed plain text m to $y^2 + h(x)y = f(x)$ to acquire divisor $D_m \in J(Fq)$. From the point forward, pick a good random integer c as a private key in the field. At that point check $E = cD$, where the public key is E.

Here the signature levels are as pursue:

1. In the finite field $GF\ (p)$, define an integer c randomly, and it is prime of p.
2. Estimate $a = k\ D$.
3. Plot this mapping function $\theta\ (Z)$ with the estimated value which is a $\overset{\theta\ (Z)}{\to}$ a'.
   It is used to estimate a' $= \theta(a)\ mod\ p$.

$$s = k^{-1}\left(m - ca'\right) \bmod p.$$

The sender generates the signature (a, s) of the message. The recipient acknowledges the following materials: the message m, public key E, the Jacobian HECC quotient group, divisor D, finite field, and finally the mapping function θ (Z). And, according to defined criteria, the recipient may authenticate the accuracy and reliability of a signature.

(i) Figure $a' = \theta(a) \bmod p$ with a and mapping function $\theta(Z)$.

(ii) $\left. sa = \begin{matrix} ak^{-1}(m-c\ a')\bmod(p-1) \\ a=kD \end{matrix} \right\}$

$\Rightarrow sa = kDk^{-1}\left(m - ca'\right) \bmod p$

$\Rightarrow as = D\left(m - ca'\right) \bmod p$

(iii) $\left. sa = \begin{matrix} (Dm-Dca')\bmod p \\ E=Dc \end{matrix} \right\}$

$\Rightarrow as + a'E = \left(a'E + Dm - Dca'\right) \bmod p$

$\Rightarrow sa + a'E = Dm$

(iv) Then, we really may assert the signature's accuracy and reliability in deciding about the equation $sa + a'E = Dm$ is being gratified.

5.3 HECC Signature Verification

Digital signatures including certification are designed to validate the signature of both parts in the CA to verify the authenticity of origins of messaging and incorporate signature authentication.

The method of signing summaries is as pursues:

1. A plain text named m_a is inscribed by the user a;
2. The signature unit produces $Sig\ (m_a)$ as a signature which is then guided to the CA.
3. Once the CA has checked the results, $CA\ (Sig\ (m_a))$ will be produced for sending it to the user b.
4. User b subsequently accepted the$CA\ (Sig\ (m_a))$, this will confirm the signature named $Sig\ (m_a)$.

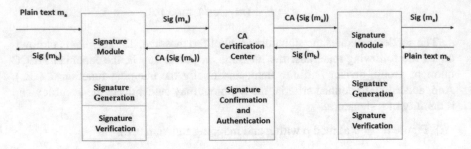

Fig. 4.3 Form and Confirmation Procedure of Signature

A signature named *Sig* (m_a) that is acquired from plain text m_a and incorporates the information of that plain text. Authentication has been managed to add to *Sig* (m_a) to get there at *CA* (*Sig* (m_a)) until the signature was sent to *CA* to verify. *CA* (*Sig* (m_a)) is an authorized signature which is from the user *a*.

The *CA* functions as just a trustee for both parties, confirming that user *a* or *b* has signed the signature. That does not need to know user *a*'s and user *b*'s contact material so far as the source is secure.

The signature *CA* (*Sig* (m_a)) is sent by the *CA* to user *b* which will unpack the signature to test its exactness. We know from the discrete logarithm problem that users can produce the *Sig* (m_a), and it cannot be retrieved by other people. This matter confirms the signature confirmation of the user on the message m_a, but this verification is not for m_a. For this reason, the content of m_a cannot be denied by the user named *a* [20].

The signature formation and validation process are demonstrated in Fig. 4.3.

6   Security of HECC

A digital signature is a part of the encryption of identification. This ensures the validity and safety of the transaction. The HECC-Elgamal integrates the HECC mechanism with the Elgamal process whose privacy arises from three parts [20]:

1. The message encryption method of the curve is exponential in the large prime finite field.
2. This model has a high computational complexity in the large prime field. For this reason, this signature security procedure is also very good.
3. Through offering encryption, defense of both parties' interests.

Depending on these defenses the hackers cannot replicate the signature. Thus the signature has no chance to be imitated. It allows for the individuality, validity, and non-repudiation of signatures.

7 Conclusion

The hyperelliptic curve cryptosystem is one of the rising cryptographic schemes of the most recent years. With the exceptional increment in the quantity and utilization of handheld gadgets, this lightweight cryptographic scheme has gone to the spotlight. This chapter suggests that to maintain safety, this curve needs a lower finite field as opposed to an elliptic curve. It is broadly acknowledged that for most cryptographic applications dependent on a hyperelliptic curve, one needs a gathering request of size at any proportion of 2^{160}. In this way, HECC over the finite field will require in any event $g.\log_2 q \approx 2^{160}$. In specific, for genus 2 hyperelliptic curves, we will require a field F_q with $|F_q| \approx 2^{80}$. Around the network, security has been a top concern in the productivity of the network operations as the threats arise. The digital signature is also a crucial factor for authentication on the Internet. In this chapter, the hyperelliptic curve's cryptosystem is acknowledged as the best characteristic of a password scheme. Organically, the two methods were merged to develop an HECC-Elgamal-based digital signature scheme. It was implemented to validate the bipolar signature in identity authentication and also evaluated the security efficiency of this strategy. This digital signature scheme inherits two systems' security technology to achieve a high-security index. This ensures credibility, verification, and authentication of identity.

References

1. Hossen, M. S. (2020). Data preprocess. In *Machine learning and big data* (pp. 71–103). John Wiley & Sons, chapter 4. https://doi.org/10.1002/9781119654834.ch4.
2. Hossen, M. S. (2017). A Java Based GUI Application for Substitution Encryption Techniques. In *Proc. International Conference on Computer, Communication, Chemical, Material and Electronic Engineering (IC4ME2–2017)*. e: University of Rajshahi.
3. Vijayakumar, P., et al. (2014). Comparative study of Hyperelliptic curve cryptosystem over prime field and its survey. *International Journal of Hybrid Information Technology, 7*, 137–146.
4. WankhedeBarsgade, M. T., & Meshram, S. A. (2014). Comparative study of elliptic and hyper-elliptic curve cryptography in the discrete logarithmic problem. *IOSR Journal of Mathematics, 10*(2), 61–63.
5. Amalraj, A. J., & Jose, J. J. R. (2016). A survey paper on cryptography techniques. *International Journal of Computer Science and Mobile Computing, 5*(8), 55–59.
6. Kahate, A. (2008). *Cryptography and network security*. Tata Mc-Graw-Hill Companies.
7. Enge, A. (1999). *Elliptic curves and their applications to cryptography* (pp. 1–165). Germany: Springer Science and Business Media, LLC, University Augsburg.
8. Blake, I., Seroussi, G., & Smart, N. (2005). Advances in Elliptic Curve Cryptography. In *London Mathematical Society Lecture Note Series*. Cambridge University Press. https://doi.org/10.1017/CBO9780511546570.
9. Cohen, H., & Frey, G. (2006). *Handbook of elliptic and HyperElliptic curve cryptography* (pp. 1–848). Chapman and Hall/CRC, Taylor and Francis Group.
10. Hailiza, K., & Jie, L. K. (2012). Elliptic curve cryptography and point counting algorithms. In *Cryptography and Security in Computing* (pp. 91–116). In Tech. Ch. 5.

11. Alimoradi, R. (2016). A study of Hyperelliptic curves in cryptography. *International Journal of Computer Network and Information Security, 8*(8), 67–72. https://doi.org/10.5815/ijcnis.2016.08.08.
12. Galbraith, S. (2012). *Hyperelliptic curves*. Cambridge University Press.
13. Van Wamelen, P. B. (2006). Computing with the analytic Jacobian of a genus 2 curve. In W. Bosma & J. Cannon (Eds.), *Discovering mathematics with magma. Algorithms and computation in mathematics* (Vol. 19). Berlin, Heidelberg: Springer. https://doi.org/10.1007/978-3-540-37634-7_5.
14. Costello, C., & Lauter, K. (2011). Group law computations on Jacobians of hyperelliptic curves. In *International Workshop on Selected Areas in Cryptography* (pp. 92–117). Berlin, Heidelberg: Springer.
15. Galbraith, S. D., Harrison, M., & Mireles Morales, D. J. (2008). Efficient Hyperelliptic arithmetic using balanced representation for divisors. In A. J. van der Poorten & A. Stein (Eds.), *Algorithmic number theory. ANTS 2008. Lecture notes in computer science* (Vol. 5011). Berlin, Heidelberg: Springer. https://doi.org/10.1007/978-3-540-79456-1_23.
16. Ruíz Duarte, E. (2017). Reduced Mumford divisors of a genus 2 curve through its Jacobian function field. In *IACR Cryptology, ePrint; no. Report 2017/006*.
17. Avanzi, R., Jacobson, M. J., Jr., & Scheidler, R. (2010). Efficient reduction of large divisors on hyperelliptic curves. *Adv. in Math. of Comm., 4*(2), 261–279.
18. Kammerer, J. G., Lercier, R., & Renault, G. (2010). Encoding points on hyperelliptic curves over finite fields in deterministic polynomial time. In *Proc. international conference on pairing-based cryptography* (pp. 278–297). Berlin, Heidelberg: Springer.
19. Jacobson, M. J., Menezes, J., & Stein, A. (2004). Hyperelliptic curves and cryptography. In *High Primes and Misdemeanors: Lectures in Honour of the 60th Birthday of Hugh Cowie Williams. Fields Institute Communications, vol. 41* (pp. 255–282). American Mathematical Society.
20. Wei, L. F. (2013). Design of hyperelliptic curve system digital signature in identity authentication. In *Fifth International Conference on Digital Image Processing (ICDIP 2013), International Society for Optics and Photonics, vol. 8878* (p. 88780X).

Chapter 5
Pairing-Based Cryptography

Ansh Riyal, Geetansh Kumar, and Deepak Kumar Sharma

Abstract Following the patterns of the modern world, it is justifiable to say that Data is one of the most valuable assets today. This change in perspective has resulted in a usefulness and popularity boost to previously neglected fields like Information security and cryptography. Cryptography, i.e. the protection of Data and messages by converting them into a senseless/unreadable format, is an age-old concept. From the Roman times where it was used for conveying covert battle plans between generals in the army, to a much later time, when it was used for sending secret messages in wars between nations, to now, when it is used to protect every strand of data in a variety of uses from social messaging and networking sites to bank accounts for the privacy of users and national secrets. Over the years, cryptography has been modified countless times and yet, each form it has taken has had the sole purpose of being nearly impossible to crack, i.e. decrypt without knowing the secret keys.

Out of the many methods/algorithms used for Encryption, each one has unique implementations, strengths and weaknesses. Pairing-based cryptography is one of the best methods known to us. It takes advantage of the Diffie–Hellman approach to make cracking the code difficult, and at the same time, it keeps computation fast. It is based on the pairing of elements from two cryptographic groups (a set based on/enveloping a binary operation which connects every two elements of the group to a third). The Diffie–Hellman Key Exchange works on the assumption that there are no secure channels, i.e. third parties (Hackers for instance) have access to every encrypted message being communicated. There are many procedures used for making groups and rings involved in the generation of our cryptographic groups like the (modified) Weil pairing, the Tate-Lichtenbaum Pairing, Eta pairing and Ate

A. Riyal · G. Kumar
Department of Computer Engineering, Netaji Subhas University of Technology, (Formerly Netaji Subhas Institute of Technology), New Delhi, India
e-mail: anshr.co18@nsut.ac.in; geetanshk.co18@nsut.ac.in

D. K. Sharma (✉)
Department of Information Technology, Netaji Subhas University of Technology, (Formerly Netaji Subhas Institute of Technology), New Delhi, India

© Springer Nature Switzerland AG 2021
K. A. B. Ahmad et al. (eds.), *Functional Encryption*, EAI/Springer Innovations in Communication and Computing, https://doi.org/10.1007/978-3-030-60890-3_5

pairing. The directions provided by the method implemented result in different sub-problems and advantages which result in different security levels of our encryption technique. The combination of these pros, cons and uniqueness acts as different methodologies for the implementation of pairing-based cryptography. Although modifications to algorithms and inventions to new approaches keep being explored every day, the backbone of a vast majority of these implementations, however, has the same concept.

This book chapter gives an introduction to pairing-based cryptography, the associated mathematical concepts, definitions and procedures and associated algorithms used for implementation. Since the main motive behind cryptography is to aid in the field of Information Security, the fulcrum of issues faced/areas of judgement for all encryption techniques to be implemented is the un-crackability/strength of the algorithm used; the reverse-engineering methods for these algorithms will also be discussed. Furthermore, there are many implementation techniques being discovered everyday which when combined with existing algorithms have scope for improvement in the future. Some of which are also mentioned.

Keywords Cryptography · Encryption · Decryption · Pairing-based cryptography · Groups (mathematical) · Rings (mathematical) · Diffie–Hellman key exchange

1 Introduction

Following the patterns of the modern world, it is justifiable to say that Data is one of the most valuable assets today. In the world of Data and data flow control, information security is one of the top priorities. This change in perspective regarding the importance of data has resulted in a usefulness and popularity boost to previously neglected fields like cryptography. Cryptography in a nutshell is the protection of information by conversion into an unreadable/seemingly senseless format having the property of being readably visible only to the intended recipient. This was originally achieved by the use of a secret key mechanism: The two parties communicating first establish a common secret agreed upon key, and then that key is used to apply a function on the message. The message then gets converted to unreadable formats which are sent by person A and received by person B who then uses the key to reverse the encryption with the use of the inverse function. Since then, secret key cryptography has come a long way, finding its implementation in every data security system throughout the world ranging from locking files while zipping to national security systems with usage access available only to the persons with highest clearance.

Secret key cryptography however has a very big issue in its very core. There needs to be an exchange of the secret key between the sender and the recipient before messages can be sent via encryption. This means that the security of the whole communication mechanism is dependent on the safe and error-free transmission of that secret key. The solution to this risk would be to have a lock and key mechanism, i.e. a set of keys which make sense only when used in combination with one another. The encryption key would encrypt the data like a lock, and the decryption key would decrypt the encrypted message like a key. It stands to reason that with the existence of such a mechanism, the private key is the fulcrum of safety. Thus for the sake of convenience, public key can be open for everyone, i.e. anyone who wants to send a message can use the encryption key that is available freely. The corresponding secret key will however only be available to the recipient. This is the exact mechanism used in public key cryptography.

Even though the idea of public key cryptography is a big hit, it is neither easy nor perfect. The biggest concern with public key cryptography was the generation of a public–private key pair which can ensure a safe data transfer pathway. The implementations of these concepts are rooted deeply into group/discrete mathematics under the concept of pairings. Implementation of different varieties are based on the various pairings defined in the set-up of our PBC-based system. These approaches are described later on in the chapter. Before that, the explanations of the discrete mathematics concepts used are very important.

2 Mathematical Terms and Concepts Used

The field of cryptography, and by extension Information Security, has deeply buried roots in the field of mathematics. The procedure of converting secret messages into unreadable format and their subsequent unauthorised decryption (unauthorised: without the access to the secret key) has been analysed and evaluated by the use of mathematical tools. This promoted the counter usage of mathematics in the overcoming of corresponding shortcomings to form an adversarial set-up. As the current scenario stands, every encryption technique in use either stems from or is heavily dependent on mathematical functionalities, understanding which lays the groundwork for understanding said cryptographic concepts and implementations. The real meaning of these terms however is not very clearly understood when placed out of context and thus is advised to be referenced when going through the main text encounters the said mathematical term.

2.1 Random Oracle

An RO is a conceptually ideal black box that acts as a randomised key-value generator which generates a uniformly distributed random output for every query that it takes as input. Two different queries never end up with the same output, and using the same query multiple times results in the same output. The implementation of this idealisation has not been a 100% possible, but there are a multitude of hashing mechanisms which offer a big realisation of the theory.

2.2 Symmetric Encryption

Symmetric encryption (SE) is the usage of a single key for all subparts of the procedure of cryptography of messages. The key, often called secret key, is transferred over safe channels between the communicating authorities along with the knowledge of its usage. The key is self-reversing and can be thus applied in a similar manner by both the sender and the receiver. Some consider the relying on safe transfer of a secret key to be the biggest flaw of symmetric encryption, but the implementation is simpler and faster than its modern counterpart and is thus preferred when the data to be transferred is very huge, especially if the safety of initial key exchange can be ensured.

2.3 Asymmetric Encryption

Asymmetric encryption (ASE) is the replacement for single secret key cryptography. ASE relies on a pair of keys, which divide the process of cryptography according to the two focal points of encryption and decryption. ASE comes in contrast to the previously prevalent SE and is considered to be a near ideal solution to the drawbacks presented by its previously used counterpart. Asymmetric key encryption is a widely explored field with multitudes of implementations available offering a wide variety of safety to speed ratios.

2.4 Public Key Encryption

Public key encryption is the implementation of asymmetric encryption which uses the existence of a pair of keys to make the encryption part of cryptography very convenient by making one key accessible to everyone while not sharing its counterpart to anyone. It is based on ignoring the security of a channel as a countermeasure to the increase in hacking skills observed.

2.5 Subexponential Algorithm

A subexponential algorithm is a comparative term used to denote any computational algorithm which has a time complexity that grows slower than b^x (base: b(>1) and no. of bits: x).

2.6 ECDLP (EC Discrete Logarithm Problem)

EC discrete logarithm problem says that given two points P and Q belonging to an EC E defined over a finite field F_q, we have to find an integer a (if it exists) such that Q = aP.

ECDLP is the bedrock of ECC and pairing-based cryptography. There are always new techniques being developed with the goal to bring the algorithm to solve the ECDLP to a subexponential algorithm, but no definitive technique has been developed yet. This is the reason ECC is considered state of the art.

2.7 Key Escrow

Splitting components of a single key into parts were viewed with scepticism due to the poor transmission environments and corresponding dangers to loss of information. To solve this, a database of key pairs is stored within a trusted central authority so that the keys can be accessed by an authorised third party in special circumstances. This database is called Key escrow.

2.8 Turing Machine

Turing machine is an abstract machine that models a computer algorithm. It is a hypothetical black box that generates/manipulates an infinite length tape on the basis of a set of rules. In terms of mathematics and cryptography, a turing machine can model/stimulate any computer algorithm, and a turing machine can be constructed for any algorithm [1].

2.9 Elliptic Curves

Elliptic curves (abbreviation: EC) by definition are smooth curves of degree 3 defined over a Finite Field F offering parameters a,b,c,d,e,f,g,h,i,j to satisfy the 3 degree EC.

$$f(x, y) = ax^3 + bx^2y + cxy^2 + dy^3 + ex^2 + fxy + gy^2 + hx + iy + j$$
(5.1)

The set of points that come under an EC satisfy the above equation defined over a fixed Finite Field F.

Apart from the generalised equation, there is a set of equations of ECs that are used in cryptography, known as Weierstrass Equation:

$$y^2 = x^3 + Ax + B \tag{5.2}$$

where A and B are the defining constants for our curve.

2.10   Jacobian of Hyper ECs

In the case of hyper ECs, there is no direct way of providing a group to the curve defined over F_q with embedding k. Instead, we introduce different objects related to the curve which when paired with each field extension k of F_q associates a group called the Jacobian of hyper ECs (Fig. 5.1).

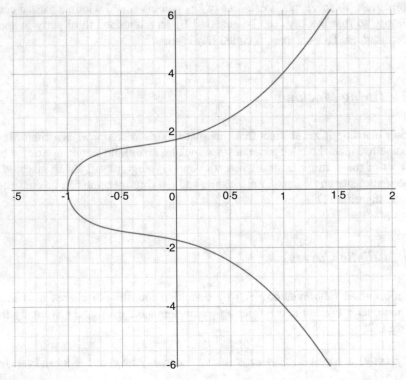

Fig. 5.1 Hyper EC $y^2 = x^5 + x^4 + 4x^3 + 4x^2 + 3x + 3$ (genus = 2)

2.11   Group [2]

Groups are symmetric algebraic structures which are basically sets combined with/equipped with binary operations (applied on the group and also having the final results in those groups) in a way such that the four conditions (called group axioms) are satisfied, i.e. the set must be closed, associative, has identity and corresponding inverses for every element.

Based on their utility, groups are selected in such a way that they perform specific operations of utility only, and those operations sometimes give the group-associated terminologies called notations. For example in ECs, we use some groups to perform addition on 2 points on the curve, thus getting the terminology of additive notations.

2.12   Field [3, 4]

A field is a fundamental algebraic structure having addition, subtraction, multiplication and division defined and acting in the same manner as with rational/real numbers.

Apart from these basic requirements, fields also must be closed, associative, having an identity under addition and multiplication and have corresponding inverse elements. Apart from this, multiplication must also be distributable over addition. These properties are often automatically satisfied when using classic sets (real number sets, rational number sets, etc.) but have to be defined for unique sets for mathematically defined theorems to be applied over them.

2.13   Finite Field

A finite field can be explained as a set of positive numbers within which each calculation must fall. In general, this acts as a range of numbers with any outliers wrapped around to make them ultimately fall within the range [5].

Finite fields have a set of pre-definitions/properties which have been proved through various higher-order mathematical methods.

Theoretically for every finite field, a number t exists having $\Sigma\, t$ equals 0, i.e.

$$1 + 1 + 1 \ldots .1 = 0.$$

There are many such numbers within a finite field, and the first number which follows this property t must belong to set of prime numbers P'' and is defined as characteristic of the field. The modulus(mod) to a number operation used as the fulcrum to modular mathematics is also an example of a finite field. Any operation

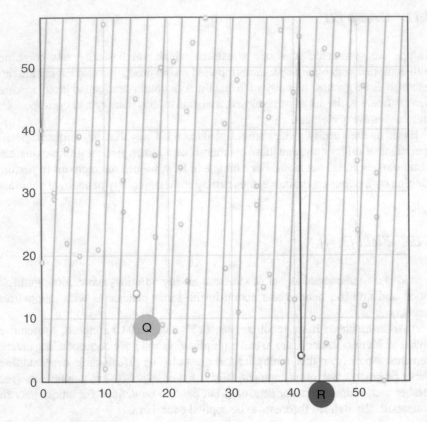

Fig. 5.2 Finite field graph of EC over F59 (mod 60)

on the members of the set results in a result which is then modded w.r.t. the number n. This finite field is represented with the $(n-1)th$ number in subscript, i.e. F_{n-1}

For example, on making the finite field as modulo 35, and all the results have to yield a result between 0 and 34. An operator of addition (+) on 20 and 30 gives 50 (mod 35) which results in 15, so according to modular mathematics, $20 + 30 = 15$ (mod 35) or F_{34}

In cryptography (especially PBC), we extensively use ECs, which are defined over (wrapped around and operate on) a finite field (Fig. 5.2).

2.14   Diffie–Hellman Key Exchange and the Problem

Diffie–Hellman algorithm is a key exchange algorithm which is used for securely exchanging keys over public communication channels [6–8]. The algorithm does not actually share keys but rather compute a secret key using which is identical for both the sender and the receiver. In this section first the algorithm will be explained and then the problems related to it with respect to pairing-based cryptography.

abbreviations: private key $\rightarrow$ PRK, public key $\rightarrow$ PBK , secret key $\rightarrow$ SK

It is an asymmetric algorithm which is used in symmetric key cryptography [9]. In this key exchange technique we have two known numbers publicly q and Pr; Pr acts as the primitive root of the prime number q, and q is a prime number. A primitive root of a prime number is a number whose power from 1 to p-1 modulo p generates all numbers from 1 to p-1. Suppose there are two users A and B. Now A randomly decides its PVK let us say Sa given that $Sa < q$, and then calculates its PBK let us say Pa where $Pa = (PR)^{Sa}$ mod q. Similarity B also independently decides its PVK Sb and calculates its PBK $Pb = (PR)^{Sb}$ mod q. Now each user makes its PBK available to other user and keeps the PVK with itself. Now both the users compute the secret key 'k' individually, for A, $k = (Pb)^{Sa}$ mod q and for B, $k = (Pa)^{Sb} \bmod q$. This SK comes to be identical for both the users, and now the key is exchanged by computation separately on both sides. The proof of why the SK 'k' come out to be identical is given as follows:

Notations:

A,B: Users

q: prime number; Pr: Primitive root of this prime number

Sa, Sb: PVKs of A and B, respectively

Pa, Pb: PBKs of A and B, respectively

$$k = (Pb)^{Sa} \ (\bmod \ q) \rightarrow \text{(calculated by A)}$$
$$= \left((PR)^{Sb} \ (\bmod \ q) \right)^{Sa} (\bmod \ q)$$

by using rules of modulo arithmetic:

$$k = \left((PR)^{Sb} \right)^{Sa} \ (\bmod \ q)$$
$$= (PR)^{Sb*Sa} \ (\bmod \ q)$$
$$= \left((PR)^{Sa} \right)^{Sb} \ (\bmod \ q)$$

using rules of modulo arithmetic

$$k = \left((Pr)^{Sa} \ (\bmod \ q) \right)^{Sb} \ (\bmod \ q)$$
$$k = (Pb)^{Sb} \bmod q \rightarrow \text{(which was calculated by B)}$$

The key exchange is based on DHP, the basic idea of DHP is that the mathematical calculations and operations are fast to compute, but it is very hard to perform the reverse of this process. Thus it is easy to perform encryption but very hard to reverse-engineer the computations. Since it is so hard to reverse-engineer the key exchange, it is hard for hackers to break the systems which contain the Diffie–Hellman problem. In this we assume that the problem cannot be solved efficiently. Diffie–Hellman is considered to be secure from hackers for if the finite cyclic group and the generator are chosen properly. If the group is large enough then it will be

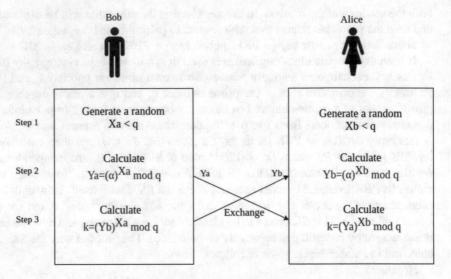

Fig. 5.3 Basic flow of Diffie–Hellman Key Exchange

really difficult to find generator to the power of both private keys multiplied. Hence with proper group size, Diffie–Hellman is secure and hard to break down (Fig. 5.3).

2.15   Miller's Algorithm

Miller's algorithm is used to map points of an EC to an element of a finite field. Miller's algorithm is basically a 'double and add'-based approach. Suppose we have two points 'a' and 'b', then miller's algorithm will calculate a value $f(a,b)$ for a particular element inside [10] the given finite field (where f is the function denoting the algorithm).

Below is the very basic loop of Miller's algorithm for better understanding [11] (here '[2]T is written as 2.T' for clarity):

Input: Two different P and Q on the EC E, i.e. $P(\in E[r])$! $= Q(\in E[r])$ *and a value s* $(\in N)$

$$\text{Output}: f_{s,p}(Q)$$

The algorithm: We take is equal to the summation of $s_j 2^j$ where j varies from 0 to L with s_j as 1 or 0 and s_j is equal to 1. Value of P is assigned to T and f = 1. Then we loop through j = L-1 to j = 0, following are the contents of the loop: $f = f^2$. $I_{T,T}(Q)/V_{[2]T}(Q)$, also T = 2.T. Check if s_j is equal to 1, then $f = f. I_{T,T}(Q)/V_{T \oplus P}$ and T = (T convolution with P). At the end of the function, we return the value of f.

One of the main properties of Miller's algorithm is bilinearity and hence is highly used in pairing-based cryptography.

2.16 Elliptic Curve Cryptography (ECC) over R

Extending from RSA using modular arithmetic, EC arithmetic has been used in multiple public key cryptosystems. Many cryptosystems which have been traditionally working with modular arithmetic, such as DSA and Diffie–Hellman, have an EC counterpart.

ECs are curves of the form:

$$y^2 = x^3 - ax + b \tag{5.3}$$

which is also called the 'short Weierstrass form', and is the general form to talk about ECs (Fig. 5.4).

The other form of EC which can be brought about in discussing is the 'Edward form':

$$x^2 + y^2 = 1 + dx^2y^2 \tag{5.4}$$

This is used to 'sign' data in a manner that third parties can authenticate the signature of the signer since the private key can only be created by the signer.

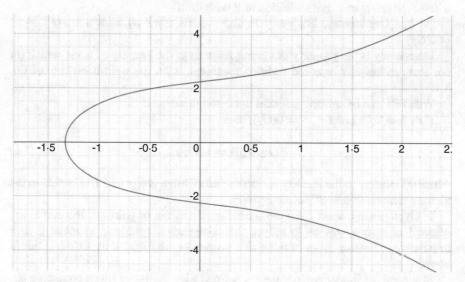

Fig. 5.4 EC $y^2 = x^3 + 2x + 5$

ECs have several useful properties which makes it usable for bitcoin: these can be studied by plotting the Weierstrass form on the graph.

Two important properties of ECDSA are used to generate repetitive addition and multiplication, those properties being point addition and point doubling.

3   Pairings and Different Pairing Methods

Pairing-based cryptography, as we know, is supposed to act on two keys: one public and one private. When these two come together, we are able to map the combination to the lock.

The mathematical equivalent of this notion is the usage of two additive notation groups: G_1 and G_2 and one multiplicative notation group G_T all three of prime order p.

A pairing E is defined as a map between $G_1 \times G_2 \to G_T$ i.e. $E(P1, P2)$ $P1 \in G1$; $P2 \in G2$ if it satisfies the following properties:

(a) Bilinearity: Bilinearity is the distributive property variant of groups, especially with ECs:

$$e(P + Q, R) = e(P, R) \times e(Q, R) \ and \ e(P, Q + R) = e(P, Q) \times e(P, R)$$

By extension, bilinearity also means that for two constants a and b,

$$e([a] P_1, [b] P_2) = e(P_1, P_2)^{ab} \tag{5.5}$$

where $[x]P$ refers to x times addition of P with itself

(b) Non-Degeneracy: For every P_1 there exists some p_2 such that $e(P_1, P_2)$! $= 1(G_T)$

Degeneracy is the property of getting reduced to the identity of a set, which (in our case) makes any repeated addition meaningless on the right-hand side of Eq. 3.1.

With PBC, an EC mapping is said to be non-degenerate if:
If P_1 ! $= 0(G_1)$ *and* P_2 ! $= 0(G_2)$, then

$$e(P_2, P_2)! = 1(G_T) \tag{5.6}$$

where 0(G) refers to the identity element of an additive group and 1(G) refers to the identity element of a multiplicative group.

In cryptography, apart from the formal requirements of pairings, the curves are selected in such a way as to reduce the computation time and to make it near impossible to inverse. Both of these side properties are fulfilled by ECs (explained above).

In general, a bilinear environment is defined with the use of seven components:

$$(p, G_1, G_2, G_T, P_1, P_2, E)$$

3.1 Weil Pairing

Weil Pairing was defined by Simone Adolphine Weil as a debut effort into the world of pairings, and even though it no longer has any practical relevance, Weil pairing and Tate pairings serve as the basis for understanding, constructing, modifying and improving other pairings.

Theorem: let there be an EC called E being defined over a finite field K, let r be an integer (greater than 1 and prime to characteristics of field K) and let P and Q be two points of r-torsion on E.

$$\text{Then, } e_{w,r} = (-1)^r x \left(f_{r,p}(Q) \right) / \left(F_{r,q}(P) \right) \tag{5.7}$$

is well defined *for P ! = Q and P, Q ! = O_E*.

The points of discontinuity can be handled to cover the domain E[r] x E[r] by selecting the bilinear environment in such a way that $e_{w,r}(P, O_E) = e_{w,r}(O_E, Q) = e_{w,r}(P, P) = 1$

This pairing obtained by extending $e_{w,r}$ over the domain to get $E[r] \times E[r] \rightarrow u_r$ is called Weil Pairing.

Weil pairing is alternate, i.e.

$$e_{w,r}(P, Q) = \left(e_{w,r}(Q, P) \right)^{-1} \tag{5.8}$$

Since Weil pairing is defined over any field K (characteristic prime to r), $E[r] \times E[r] \rightarrow u_r$, u_r is a subset of ~K. Since K is a finite field, for cryptography, we assume $K = F_q$ (for some prime number q). Now, ~$K = F_q k$ is defined for the minimum value of integer k such that u_r is a subset of $F_q k$ and k is called the embedding degree of the EC.

From the above relation, we can also say that $q^k = 1 \ (mod \ r)$.

3.2 Tate-Lichtenbaum Pairing

The Tate-Lichtenbaum pairing was introduced by Tate by his duality pairings and was later extended by Lichtenbaum.

Like Weil pairing, Tate-Lichtenbaum pairing (or simply Tate pairing) was introduced into the world of cryptography as a debut mechanism for the introduction of duality-based pairings and holds very little practical usage. The concept of duality pairings introduced by Tate was however adopted and improved in successive attempts at pairings [12–16].

Theorem: let there be E (an EC), r (a prime number dividing the order of $E(F_q)$), P (a point of r torsion defined over F_qk ($P \in E[r](F_qk)$)) and Q (a point of the EC defined over F_qk ($Q \in E(F_qk)$)).

A point R is selected from $E(F_qk)$ such that neither R nor Q + R is equal to P or O_e, Then,

$$e_{T,r}(P, Q) = \left(f_{r,P}(Q+R) / f_{r,P}(R) \right)^{((q^k-1)/r)} \tag{5.9}$$

is well defined and independent of R and the pairing:

$E[r](F_qk) * \frac{E(F_qk)}{rE(F_qk)} \rightarrow u_r$ (or $(P, Q) \rightarrow E_{T,r}(P, Q)$) is called Tate pairing.

For a cryptographic system, the value of r is sufficiently high (as r is a large prime number) so that there are no points of r^2 - torsion in $E(F_qk)$ as embedding k (the smallest possible) is found to be much smaller than required to make the EC defined over finite field F_qk has r^2 torsion points.

Thus such a Tate pairing restricted over r-torsion points is also non-degenerate.

Since there is only one restriction on the selection of R, for practical purposes, we can select $R = O_E$ and thus the pairing gets reduced to $e_{T,r}(P, Q) = (f_{r,P}(Q)^{((q^k-1)/r)}$, then we can apply fast exponential algorithm to do the final exponential.

3.3 Hyperelliptic Tate-Lichtenbaum Pairing [12, 17]

In practical application, with the freedom of selecting R in Tate pairing combined with efficient exponential algorithms, it was observed that Tate pairing takes half the computational time as compared to Weil pairing, so for an efficient implementation, modifications to the Tate pairing were made.

Hyperelliptic curves are generally represented in the form $Y^2 + M(x)y = N(x)$.

Here, M(x) and N(x) are polynomials with their coefficients in field F_q with degree of M(x) < = genus of hyperelliptic curve and degree of $N(x) = 2 \, x \, genus + 1$

Using the Jacobian of hyperelliptical curves in place of direct groups as in the case of ECs was one such suggestion originally given by Koblitz and subsequently combined with existing pairings.

Hyperelliptic Tate-Lichtenbaum pairing is a class of very new pairings being defined over a large experimental set of hyperelliptic curves like $y^2 = x^p - x + d$ (in characteristic p) to give rise to a class of hyperelliptic curves which show better encryption and decryption speeds and more cumbersome addition operations in a constrained atmosphere due to the small genus size making hyperelliptic curves a very viable option for researchers to explore.

3.4   Eta Pairing

In Tate pairing, the length of the Miller's loop determines the speed of the pairing taking place. So in Eta pairing that is exactly what we are trying to reduce, i.e. the length of Miller's loop.

In Duursma and Lee [18], the author introduced some findings and techniques which laid the ground stone for the development of Eta Pairing [19]. In other words, we can say that Eta pairing combines or generalises the outcome of Duursma and Lee for Supersingular curves. So in this section we will talk about the findings of Duursma–Lee and on the basics of Eta pairing. Duursma–Lee techniques provided considerable enhancement to the computation for the curves of type

$$y^2 = x^p - x + d \tag{5.10}$$

applied on $f_q k$ where q is taken as greater than or equal to 3 (q > =3) and gcd of k and $2q = 1$. Using the value $p^{mp} + 1$, with hamming weight = 2 (base q), instead of r is one of the critical features of the Duursma–Lee technique. The final exponent is raised to the power $\frac{q^{2kq}-1}{q^{qk}+1} = q^{kq} - 1$, which if we say in words is just performing a division and calculating a Frobenius [11]. This simplifies the central part of the Miller's algorithm but by sacrificing the number of iterations. One of the major benefits of Duursma–Lee was they showed that the iterations can be lowered to m iterations only from mp.

One of the ways to define Eta pairing given by Barreto et al. is the following:
Definition
Eta pairing can be defined as the relationship $\eta T(D, D) = fT, D(\psi(D))$, where T belongs to Z. The equation using points P,Q is as follows:

$$\eta T\,(P,\,Q) = \eta T\,((P) - (\infty),\,(Q) - (\infty)). \tag{5.11}$$

The main goal is to take values of T such that N is greater than T. This definition is basically a generalisation of the idea for shortening the loop.

If we summarise Duursma–Lee, there are a few major independent contributions we can list out; three of them are the following:

1. Introduced the use of degenerate divisors instead of simply general divisors;
2. Shortening of the loop for certain group orders;
3. Introducing Frobenius operations directly into the formulae.

Eta pairing is basically a generalisation of these which results in a simpler and unified approach.

3.5 Ate Pairing

Ate pairing is one of the fastest technique available to us today. It is called Ate pairing because it is very similar to Eta pairing, but the order of arguments is reversed as compared to Eta pairing and also if we notice Ate is exactly spelled backwards of Eta [20]. It is like Tate pairing as well but obviously a much more improved version and faster, so the 'T' is absent in Ate.

Ate pairing has similar aim as Eta pairing, i.e. to reduce the length of the loop and make the pairing faster. Ate pairing can be applied to ECs as well as hyperelliptic curves [21]. Thus, generalising Ate pairing for most of the ordinary curves. Ate pairing abbreviates the range of Miller's loop, and it is dependent on the value of Frobenius t modulo r (where the order is taken as r for the subgroup). There are certain pairing friendly ECs in which the value of t can be as low as $r^{1/\phi k}$ (where k is the embedding degree). The best forms of Ate pairing have the Miller's loop range as worst-case scenario to be $log2(r/\phi(k))$. We get the most efficient results when $(t - 1)\ mod\ r$ has a very low value.

If we generalise the optimal ate pairings, Miller's loop has the lower bound of $r^{1/\phi k}$ and ultimately helps in faster computation .The optimal Ate and twisted optimal Ate pairing are faster than or as fast as the Tate pairing in any conditions. Hence, Ate pairing is the fastest pairing we have as of now.

4 Drawbacks/Vulnerabilities of PBC

Security of modern cryptosystems is based on their mathematical implementations. A good cryptosystem is created in a way such that illegally breaking the encryption theoretically takes the same time as a serial exhaustive search of all possible combinations and thus is computationally expensive [22–25]. The time graph of a modern cryptosystem grows exponentially with length of the encryption keys. Thus reverse engineering of pairings is also based on their mathematical implementations.

Pairing-based cryptography emerged as a system which decreases the bit size needed while maintaining the security offered by ECs. This notion, however, is a little misleading. ECC is defined over the base finite field F_q and so its parameters have size $O(log(q))$ bits; however, PBC is defined over the extension of the field F_qK and thus its members have the parameters of bit order $O(klog(q))$.

This means that PBC security will depend directly on the intractable levels (finding the complete key pair given one-half and final transmitted message) either on the ECDLP of F_q or DLP of F_qK. This is the point of inconsistency as ensuring the immunity of Weil and Tate pairings needs the embedding degree k to be large (>20) but that takes away the small bit size computation time advantage by a large factor [26].

The way PBC works in implementation is by the use of a master key which generates the user's private keys. This generation is done with the help of a central trusted agency. These central agencies handle a large number of users and are practically very few in number. So, for any system, theoretically each user's private key is not truly exclusive and this fact can be exploited [26].

The biggest implementation of pairing-based cryptography is found in identity-based encryption which uses a public key (derived from the user identity). This combined with the pseudo-exclusivity of generated keys in a system can also be exploited as is the case with a subclass of side channel attacks [27].

The usage of pairing keys in cryptography often increases the collision probability and thus the birthday paradox gets a special boost in the mathematics of theory of probability. A branch of brute force attacks focusing on the collision of keys in a network called Birthday attacks are also statistically very dangerous to pairing-based cryptography systems [28–30].

Weil and Tate pairings have a specific weakness because of their simplistic structure. Because of the generation of additive (and multiplicative points) of point P belonging to Group, and the hinging of security on finding them, these systems can convert the discrete logarithm problem on ECs to discrete logarithm problem on finite fields where subexponential attacks are extensively researched and discovered. A class of attacks called the MOV attacks utilise this weakness by taking a point Q of the same order as P such that no n exists such that $Q = nP$ (i.e. linearly independent of P). By doing so, we can calculate $e(P, Q)$ and $e([x]P, Q)$ $(=e(P, Q)^x)$ since they are both mth roots of unity and members of the field. Since P and Q are linearly independent, the non-degeneracy of the Weil pairing dictates that e(P,Q) cannot be unity. Thus DLP on EC is reduced to DLP on finite fields [31–33].

Hyperelliptic Tate works on the Jacobian of Hyperelliptic curves instead of a curve defined over F_q and thus some of the computational requirements are different and curve-dependant in the case of hyperelliptic Tate and thus sometimes do not display the advantages of other pairing methods being implemented.

Eta and Ate have been designed to take care of the weaknesses of Weil and Tate pairing and thus do not have many specific weaknesses.

5 Security of Pairing-Based Cryptography

Security is the most important aspect of pairing-based cryptography since the main objective of cryptography is safety of information from the intruders. So, it becomes increasingly important that the technique we are using for encryption and decryption is also secure in itself. The safety of pairing-based cryptography techniques depends on the fact that the reverse engineering of the methods used is very difficult to figure out to crack the cryptosystem [34]. Since pairing-based techniques cannot be cracked, so, we focus on other kinds of attacks that can be used against these kinds of systems, and these attacks are called physical attacks.

Physical attacks include techniques like electromagnetic waves, power, clock pulses, timing, and using change in temperature. Physical attacks are further divided into two major parts called side channel analysis and fault attacks. These two attacks are further divided into more subsections [27]. In this section, we will take up these attacks one-by-one and give a brief description about how they are used to alter or break the cryptography system. Since we have already stressed enough on the importance of Miller's algorithm, so, most of the attacks are focused to disrupt the Miller's algorithm.

5.1   Fault Attacks

In this class of attacks, we mainly focus on disrupting the algorithm first and then retrieving the information. The basic idea of fault attack is to introduce some errors in the code which can eventually lead to leak of some sensitive information. Also the hackers can change the source code as well which can help them identify some of the key values of the cryptosystem. Fault attacks are performed with various techniques such as errors in clock pulse and electromagnetic pulses. We will give a brief overview of a few fault attacks here (Fig. 5.5).

5.1.1   Fault against Duursma–Lee

Since Duursma–Lee techniques hold high importance in eta and ate pairing, attacks against it have been developed. The fault attack on Duursma–Lee is done by altering the number of times the algorithm is executed. Some assumptions are in order to perform this attack; some of which are given below:

1. Pairing is done publicly.
2. Constant input parameters are taken.
3. Two computations are done.

Using the results of these two computations, the secret is calculated.

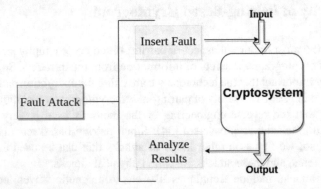

Fig. 5.5 Basic fault attack mechanism

5.1.2 Attacks against the Miller's Algorithm

As mentioned earlier, Miller's algorithm plays one of the most crucial parts in pairing-based cryptography and that is why many attacks have been proposed for it. There are two major ways in which fault attacks can be implemented against the Miller's algorithm. First is by trying to determine the no. of repetitions implemented in the Miller's loop, this can be achieved by keeping track of the timing and after some information can be retrieved. Second is somewhat similar to what we did against the Duursma–Lee techniques by performing two computations, one authentic and one with the fault, and then use the results of these to get the secret. In this method we do not alter the number of iterations; rather we inject fault in Miller's variables.

5.2 Side Channel Analysis

Second type of attack is the side channel analysis or the study of side channel information which is leaked at times. Side channel information can be defined as the information which is leaked due to certain physical characteristics of the cryptosystem or certain physical faults in the circuit (either design or some faulty component). Here we will discuss two types of side channel analysis.

5.2.1 Timing Attacks

Timing attack is one of the oldest known side channel analysis techniques. This is primarily based upon the idea that the circuit behaves differently and shows various patterns depending upon certain inputs (sensitive information). Due to the timing attack, we can tell which kind of operation is being executed.

5.2.2 Power Analysis

In power analysis, first we take two basic assumptions: first, that the attackers know the type of algorithms and techniques available, and second, public key cryptography is implemented. Now since we know what kind of algorithms are present, one can try to figure out patterns and variations by analysing power consumption, electromagnetic waves or execution traces of the machine.

One of the main points and a drawback of side channel analysis is that in order to implement it, we need to place the device on the circuit, transistor or any other component of the cryptosystem in order to get the information required for performing side channel analysis.

6 Functional Encryption and its Impact on Cryptography

Functional encryption (FE) is the conceptual generalisation of the concept of PBC and public key cryptography. FE's mathematical definition is very similar to that of public key cryptography, but in an abstracted manner, FE is the usage of some mathematical function to generate a pair of interrelated keys which are used to lock and unlock the encryption of mechanism of a cryptography system. One of the fulcrum points of the generation of these keys is the lack of interrelation between them. The pair of keys are generated and related in such a way that even though both combine to form the encryption mechanism and thus have mathematical cooperation, there is no interrelation through which one can be derived from the other. FE revolutionized encryption standards, offering ease of usage (by making the public key free for all), while increasing security (by eliminating the need to share the second part of the puzzle) [35].

Functional encryptions are defined with a 'functionality' at its centre which is basically a deterministic turing machine mapping a function between two sets (generally groups).

FE is still far from developed and has realisation problems ranging from constructing robust environments for integration with existing set-ups to constructing secure functional encryption schemes for all polynomial time predicates. These areas of research however are not all drawbacks; in reality, FE has a huge scope for improvement. For example, the fulcrum of functional encryption, i.e. the functionality which maps sets of points by a turing machine can be converted into a cleartext model (decision tree), but the challenge/scope for improvement is to make one in a way that reveals nothing about the real data. Other scopes for improvement include figuring out the correlation between functionalities so as to create black boxes to separate them and thus avoid the collision problems observed with functionalities [36].

With the introduction of functional encryption, we are no longer dependant on designing direct operations on transmission data, and introduction of division and subdivision with abstracted correlation among both sides of a mapping gives us a very amicable way of increasing computational complexity of unauthorised access while reducing the time taken for internal computation (by compartmentalisation).

7 Conclusion and Future Advancements

Pairing-based cryptography is one of the best implementations of functional encryption, and the advancement of it will only lead to betterment of functional encryption in general. In recent years, pairing-based cryptography has grabbed a lot of attention in the research field. Pairing-based cryptography because of its novel properties improved the cryptosystems significantly and made things possible which were really difficult to implement before the introduction of pairings. There are various types of pairing techniques available to us in the present day [37–40]; some of which have been mentioned in this chapter. With time and more research the cryptography community has developed better pairing techniques as well. Even though pairing-based cryptography is very effective and secure, it is still vulnerable to some kinds to physical attacks and techniques (such as fault attacks and side channel), but counter measures to some of these attacks are available as well, and it is hard to implement these kinds of attacks in a secure environment. Many major companies have adopted pairing-based cryptography, and its commercial importance is growing rapidly.

Although it is a very fast-growing field, there are still things that can be improved or introduced to make pairing-based cryptography better. Counter measures to more attacks should be devised to make the cryptosystems much secure since it should be taken care first in order to protect the information. Less expensive cryptosystems or circuits can be developed. With further research more drawbacks are introduced, but the advantages outweigh the disadvantages and hence with more research the future of pairing-based cryptography looks very bright.

References

1. https://en.wikipedia.org/wiki/Turing_machine#:~:text=A%20Turing%20machine%20is%20a, algorithm's%20logic%20can%20be%20constructed. Last visited 11 June 2020.
2. https://en.wikipedia.org/wiki/Group_(mathematics)#:~:text=In%20mathematics%2C%20a %20group%20is,%2C%20associativity%2C%20identity%20and%20invertibility.&text= Groups%20share%20a%20fundamental%20kinship%20with%20the%20notion%20of%20 symmetry. Last visited 11 June 2020.
3. https://en.wikipedia.org/wiki/Field_(mathematics) Last visited 11 June 2020.
4. https://study.com/academy/lesson/field-theory-definition-examples.html Last visited 11 June 2020.
5. https://blog.cloudflare.com/a-relatively-easy-to-understand-primer-on-elliptic-curve-cryptography/ Last visited 11 June 2020.
6. Azim, M. A., & Jamalipour, A. (2005). An efficient elliptic curve cryptography based authenticated key agreement protocol for wireless LAN security. In *IEEE International Conference on High Performance Switching and Routing.*
7. Wang, Y., Ramamurthy, B., & Zou, X. (2006). The performance of elliptic curve based group Diffie-Hellman protocols for secure group communication over ad hoc networks. In *IEEE International Conference on Communication.*
8. Rahman, M. M., & El-Khatib, K. (2010). Private key agreement and secure communication for heterogeneous sensor networks. *J. Parallel and Distributed Computing, 70,* 858–870.
9. https://www.ques10.com/p/7533/explain-diffie-hellman-key-exchange-algorithm-wi-1/ Last visited 11 June 2020.
10. https://crypto.stackexchange.com/questions/61930/simple-explanation-of-millers-algorithm Last visited 11 June 2020.
11. Vercauteren, F. (2010). Optimal Pairings. *IEEE Transactions on Information Theory, 56*(1), 455–461.
12. Duursma, I., & Lee, H. S. (2003). Tate pairing implementation for Hyperelliptic curves $y2 = xp - x + d$. In C. S. Laih (Ed.), *Advances in cryptology - ASIACRYPT 2003. ASIACRYPT 2003. Lecture notes in computer science* (Vol. 2894). Berlin, Heidelberg: Springer.
13. Juang, W. S., Chen, S. T., & Liaw, H. T. (2008). Robust and efficient password –authenticated key agreement using Smart cards. *IEEE Transactions on Industrial Electronics, 55*(6), 2551.

14. Yang, J. H., & Chang, C. C. (2009). An ID-based remote mutual authentication with key agreement scheme for mobile devices on elliptic curve cryptosystems. *J Computer Security, 28*, 138–143.
15. Yang, J. H., & Chang, C. C. (2009). An efficient three-party authenticated key exchange protocol using elliptic curve cryptography for mobile-commerce environments. *J Systems Software, 82*, 1497–1502.
16. Tzeng, S. F., & Hwang, M. S. (2004). Digital signatures with message recovery and its variants based on elliptic curve discrete logarithm problem. *J Computer Standards Interface, 26*, 61–71.
17. Wankhede-Barsgade, Meshram, & Suchitra. (2014). Comparative study of elliptic and hyper elliptic curve cryptography in discrete logarithmic problem. *IOSR Journal of Mathematics, 10*, 61–63. https://doi.org/10.9790/5728-10256163.
18. Barreto, P. S. L. M., Galbraith, S. D., hÉigeartaigh, C. Ó., & Scott, M. (2007). Efficient pairing computation on supersingular abelian varieties. *Designs, Codes and Cryptography, 42*(3), 239–271. https://doi.org/10.1007/s10623-006-9033-6.
19. Nanjo, Y., Khandaker, M. A. A., Kusaka, T., & Nogami, Y. (2018). Efficient pairing-based cryptography on raspberry Pi. *Journal of Communications, 13*(2), 88–93. https://doi.org/10.12720/jcm.13.2.88-93.
20. Zhao, C.-A., Zhang, F., & Huang, J. (2008). A note on the ate pairing. *International Journal of Information Security, 7*(6), 379–382. https://doi.org/10.1007/s10207-008-0054-1.
21. Hess, F., Smart, N. P., & Vercauteren, F. (2006). The eta pairing revisited. *IEEE Transactions on Information Theory, 52*(10), 4595–4602. https://doi.org/10.1109/tit.2006.881709.
22. Chen, T. S., Chung, Y. F., & Huang, G. S. (2003). Efficient proxy multisignature scheme based on the elliptic curve cryptosystem. *Computer & Society, 22*(6), 527–534.
23. Hwang, M. S., Tzeng, S. F., & Tsai, C. S. (2004). Generalization of proxy signature based on elliptic curves. *J. Computer Standards & Interface, 26*, 73–84.
24. Sun, X., & Xia, M. (2009). An improved proxy signature scheme based on elliptic curve cryptography. In *International Conference on Computer and Communications Security*. Los Alamitos: IEEE Computer Society.
25. Zuhua, S. (2004). Improvement of digital signatures with message recovery and its variants based on elliptic curve discrete logarithm problem. *J. Computer Standards & Interface, 27*, 61–69.
26. Cao, Z., & Liu, L. (2015). On the disadvantages of pairing-based cryptography. In *IACR Cryptology ePrint Archive* (p. 84).
27. El Mrabet, N., & Joye, M. (2017). Nadia. In *Guide to Pairing-Based Cryptography*. New York: Chapman and Hall/CRC. https://doi.org/10.1201/9781315370170.
28. https://thisismyclassnotes.blogspot.com/2017/07/cryptography-birthday-problem.html#:~:text=%C2%A7A%20birthday%20attack%20is,birthday%20problem%20in%20probability%20theory.&text=Such%20a%20result%20is%20called,find%20collisions%20of%20hash%20functions. Last visited 11 June 2020.
29. Chen, T. S. (2004). A specifiable verifier group-oriented threshold signature scheme based on the elliptic curve cryptosystem. *J Computer Standards Interface, 27*, 33–38.
30. Jianfen, P., Yajian, Z., Cong, W., & Yixian, Y. (2010). An application of modified optimal – type elliptic curve blind signature scheme to threshold signature. In *International Conference on Networking and Digital Society*. Los Alamitos: IEEE.
31. Chen, T. S., Huang, K. H., & Chung, Y. F. (2004). A practical authenticated encryption scheme based on the elliptic curve cryptosystems. *Computer Standards & Interface, 26*, 461–469.
32. Boneh, D., Goh, E., & Nissim, K. (2005). Evaluating 2-dnf formulas on ciphertexts. In J. Kilian (Ed.), *TCC 2005. LNCS, vol. 3378* (pp. 325–341). Heidelberg: Springer.
33. https://crypto.stanford.edu/pbc/notes/elliptic/movattack.html Last visited 11 June 2020.
34. Blomer, J., Gunther, P., & Liske, G. (2014). Tampering Attacks in Pairing-Based Cryptography. In *2014 Workshop on Fault Diagnosis and Tolerance in Cryptography*. https://doi.org/10.1109/fdtc.2014.10.

35. https://en.wikipedia.org/wiki/Functional_encryption#Formal_definition Last visited 11 June 2020.
36. Boneh, D., Sahai, A., & Waters, B. (2011). Functional encryption: Definitions and challenges. In *Proceedings of Theory Cryptogr* (pp. 253–273).
37. Boneh, D., & Franklin, M. (2001). Identity-Based Encryption from the Weil Pairing. In J. Kilian (Ed.), *CRYPTO'2001. LNCS, vol. 2139* (pp. 213–229). Heidelberg: Springer.
38. Hankerson, D., Menezes, A., & Vanstone, S. (2004). *Guide to elliptic curve cryptography*. Heidelberg: Springer.
39. Koblitz, N. (1987). Elliptic curve cryptosystems. *Mathematics of Computation, 48*(177), 203–209.
40. Liu, J., Yuen, T., & Zhou, J. (2011). Forward secure ring signature without random oracles. In S. Qian et al. (Eds.), *ICICS'2011. LNCS, vol.7043* (pp. 1–14). Heidelberg: Springer.

Chapter 6
NTRU Algorithm: N^{th} Degree Truncated Polynomial Ring Units

Afsar Kamal, Khaleel Ahmad, Rosilah Hassan, and Khujamatov Khalim

Abstract NTRU is a public key cryptosystem designed over a polynomial ring. It is based on the polynomial algebra. NTRU operations are based on addition, modular inverse, convolutional product, etc. The modular inverse plays an important role in generating the public/private keys. It provides low memory use and high speed compared to other cryptosystems. It is a lattice-based shortest vector problem. Its security is based on the product of polynomials and reducing the coefficients using two co-prime numbers p and q. Its smallest key size grants it better performance over other numerical based cryptosystems. It is the first asymmetric cryptosystem that is independent of the discrete algorithmic problem (ECC and Elgamal cryptosystem) or factorization (RSA cryptosystem).

Keywords NTRU cryptosystem · Lattice · Polynomial · Modular inverse · Convolutional multiplication · Low-Hamming-weight product

1 Introduction

The origin of the word "cryptography" comes from the Greek words "Kryptos and Graphein" which means hiding and writing, respectively. It was used for the first time by Spartans for sending secret messages to the military operations dated

A. Kamal (✉) · K. Ahmad
Department of Computer Science and Information Technology, Maulana Azad National Urdu University, Hyderabad, India
e-mail: kamal786lari@gmail.com; khaleelahmad@manuu.edu.in

R. Hassan
Centre for Cyber Security, Faculty of Information Science and Technology, Universiti Kebangsaan Malaysia (UKM), Bangi, Malaysia
e-mail: rosilah@ukm.edu.my

K. Khalim
Department of Data Communication Networks and Systems, Tashkent University of Information Technologies named after Muhammad al-Khwarizmi, Tashkent, Uzbekistan

© Springer Nature Switzerland AG 2021
K. A. B. Ahmad et al. (eds.), *Functional Encryption*, EAI/Springer Innovations in Communication and Computing, https://doi.org/10.1007/978-3-030-60890-3_6

back as 400 BC. The latest cryptography is based on the mathematics on computer science designed to make the hard computational process such that no adversary can break the secret communication. In this modern era, the quantum computer is able to break any cryptosystem that is based on integer factorization or discrete logarithm problem in polynomial time. It can affect the most commonly used asymmetric key cryptosystem such as RSA, DSA, and ECC. RSA is the most familiar which was first described in 1977. Cryptography algorithms have the purpose of altering in a way that just trusted users may access the keys to read transmitted data. Cryptography techniques are categorized into private and public as well [1–3].

The public key cryptosystem uses different keys for encryption and decryption processes. It is different from a private key cryptosystem that has only one key for encryption and decryption. In public key cryptosystems, the sender first encrypts his/her messages using the receiver's public key and then forwards it to the recipient. After receiving the message, the recipient decrypts the messages using his/her private key to extract the original message. This cryptosystem is very powerful, gives more flexibility, is very different from LUC cryptosystem which is based on Lucas functions [4], is a special form of second-order linear recurrence relation, and uses a public key for encryption and a private key for decryption [5], as well as a large public integer as a modulus [6]; RSA cryptosystem depends on the integer factorization problem [7], and Elgamal cryptosystem is based on discrete logarithm problem [7].

NTRU asymmetric key cryptosystem was developed in 1996 by three professors J. Hoffstein, J. Pipher, and J. H. Silverman from the mathematics department of Brown University in the United States [6]. It is a fully safe and secure cryptosystem against any known and unknown attacks. It can resist the attacks against the quantum calculation while ECC and RSA cannot. It is very fast and unaffected by quantum computers which represent the powerhouses of the computer world, and which outperform all the other classes and mainly silicon computers [8]. The institution of NTRU declared the speed of this algorithm to be 200 times faster than other public key cryptosystems and it is widely used for the safety of wireless communication. Additionally, the smaller key size O(N) for the message block of length N makes the system attractive and smart. It is named on the abbreviation for "Nth Degree Truncated Polynomial Ring" and is also called "Number Theory Research Unit." It requires only $O(N^2)$ time to encrypt/decrypt the message block of length N while RSA takes $O(N^3)$ time. It fulfills the IEEE standard as P1363 under the lattice-based cryptography (IEEE P1363.1). It is already accepted by many institutions such as the IEEE 802.15.3, IETF (Internet Engineering Task Force), and CEES (Consortium for Efficient Embedded Security) [9]. It is easy for coding and building into the hardware. It takes a little bit of memory for the software and fewer gates for hardware. It is easily embedded in the cellular phone, smart card with low power and low cost, radio-frequency identification device (RFID) and handheld devices, etc. Table 6.1 concludes the complexity of the NTRU algorithm.

Table 6.1 Complexity

Block of original message in bits	$N\log_2(p)$
Message expansion	Log_p q-to-1
Public key size in bits	$N\log_2(q)$
Block of encrypted message in bits	$N\log_2(q)$
Encryption speed	$O(N^2)$
Private key size	$2N\log_2(p)$
Decryption speed	$O(N^2)$

2 NTRU Cryptosystem

The NTRU cryptosystem is defined by the following symbols and mathematical notations which are used in encryption, decryption, and key generation process.

2.1 Symbols and Notations

Symbols	Descriptions
f_p	Polynomial inverse of f(y) modulo p
f_q	Polynomial inverse of f(y) modulo q
P_k	Public key
M_{sg}	Original message (polynomial set in ring R having coefficients between $-(p-1)/2$ and $(p-1)/2$
f(y)	Private key (polynomial set in ring R having coefficients 1 s > −1 s
g(y)	Generator polynomial set in ring R having equal coefficients 1 s = −1 s
r(y)	Blinding message (polynomial set in ring R having equal coefficients 1 s = −1 s
M_e	Encrypted message
M_d	Decrypted message
M_{d1}	Partially decrypted message 1
M_{d2}	Partially decrypted message 2

The whole operation in this cryptosystem is based on a truncated polynomial ring:

$R = Z[Y]/(Y^N\text{-}1)$.

The polynomial in this ring includes only integer coefficients having the highest degree N-1:

$$a = a_0 + a_1 Y + a_2 Y^2 + a_3 Y^3 + a_4 Y^4 + a_5 Y^5 + \cdots + a_{N-3} Y^{N-3}$$

$$+ a_{N-2} Y^{N-2} + a_{N-1} Y^{N-1}$$

$$= \sum_{i=0}^{N-1} a_i Y_i \ldots$$

(6.1)

The polynomial can be represented using the vector form as.

$$a = (a_0, a_1, a_2, a_3, a_4, a_5, \ldots, a_{N-3}, a_{N-2}, a_{N-1}) \tag{6.2}$$

It represents the polynomial f as

$$f = a_0 + a_1 y + a_2 y^2 + a_3 y^3 + a_4 y^4 + a_5 y^5, \ldots, a_{N-3} y^{N-3} + a_{N-2} y^{N-2} + a_{N-1} y^{N-1}$$

The summation (+) of the polynomials is same as the ordinary addition while the product of polynomials is a little bit different to the general multiplication. It is also called convolutional multiplication, in which the highest degree is the degree replaced by 1. Y^N is replaced by 1, Y^{N+1} by Y, Y^{N+2} by Y^2, and so on. For example: N = 7:

$$\begin{aligned}
\left(y^5 + 5y^4 + 2y^3 + 8y^2\right) &* \left(3y^4 + y^2 + 2y\right) \\
&= 3y^9 + 15y^8 + 7y^7 + 31y^6 + 12y^5 + 12y^4 + 16y^3 \\
&= 3y^2 + 15y + 7 + 31y^6 + 12y^5 + 12y^4 + 16y^3 \\
&= 31y^6 + 12y^5 + 12y^4 + 16y^3 + 3y^2 + 15y + 7
\end{aligned}$$

The operations on polynomials in the ring are calculated as follows:

$$a + b = \left(\sum_{i=0}^{N-1} a_i Y_i\right) + \left(\sum_{i=0}^{N-1} b_i Y_i\right) = \sum_{i=0}^{N-1} (a_i + b_i)\, Y_i \tag{6.3}$$

$$a * b = \left(\sum_{i=0}^{N-1} a_i Y_i\right) * \left(\sum_{i=0}^{N-1} b_i Y_i\right) = \sum_{i=0}^{N-1} \left(\sum_{i+j=k (\mathrm{mod}\, N)} a_i b_j\right) Y^k \tag{6.4}$$

The basic modular arithmetic operation is also applied on polynomials which have the following properties:

$$(a \bmod m) \pm (b \bmod m) = (a \pm b) \bmod m \tag{6.5}$$

$$(a \bmod m) * (b \bmod m) = (a * b) \bmod m \text{ where} \tag{6.6}$$

$a = (b \bmod m)$ describes that a and b both have the same remainder if they are divided by m.

To perform the modular arithmetic operation on polynomial, the integer modulus is used to divide each coefficient and the remainders are treated as the new coefficients. Applying a multiplication modulo (let) m means to reduce the polynomial coefficient modulo m.

Table 6.2 Recommended parameters

	Security levels			
Parameters	Highest	High	Standard	Moderate
p	3	3	3	3
q	256	128	128	128
N	503	347	251	167

2.2 Parameters

There are three integer parameters, N, p, and q, in the NTRU algorithm. N restricts the polynomial degree to be at most N-1. N is highly preferred to be the prime number to improve the security while p and q are co-prime and used to decrease the polynomial coefficients. The modulus p has to be less than modulus q, and the modulus q is generally less than N. The recommended parameters of NTRU for different security levels are shown in Table 6.2.

2.3 Algorithm

It describes the step-by-step procedure for the mathematical calculation to achieve the key generation, encryption, and decryption process.

1. Select the parameters: N, p, and q.
2. Choose random polynomials: $f(y)$ and $g(y)$.
3. Compute the polynomial inverse: $f(y) * f_p = 1 \mod p$ and $f(y) * f_q = 1 \mod q$.
4. Calculate public key: $P_k = p * (f_q * g(y)) \pmod q$.
5. Generate a blinding polynomial: $r(y)$.
6. Convert the message into binary polynomial: $m(y)$ or M_{sg}.
7. Encrypt the message: $M_e = \{r(y) * P_k + Msg\} \pmod q$.
8. Decrypt the ciphertext as follows:

   - $M_{d1} = f(y) * M_e \pmod q$, coefficient between $-q/2$ and $q/2$.
   - $M_{d2} = M_{d1} \pmod p$, coefficient between $-p/2$ and $p/2$.
   - Final decrypted original message: $M_{sg} = f_p * M_{d2} \pmod p$.

2.4 Key Generation

Two random polynomials $f(y)$ and $g(y)$ in a ring are chosen to generate the public key. The integer coefficients of these two random polynomials must be in $-1 \le Int.Coif \le 1$. The no. of 1 s coefficients in polynomial $f(y)$ is greater than the negative coefficients and the rest of the coefficients become 0. For the polynomial $g(y)$, the no. of both coefficients 1 s and -1 s should be equal and the rest 0. It does not need to be invertible or not by p and q [10]. The next step is to compute the inverse of f(y) using modulo q and p with the given properties:

$$f(y) * f_q = 1 \bmod q \tag{6.7}$$

and

$$f(y) * f_p = 1 \bmod p \tag{6.8}$$

If f(y) does not have the exact inverse of modulo q and p then another polynomial f(y) is randomly chosen that satisfies the above Eqs. (6.7) and (6.8). The pair of polynomials f(y) and f_p is treated as the private key while the public key P_k is calculated as follows:

$$P_k = p * \left(f_q * g(y)\right) \pmod{q} \tag{6.9}$$

Example of a public and private key:

1. The public parameters N, p, and q are, respectively, 7, 3, and 41. The sender S chooses a random polynomial f(y) that is the private key as

$$f(y) = -1 + y^2 + y^3 - y^4 + y^6 \text{ and}$$

$$g(y) = -y - y^2 + y^4 + y^6$$

2. Compute the inverse of a polynomial using modulo p and q that gives out the exact inverse. Here we are using Wolfram Mathematica 10.4 for calculation:

$f_p = f(y)^{-1}$ (modulo p)

$$= \left(-1 + y^2 + y^3 - y^4 + y^6\right)^{-1} \pmod{3}$$

PolynomialMod[Algebra'PolynomialPowerMod'PolynomialPowerMod[f(y), −1, y, y^7−1], 3] that gives the result as

$$f_p = 1 + y + y^2 + y^3 + 2y^5 + y^6 \text{ and}$$

$f_q = f(y)^{-1}$ *(modulo q)*

$$= \left(-1 + y^2 + y^3 - y^4 + y^6\right)^{-1} \pmod{41}$$

PolynomialMod[Algebra'PolynomialPowerMod'PolynomialPowerMod[f(y), −1, y, y^7−1], 41] that gives the result as

$$f_q = 32 + 2y + 40y^2 + 21y^3 + 31y^4 + 26y^5 + 8y^6$$

3. The pair of f(y) and f_p is stored as the private keys. The public key is calculated as follows:

$$
\begin{aligned}
P_k &= p * \left(f_q * g(y)\right) \pmod{q} \\
&= 3 * \{(37 + 2y + 40\,y^2 + 21\,y^3 + 31\,y^4 + 26\,y^5 + 8\,y^6) \\
&\quad \times (-y - y^2 + y^4 + y^6)\} \pmod{41} \\
&= 3 * (-37y - 39y^2 - 42y^3 - 24y^4 - 50y^5 + 20y6 - 11y^7 + 63y^8 \\
&\quad + 47y^9 + 39y^{10} + 26y^{11} + 8y^{12}) \pmod{41} \\
&= 3 * (-11 + 26y + 8y^2 - 3y^3 + 2y^4 - 42y^5 + 20y^6) \pmod{41} \text{ [Truncated]} \\
&= (-33 + 78y + 24y^2 - 9y^3 + 6y^4 - 126y^5 + 60y^6) \pmod{41} \\
&= 8 + 37y + 24y^2 + 32y^3 + 6y^4 + 38y^5 + 19y^6
\end{aligned}
$$

The computation of $f_q * g(y)$ needs N^2 multiplication. If any one of them has small coefficients, then this computation is very fast.

2.5 Encryption

The message M_{sg} is converted into a form of polynomial in which the coefficients must be in the range of $[-(p-1)/2, (p-1)/2]$ and the degree should not be more than $N-1$. A blinding polynomial r(y) in the ring is also chosen which has equal positive and negative coefficients. The encrypted message M_e is computed as.

$$M_e = \{r(y) * P_k + M_{sg}\} \pmod{q} \tag{6.10}$$

Example of encryption:

- Sender wants to send a message $M_{sg} = 1 - y + y^2 + y^3 - y^5$.
- Using a bling random polynomial $r(y) = -1 + y - y^5 + y^6$ will give

$$
\begin{aligned}
M_e &= \{\left(-1 + y - y^5 + y^6\right) * \left(8 + 37y + 24y^2 + 32y^3 + 6y^4 \right. \\
&\quad \left. + 38y^5 + 19y^6\right) + Msg\} \pmod{q} \\
&= \left(-8 - 29y + 13y^2 - 8y^3 + 26y^4 - 40y^5 - 10y^6 + 32y^7 - 8y^8 + 26y^9 - 32y^{10} \right. \\
&\quad \left. + 19y^{11} + 19y^{12} + Msg\right) \pmod{q} \\
&\quad - \left(24 - 37y + 39y^2 - 40y^3 + 45y^4 - 21y^5 - 10y^6 + M\varepsilon g\right) \pmod{q} \text{ [Truncated]} \\
&= \left(24 - 37y + 39y^2 - 40y^3 + 45y^4 - 21y^5 - 10y^6 + 1 - y + y^2 + y^3 - y^5\right) \pmod{41} \\
&= \left(25 - 38y + 40y^2 - 39y^3 + 45y^4 - 22y^5 - 10y^6\right) \pmod{41}
\end{aligned}
$$

Then $M_e = 25 + 3y + 40y^2 + 2y^3 + 4y^4 + 19y^5 + 31y^6$

2.6 *Decryption*

The encrypted message M_e is partially decrypted using the private key $f(y)$ as follows:

$$M_{d1} = f(y) * M_e \ (\text{mod } q) \tag{6.11}$$

The coefficients of M_{d1} is centered between $-q/2$ and $q/2$. Modulo q is also applied on the partially decrypted message M_{d1} to reduce the coefficients of the polynomial to take place between $-p/2$ and $p/2$ as.

$$M_{d2} = M_{d1} \ (\text{mod } q) \tag{6.12}$$

The final step is to compute the original message M_{sg} using the other part of the private key f_p as follows:

$$M_{sg} = f_p * M_{d2} \ (\text{mod } p) \tag{6.13}$$

Example of decryption:

1. Once the encrypted M_e is received, the partial decryption process starts using the private key $f(y)$ as the following:

$$
\begin{aligned}
M_{d1} &= f(y) * M_e \ (\text{mod } q) \\
&= \left(-1 + y^2 + y^3 - y^4 + y^6 \right) * \left(25 + 3y + 40y^2 + 2y^3 + 4y^4 + 19y^5 + 31y^6 \right) \\
&\quad (\text{mod } 41) \\
&= \left(-25 - 3y - 15y^2 + 26y^3 + 14y^4 + 20y^5 - 40y^6 + 24y^7 + 86y^8 \right. \\
&\quad \left. + 14y^9 - 27y^{10} + 19y^{11} + 31y^{12} \right) (\text{mod } 41) \\
&= \left(-1 + 83y - y^2 - y^3 + 33y^4 + 51y^5 - 40y^6 \right) (\text{mod } 41) \ [\text{Truncated concept}] \\
&= 40 + y + 40y^2 + 40y^3 + 33y^4 + 10y^5 + y^6
\end{aligned}
$$

2. Choosing of the coefficients lying between $-q/2$ and $q/2$ or $[-20,20]$ modulo q is done to obtain the partial message M_{d2} as mentioned below:

$$
\begin{aligned}
M_{d2} &= M_{d1} \ (\text{mod } p) \\
&= \left(40 + y + 40y^2 + 40y^3 + 33y^4 + 10y^5 + y^6 \right) (\text{mod } 41) \\
&= -1 + y - y^2 - y^3 - 8y^4 + 10y^5 + y^6
\end{aligned}
$$

3. Finally, the original message M_{sg} is calculated as follows:

$$
\begin{aligned}
M_{sg} &= f_p * M_{d2} \ (\text{mod } p) \\
&= \left(1 + y + y^2 + y^3 + 2y^5 + y^6 \right) \left(-1 + y - y^2 - y^3 - 8y^4 + 10y^5 + y^6 \right) (\text{mod } 3) \\
&= \left(-1 - y^2 - 2y^3 - 9y^4 - 2y^5 + 3y^6 + 2y^7 + 8y^8 - 16y^9 + 12y^{10} + 12y^{11} + y^{12} \right) (\text{mod } 3) \\
&= \left(1 + 8y - 17y^2 + 10y^3 + 3y^4 - y^5 - 3y^6 \right) (\text{mod } 3) \ [\text{Applying truncated concept}] \\
&= 1 + 2y + y^2 + y^3 - y^5
\end{aligned}
$$

Choose the coefficients centering between $[-1,1]$ or $-p/2$ and $p/2$. Hence, the original message after decryption is $M_{sg} = 1 - y + y^2 + y^3 - y^5$.

2.7 Working Method

In this process, the partially decrypted message M_{d1} satisfies the following:

$$
\begin{aligned}
M_{d1} &= f(y) * M_e \ (\text{mod } q) \\
&= f(y) * \big(r(y) * P_k + M_{sg}\big) \ (\text{mod } q) && \big[\text{using (10)}\big] \\
&= f(y) * \big(r(y) * p * \big(f_p * g(y)\big) + M_{sg}\big) \ (\text{mod } q) && \big[\text{using (9)}\big] \\
&= p * r(y) * g(y) + f(y) * M_{sg} \ (\text{mod } q) && \big[\text{using (8)}\big]
\end{aligned}
$$

All the coefficients of the $f(y)$, $g(y)$, $r(y)$, M_{sg}, and p are much smaller than q. It is to be ensured that the coefficients are in $-q/2$ and $q/2$. Also

$$
\begin{aligned}
M_{d2} &= M_{d1} \ (\text{mod } p) \\
&= \big(p * r(y) * g(y) + f(y) * M_{sg}\big) (\text{mod } p) \\
&= \big(f(y) * M_{sg}\big) (\text{mod } p) \\
M_d &= f_p * M_{d2} \ (\text{mod } p) \\
&= f_p * f(y) * M_{sg} (\text{mod } p) \\
&= M_{sg} \ (\text{mod } p) && \big[\text{using (7)}\big]
\end{aligned}
$$

3 NTRU Optimization

To enhance the speed of NTRU algorithm, some optimization techniques may be applied as follows:

1. *Formation changing of p and f:*

    The random polynomial $f(y)$ must be invertible by modulo q and p with small coefficients. Hence it can be chosen in the form of $f(y) = 1 + p * F(y)$. Here $F(y)$ is a random polynomial of message space with small coefficients. As described in Eq. (6.7), $f(y)$ should have inverse modulo p because the inverse of f_p is equal to 1. This process decreases the time complexity in key generation because there is no need to calculate the inverse modulo p. It also saves the time in decryption process, since there is no requirement of last step (13) to multiply $f_p * M_{d2}$. It means no need to save f_p as the private key [11].

2. *Polynomial Form of P*

    p can be taken as a polynomial instead of an integer type. It needs to have small coefficients in ring R and the principles produced by q and p must be relatively prime. The most suitable form of polynomial p is $p = 2 + Y$. The

binary polynomial [0, 1] is used for p instead of a trinary polynomial [−1, 0, 1].
It makes the encoding of messages simple to use.

3. *Product of Low Hamming Weight*

The highly time-consuming process in the NTRU algorithm involves a product of $r(y) * P_k$ and $f(y) * M_e$ to encrypt and decrypt the messages, respectively.
Low-Hamming-weight products can be used to increase the speed of encryption
and decryption. Because the private key $f(y)$ and the bling message $r(y)$ both have
small coefficients, the time consumptions for these product calculations may be
reduced by the low-Hamming-weight product and the following operations:

$$f = 1 + p\,(f_1 * f_2 + f_3) \tag{6.14}$$

$$r = r_1 * r_2 \tag{6.15}$$

f_1, f_2, and f_3 are chosen carefully to convert $f = 1 + p\,(f_1 * f_2 + f_3)$ into
pure binary polynomial. If the polynomials r_1 and r_2 both have equal binary
coefficients 1 s then r will also be a binary polynomial having coefficients 1 s
[12].

4 Security

The security of the NTRU algorithm comes from the mixing of polynomial
multiplication in reduction with modulo q and p. Security also depends on the
lattices wherein finding out the extremely short vector is very difficult. It is an
important step to consider gcd(q,p) = 1 rather than gcd(q,p) > 1 because it will
decrease the security level [13]. Silverman and Hoffstein provided different ways to
counter the attacks [14]:

- Repetition of key changing: Each and every time the sender uses a new public
  key with a digital certificate along his/her message to the recipients verifying
  the key origin. This process restricts communication via off-line and secures the
  information.
- Distribution analysis of coefficient: The total number of coefficients in the
  resultant polynomial $p * r(y) * g(y) + f(y) * M_{sg}$ lying to the range $(−q/2, q/2)$
  is larger than usual. Hence it is easy to identify the attacker counting the number
  of coefficients between the ranges.
- Failure of decryption track: Decryption failure is possible in a normal situation.
  However it is different when an attack occurs. Anyone can detect the ongoing
  attack and be forced to change the public key.
- Message concealing: This is useful to hide the message by adding a random
  polynomial to the plain text. This process produces a different message and
  misleads the attacker when optimal asymmetric encryption padding is used. It
  is also able to produce errors in attack.

5   Improvements

Many researchers tried to enhance its performance during the past 15 years and many good alternatives were introduced based on a similar configuration of NTRU. They are designed to enhance the performance by changing the original polynomial ring over Z [15] and by many variant polynomial rings rather than Z. All of them have an aim to design a secure cryptosystem like NTRU using a small key size and provide security against lattice-based attacks [16]. P. Gaborit et al. presented a CTRU cryptosystem in 2002 that is based on the polynomial ring over a finite field [17]. M. Coglianes et al. introduced MaTRU in 2005 that depends on the ring of (m × m) matrices of a polynomial on order n [18]. In 2009, the QTRU cryptosystem was presented by Malekian et al., which depends on the quaternion algebra [19]. They have also presented OTRU in 2010 based on the octonion algebra [20]. In the same year, NNRU was presented by Vats, which is basically a noncommutative operation on the noncommutative ring $R = R_k (Z[x])/(X^n - I_{k \times k})$. He claimed that his recommended system is fully protected against any lattice attack [21]. K. Jarvis proposed ETRU in 2011 based on the Eisenstein integers [22]. Alsaidi et al. proposed a cryptosystem in 2015 based on commutative quaternion algebra that is called CQTRU [23]. In this same year, Karbasi et al. also introduced a public key cryptosystem like NTRU with provable security based on the hardness worst case of the approximate both closest vector problem (CVP) and shortest vector problem (SVP) and named it as ILTRU [24]. In 2015, Yasuda et al. also proposed an NTRU variant using the group ring called GR-NTRU [25]. In 2016, Thakur et al. presented BTRU, in which Z is replaced by a polynomial ring with a single variable α over a rational field to make it faster than NTRU [26]. The same year, Al-Saidi et al. proposed a new alternative of an asymmetric key cryptosystem that is called BITRU. It is also based on a new algebraic structure like the mathematical structure of NTRU that is called binary algebra. It is associative as well as commutative. This new algebraic structure provides higher security and complexity for BITRU [27]. In 2017, they also presented a new look of the NTRU cryptosystem in the name of HXDTRU that depends on hexadecnion algebra. It is used to create a highly performed public key with higher security among a small key size [28]. In 2018, Atani et al. presented a new public key cryptosystem based on the finite field. It is a noncommutative variant of CTRU. In this cryptosystem, the encryption and decryption both are noncommutative that makes the system secure against any linear algebra attack such as the lattice-based attack. It provides a higher level of security using the two-sided matrix multiplication [29]. In that year, Karbasi presented a new form of public key cryptosystem PairTRU based on the lattice [30].

6 Conclusion

The NTRU algorithm is simple to use with little resources and very fast in key generation, encryption, and decryption processes. All the operations performed on this algorithm are modular arithmetic, convolutional multiplication, and addition. All the polynomial coefficients are reduced by modulo q that results mostly in 11-bit integers. Hence NTRU computation does not require any additional multi-precision library. It requires low memory use compared to other cryptosystems. Decryption failure is extremely low. It is more safe and secure for known and unknown attacks. This algorithm proves that it is highly secure against different types of attacks such as a meet-in-the-middle attack, lattice-based attack, and brute force attack.

Acknowledgement This work was supported by the Universiti Kebangsaan Malaysia under the grant DIP-2018-040.

References

1. Samar, K., Masri, A., Siti, N., Huda, S., Abdullah, & Zulkifli, A. (2018). Challenges in multi-layer data security for video steganography revisited. *Asia Pacific Journal of Information Technology and Multimedia, 7*(2), 53–62.
2. Nguyen, H. B. (2014). An overview on the Ntru cryptographic system. PhD diss. *Sciences*.
3. Zhao, N., & Shenghui, S. (2011). An improvement and a new design of algorithms for seeking the inverse of an NTRU polynomial. In *2011 seventh international conference on computational intelligence and security, pp. 891–895. IEEE*.
4. Ali, Z. M., Othman, M., Said, M. R. M., & Sulaiman, M. N. (2008). An efficient computation technique for cryptosystems based on Lucas functions. In *Proceedings of the international conference on computer and communication engineering, ICCCE08: Global links for Human Development*.
5. Md Ali, Z., & Makhzoum, N. M. A. (2012). Computation of private key based on divide-by-prime for Luc cryptosystems. *Journal of Computer Science, 8*(4), 523–527.
6. Ali, Z. M., Othman, M., Said, M. R. M., & Sulaiman, M. N. (2008). Parallel computation for LUC cryptosystems on distributed memory multiprocessor machine. In *Proceedings of the 4thIASTED international conference on advances in computer science and technology, ACST*.
7. Ahmed, J. M., & Md Ali, Z. (2011). The enhancement of computation technique by combining RSA and El-Gamal cryptosystems. In *International conference on electrical engineering and informatics, Bandung, Indonesia*.
8. Aisar, M., MMI, Fauzi, S. S. M., Baharin, H., Sobri, W. A. W. M., Suali, A. J., Gining, R. A. J. M., & Jamaluddin, M. N. F. (2018). Performance analysis between quantum computers and silicon computers: A preliminary investigation? In *IOP conferences series, journal of physics*.
9. Bu, S. Y., & Zhang, H. (2009). Research on the method of choosing parameters for NTRU. In *2009 international conference on multimedia information networking and security, vol. 2, pp. 334–337. IEEE*.
10. Pipher, J. (2002). Lectures on the ntru encryption algorithm and digital signature scheme: Grenoble june 2002. In *Brown University, Providence RI 02912, report*.
11. Shen, X., Zhenjun, D., & Chen, R. (2009). Research on NTRU algorithm for mobile java security. In *2009 international conference on scalable computing and communications; eighth international conference on embedded computing, pp. 366–369. IEEE*.

12. Jha, R., & Saini, A. K. (2011). A Comparative Analysis & Enhancement of NTRU algorithm for network security and performance improvement. In *2011 international conference on communication systems and network technologies, pp. 80–84. IEEE.*
13. Hoffstein, J., Pipher, J., Joseph, H., & Silverman. (1998). NTRU: A ring-based public key cryptosystem. In *International algorithmic number theory symposium* (pp. 267–288). Berlin, Heidelberg: Springer.
14. Jaulmes, Éliane, and Antoine Joux. "A chosen-ciphertext attack against NTRU." In Annual international cryptology conference, pp. 20–35. Springer, Berlin, Heidelberg, 2000.
15. Nevins, M., Karimianpour, C., & Miri, A. (2010). NTRU over rings beyond $\mathbb{Z}$. *Designs, Codes and Cryptography, 56*(1), 65–78.
16. Coppersmith, D., & Shamir, A. (1997). Lattice attacks on NTRU. In *International conference on the theory and applications of cryptographic techniques* (pp. 52–61). Berlin, Heidelberg: Springer.
17. Gaborit, Philippe, Julien Ohler, and Patrick Solé. "CTRU, a polynomial analogue of NTRU." (2002).
18. Coglianese, M., & Goi, B.-M. (2005). MaTRU: A new NTRU-based cryptosystem. In *International conference on cryptology in India* (pp. 232–243). Berlin, Heidelberg: Springer.
19. Malekian, Ehsan, Ali Zakerolhosseini, and Atefeh Mashatan. "QTRU: a lattice attack resistant version of NTRU PKCS based on quaternion algebra." preprint, Available from the Cryptology ePrint Archive: http://eprint. iacr. org/2009/386. pdf (2009).
20. Malekian, E., & Zakerolhosseini, A. (2010). OTRU: A non-associative and high speed public key cryptosystem. In *2010 15th CSI international symposium on computer architecture and digital systems, pp. 83–90. IEEE.*
21. Vats, N. (2009). *NNRU, a noncommutative analogue of NTRU.*" arXiv preprint arXiv:0902 (p. 1891).
22. Jarvis, K. (2011). *NTRU over the Eisenstein integers.* Ottawa: University of Ottawa.
23. Alsaidi, N., Saed, M., Sadiq, A., & Majeed, A. A. (2015). An improved NTRU cryptosystem via commutative quaternions algebra. In *Proceedings of the international conference on security and management (SAM)* (p. 198). The Steering Committee of The World Congress in Computer Science, Computer Engineering and Applied Computing (WorldComp).
24. Karbasi, A. H., & Atani, R. E. (2015). ILTRU: An NTRU-like public key cryptosystem over ideal lattices. *IACR Cryptology ePrint Archive, 2015,* 549.
25. Yasuda, T., Dahan, X., & Sakurai, K. (2015). Characterizing NTRU-variants using group ring and evaluating their lattice security. *IACR Cryptology ePrint Archive, 2015,* 1170.
26. Thakur, K., & Tripathi, B. P. (2016). BTRU, a rational polynomial analogue of NTRU cryptosystem. *International Journal of Computer Applications, 12,* 145.
27. Alsaidi, N. M., & Yassein, H. R. (2016). BITRU: Binary version of the NTRU public key cryptosystem via binary algebra. *International Journal of Advanced Computer Science & Applications, 1*(7), 1–6.
28. Al-Saidi, N. M. G., & Hassan, R. (2017). Yassein. "a new alternative to NTRU cryptosystem based on highly dimensional algebra with dense lattice structure." Malaysian. *Journal of Mathematical Sciences, 11,* 29–43.
29. Atani, R. E., Atani, S. E., & Karbasi, A. H. (2018). NETRU: A non-commutative and secure variant of CTRU cryptosystem. *ISeCure, 1,* 10.
30. Karbasi, A. H., Atani, R. E., & Atani, S. E. (2018). PairTRU: Pairwise non-commutative extension of the NTRU public key cryptosystem. *International Journal of Information Security Science, 7*(1), 11–19.

Chapter 7
Cocks IBE Scheme

Deepak Kumar Sharma, Bhanu Tokas, Venkata Rohit Jakkinapalli, and Ritvik Nagpal

Abstract Until the 1970s, the symmetric key ciphers were the golden standard of encryption. But all this changed with the rise of the asymmetric key ciphers (i.e., R.S.A.) which provided a level of security unimaginable by symmetric key ciphers. Yet, these ciphers could only be utilized by the military as it was not practical for the general public to keep records of public keys of various individuals they communicate with. Apart from the extreme complexity of such a system, it also faced another challenge in the form of the cost of storage of the said collection of keys. This led to the call for IBE (identity-based encryption) wherein a public identity of a user could be converted into a public key by an appropriate hashing function.

Adi Shamir proposed an asymmetric scheme in 1984 but an effective implementation of this scheme could only be formed in 2001. These included the Boneh-Franklin scheme and Cocks scheme, both of which were introduced in 2001. This scheme that was proposed by Adi Shamir had a lot of advantages over traditional public key-based systems. These systems eliminated the problem of prior distribution of keys. It thus reduced the complexity significantly and also reduced the cost relative to public key-based schemes.

Through this chapter we explore the Cocks IBE, the mathematical working of the algorithm, its correctness, and its security. This will not only be limited to the working of the Cocks scheme but we shall also discuss the mathematical proof and security of this scheme.

Keywords Cocks scheme · IBE · Quadratic residuosity problem · Asymmetric key cryptography

D. K. Sharma (✉)
Department of Information Technology, Netaji Subhas University of Technology, (Formerly Netaji Subhas Institute of Technology), New Delhi, India

B. Tokas · V. R. Jakkinapalli · R. Nagpal
Department of Computer Engineering, Netaji Subhas University of Technology, New Delhi, India

© Springer Nature Switzerland AG 2021
K. A. B. Ahmad et al. (eds.), *Functional Encryption*, EAI/Springer Innovations in Communication and Computing, https://doi.org/10.1007/978-3-030-60890-3_7

1   Introduction

1.1   Identity-Based Encryption (IBE)

In IBE technology, the public address of a user is used in the encryption technique. It works by generating the public key of the user from their public address. The public address that is used in IBE techniques is always in the form of a string. In IBE technique private key generator (PKG), a trusted third-party system, is used to generate the private key of the user. The private key generator gives the private key only to authorized parties. After getting the private key, the message is obtained by using it. This is a basic introduction to working with IBE [2]. It can be better understood through an example:

1. Ritvik performs the encryption of the message and sends it to Rohit with his public address as the public key.
2. Rohit requests the private key from the key server after receiving the message from Ritvik.
3. The key server authenticates Rohit's identity and allows him to access the private key.
4. Rohit decrypts the ciphertext after receiving the private key from the key server, and for future messages the same private key can be used between Ritvik and Rohit [3].

1.2   Cocks IBE

In the modern world the pace with which technology is progressing humans are generating more and more data. Security of information is becoming more and more relevant. Considering the volume of data being produced and recent global events, the security of a user's data needs to be ensured. One way to protect the user's data is to use encryption techniques. The Cocks IBE scheme was one of the first secure IBE schemes developed and it was proposed in 2001 by Clifford Cocks. Cocks IBE scheme is based on two mathematical concepts: quadratic residuosity problem and integer factorization [1]. Before discussing the Cocks IBE scheme we need to understand the IBE.

1.3   Working of Cocks IBE

Cocks IBE scheme is done in four steps: *setting up parameters, extraction of private key, encryption of the plaintext, and decryption of the ciphertext* [4]:

- Setting up of parameters in the Cocks IBE scheme is used to initialize the parameters that are required for the algorithm.
- In the extraction phase the private key of the user is generated.
- Encryption phase is used to encrypt the plaintext bit by bit. After this phase a ciphertext is generated. Ciphertext is another name for the encrypted message.
- Decryption is the last phase of the Cocks IBE scheme. This phase is executed at the receiver end. In this phase the message is obtained from the ciphertext through the private key [3–5].

1.4   Features of Cocks IBE

Cocks IBE scheme does not use bilinear pairings [5, 6]. This scheme does not have much of a practical use however due to high degree of ciphertext expansion. Ciphertext expansion means there is an increase in the number of bits when a message is encrypted. The number of bits is doubled in Cocks IBE scheme; how it does will be explored in the encryption phase [7].

2   Mathematical Concepts

Before progressing into its working there is a need to understand some of the mathematical concepts that are required to understand the working of all the four phases.

1. **Number theory:** Number theory is a mathematical concept that is mainly concerned about the properties of numbers, especially integers.
2. **Congruence modulo:** Congruence modulo is a mathematical concept that is represented as $x \equiv y \pmod{z}$. It means that x and y when divided by z give the same remainder [8].

   For example:

$$8 \equiv 18 \equiv -2 \pmod{10}$$

   These three have the same remainder -2 when divided by the value 10.
3. **Relative Prime Numbers:** Relative Prime Numbers Are Two Numbers whose HCF or Highest Common Factor Is Equal to One [9–12].

   For example:

   HCF of 98 and 95 is 1.
4. **Euler phi:** It denotes the total number of integers that are relative prime numbers of n and are less than the value of n with a condition that n is a positive number. It is represented by $\phi(n)$ [13, 14].

5. **Quadratic residue:** It is represented by $x^2 \equiv a(mod\,n)$ where a belongs to a set of positive integers and a is not equal to 0. The value a is called a quadratic nonresidue when there exists no such value of x [7, 15, 16].

6. **Legendre symbol:** Legendre symbol can be represented as a/p where a is represented by an integer while p is represented by an odd prime number [17, 18]. Its value can be defined as follows:

   - 0 if p is divisible by a.
   - If the value of a is a quadratic residue modulo of p then $(a/p) = +1$.
   - If the value of a is a quadratic nonresidue modulo of p then $(a/p) = -1$.

7. **Jacobi symbol:** It is a generalized form of Legendre symbol that is used in the case of non-prime numbers [19–21].

   In Jacobi symbol we take the value of n as

$$n = \prod_{i=1}^{k} p_i^{a_i}$$

Here, n is a positive odd number and a is an integer. Jacobi symbol can be defined as

$$\frac{a}{n} = \prod_{i=1}^{k} (a/p_i)^{a_i}$$

This algorithm is helpful as the factors of n are not required to be known to calculate the value. Jacobian symbol (a/n) value can be calculated using this algorithm:

   - The input constraints are that n should be an odd integer that is ≥ 3 and a should be an integer between 0 and n.
   - If the value of a is equal to zero, return zero.
   - If the value of a is equal to one, return one.
   - Else, write a in the form of $a = 2^k a_1$ such that a_1 is odd.
   - Case 1: value of k is even, then $s \leftarrow 1$.
   - Case 2: $n \equiv 1(mod\,8)$ or $n \equiv 7(mod\,8)$, $s \leftarrow 1$.
   - Case 3: $n \equiv 3(mod\,8)$ or $n \equiv 5(mod\,8)$, $s \leftarrow -1$.
   - Then compute the value n_1 as $n_1 \leftarrow n\,mod\,a_1$.
   - Return $s \cdot$ Jacobi symbol (n_1, a_1).
   - The output that will be returned by this recursive function will be the Jacobi symbol value for (a/n).

3 Setting up of Parameters

The setting up phase in Cocks IBE algorithm is the first step in the entire IBE framework. The algorithm setup phase is when initialization of the parameter required for the various calculations is done. In this phase, the master secret (the PKG uses this to calculate the private keys) is also initialized. It is run only one time and it is used to set up the entire IBE environment. Both the public and the private keys are generated in this phase. Only the PKG has the private key in this phase [4, 22].

Table 7.1 describes all the parameters in Cocks IBE scheme.

3.1 How it Works

In the setup phase of Cocks IBE there are a few requirements. Two prime numbers, p and q are required, such that they are congruent to $3 \bmod ulo\, 4$. Then the value of n is calculated as a product of p and q. The obtained value of n is public while the values of p and q are known only to the private key generator. In this phase the identity ID of the user is sent through a hash function to obtain a value a. After this, we obtain the Jacobi symbol value as $\left(\frac{a}{n}\right)$ is equal to $+1$. When the Jacobi symbol is equal to $+1$ it guarantees that the value a (positive or negative) is the square modulo n. To achieve the above condition, we use a hash function on the identity to obtain the integer a and we keep adding 1 to a till we obtain the Jacobi symbol value as $+1$.

The explanation is as follows:

First case: a is positive.

By the properties of Jacobi symbol and that n is equal to p times q, we know that

$$(a/n) = (a/p) \cdot (a/q) \tag{7.1}$$

As a square modulo n has to be $+1$ we have two possibilities:

- Values of both a square modulo p and a square modulo q have to be $+1$ or
- Values of both a square modulo p and a square modulo q have to be $.-1$

Table 7.1 Cocks IBE parameters

Parameter type	Parameter	Properties
Private global parameters	p, q	$primes = 3(mod\,4)$
Public global parameter	n	$n = p.q$
Public hash function	H_1	$H_1 : \{0,1\}^* -> Z_n; (H_1(ID)/n) = +1$
Per-user public key	a	$\left(\frac{a}{n}\right) = +1$
Per-user private key	r	$r^2 = a(mod\,n)$

Substitute the values of (a/p) and (a/q) into Eq. 7.1 for both the cases.

Case 1:

$$(a/n) = (a/p) \cdot (a/q) = (+1)(+1) = +1$$

Case 2:

$$(a/n) = (a/p) \cdot (a/q) = (-1)(-1) = +1$$

Second case: a is negative.

We know that p and q are congruent to 3 *mod ulo* 4; thus we can write

$$(-1/p) = (-1/q) = -1 \tag{7.2}$$

Now the same set of equations should be written for $-a$ being the square modulo of n. This implies

$$(-a/n) = (-a/p) \cdot (-a/q)$$
$$= (a/p) \cdot (-1/p) \cdot (a/q) \cdot (-1/q)$$

Using Eq. (7.2)

$$= (a/p) \cdot (-1) \cdot (a/q) \cdot (-1)$$

$$= (a/p) \cdot (a/q) = (+1)(+1) = +1$$

From this we can infer that $-a$ is a square because it is the product of two numbers that are squares.

The two possible values of a can lead to a lot of confusion, as in which value to consider as it is the user public key. As it will be seen later in this chapter, it will lead to extra computation in the encryption section. Also, this is the reason for ciphertext carrying two values in order to account for each of these two cases.

The resulting value in either case is the value a which will become the corresponding public key to identity ID [3, 7, 23].

4 Extraction of Private Key

In the phase two, the user obtains the private key from the PKG. PKG uses a user's identity, and the private variables p and q for the extraction of private keys (mentioned in Table 7.1). The private key is usually delivered through a secure channel [4, 22].

4.1 How it Works

Extraction phase involves the calculation of private key, in correspondence to the public key a, by calculation of either square root a or $-a$ modulo of n. Now, we know that both p and q are equivalent to 3 *mod ulo* 4 OR both $p - 1$ and $q - 1$ are equivalent to 2 *mod ulo* 4. Thus,

$$p = 4k_1 + 2 \tag{7.3}$$

and

$$q = 4k_2 + 2 \tag{7.4}$$

Now, we have $n = p \cdot q$, and $\emptyset(n) = (p - 1) \cdot (q - 1)$; thus

$$\emptyset(n) + 4 = (p - 1)(q - 1) + 4$$

Using the values from Eqs. (7.3) and (7.4)

$$= (4k_1 + 2) \cdot (4k_2 + 2) + 4$$

$$= (2k_1 k_2 + k_1 + k_2 + 1) \cdot 8$$

such that $\emptyset(n)$ is divisible by 8.

From the above equation, we can calculate a square root modulo of n as

$$r = a^{((\emptyset(n)+4)/8)} \bmod n \tag{7.5}$$

Now, we have a square root modulo of n, as squaring Eq. (7.5) on both sides:

$$r^2 = a^{2((\emptyset(n)+4)/8)}$$
$$= a^{((\emptyset(n)+4)/4)}$$
$$= \left(a^{\emptyset(n)}\right)^{1/4} . a$$
$$= \pm a \pmod{n} \text{ (Using Euler Theorem)}$$

So, if a is the square root modulo of n, then r^2 is congruent to $a(\bmod n)$, and if $-a$ is a square root modulo of n, then r^2 is congruent to $-a(\bmod n)$ [3, 7, 23].

5 Encryption

This is the third phase of the Cocks IBE scheme. The main aim of this phase is to encrypt and obtain the ciphertext. In this phase, the message is encrypted bit by bit and for each bit we obtain two integers. The reason for getting two integers is that we need to know if either a or $-a$ is the square root modulo of n [4, 22].

5.1 How it Works

If we take a bit from the message, say m, to encode it as $x = (-1)^m$, then bit 0 will become $+1$ and bit 1 will become -1. After the encoding step is completed two random integers t_1 and t_2 are picked such that both

$$(t_1/n) = x$$

and

$$(t_2/n) = x$$

The values of t_1 and t_2 so obtained are used to calculate the ciphertext which is represented by (s_1, s_2). The ciphertext so obtained is sent to the recipient. The values s_1 and s_2 are calculated using the following formula:

$$s_1 = (t_1 + a/t_1) \bmod n \tag{7.6}$$

and

$$s_2 = (t_2 + a/t_2) \bmod n \tag{7.7}$$

Out of these two ciphertext values the recipient will decrypt s_1 or s_2 depending on the value of a. The value s_1 will be chosen if a is the square root modulo of n; else s_2 will be chosen.

We need to make sure that we take two separate values t_1 and t_2 instead of just one value t. The reason for this is that it becomes easier for someone to decrypt the message from the ciphertext if we use only one integer. This is how it works: replacing t_1 and t_2 with t in Eqs. (7.6) and (7.7)

$$s_1 = (t + a/t) \bmod n$$

and

$$s_2 = (t - a/t) \bmod n$$

So, for a hacker to hack the ciphertext and obtain the message he/she has to calculate

$$\frac{s_1 + s_2}{2} = \frac{1}{2} \left((t + (a/t)) + (t - a/t) \right) \bmod n = t \bmod n$$

After this, the hacker can compute x using the value of t obtained:

$$t/n = x$$

Therefore, the message can be decrypted in this way [3, 7, 23].

6 Decryption

This is the last phase of Cocks IBE scheme; it happens on the recipient side. We decrypt the encrypted message with the help of the private key which is given to the recipient after authentication by the key server. The decryption phase is dependent on the value of r as mentioned in encryption. Based on the two conditions the method for derivation of the plaintext varies [4, 22].

6.1 How it Works

Now, after the recipient receives the pair of s_1 and s_2, he/she decides upon the two choices that he/she needs to decrypt:

Making

$$s = s_1 \; if \; r^2 = a \, (\bmod \; n) \, , and$$

$$s = s_2, if \; r^2 = -a \, (\bmod \; n)$$

1. In the first case (where $r^2 = a(\bmod n)$), the user calculates

$$x = ((s + 2r)/n) \tag{7.8}$$

In this case, we must note that using Eq. (7.6) we get

$$\begin{aligned} s + 2r &= (t_1 - a/t_1) + 2r \\ &= t_1 + 2r - a/t_1 \\ &= t_1 \left(1 + 2\,r/t_1 - a/(t_1)^2 \right) \\ &\equiv t_1 \left(1 + 2\,r/t_1 + (r)^2/(t_1)^2 \right) \bmod n \\ &\equiv t_1 (1 + r/t_1)^2 \bmod n \end{aligned}$$

So, we have $s + 2r$ as a square modulo n whenever t_1 is such that

$$(s + 2r)/n = t_1/n = x$$

So using Eq. (7.8), we can recover the plaintext bit x

2. In the second case (where $r^2 = -a(mod\,n)$), we must note that using Eq. (7.7) we get

$$\begin{aligned}
s + 2r &= (t_2 - a/t_2) + 2r \\
&= t_2 + 2r - a/t_2 \\
&= t_2 \left(1 + 2\,r/t_2 - a/(t_2)^2\right) \\
&= t_2 \left(1 + 2\,r/t_2 + (r)^2/(t_2)^2\right) \bmod n \\
&\equiv t_1 (1 + r/t_2)^2 \bmod n
\end{aligned}$$

So, we have $s + 2r$ as a square modulo n whenever t_2 is such that

$$(s + 2r)/n = t_2/n = x$$

So, using Eq. (7.8), we can recover the plaintext bit x [3, 7, 23].

7 Examples

Now, we have seen the four steps of the algorithm. There are three possible cases that can emerge considering the conditions mentioned in the Cocks IBE algorithm. The first case is when $r^2 \equiv a(mod\,n)$.

We follow the four steps of the algorithm in the following order:

- In the setting up phase we assume two prime integers $p = 7$ and $q = 11$. Both p and q satisfy the condition $3(mod\,4)$. The value $n = p \cdot q = 7 \cdot 11 =$. Let us assume that the public key value is $a = 9$.
- We have to calculate the value of the private key r as mentioned in the extraction phase. It is given by the equation $r = a^{((\phi(n) + 4)/8)} \, mod\,n$.

$$\emptyset(77) = \emptyset(7) \cdot \emptyset(11).$$

  - $\phi(77) = 6 \cdot 10 = 60$ (using Euler function properties for prime numbers).
  - So $r = 9^{((60 + 4)/8)} \, mod\,77$.

$$r = 9^8 \bmod 77.$$

$$r = 25$$

- In this case $r^2 \equiv a(mod\,n)$.

- Let us assume that the bit "0" is to be encrypted with the public key; the bit "0" has to be encoded as $+1$. We then pick two values t_1 and t_2 which satisfy the condition mentioned in the encryption phase.
- We choose t_1 and t_2 as 4 and 6 as

  - The Jacobi value $(4/77) = +1$
  - The Jacobi value $(6/77) = +1$

- Now the ciphertext value is calculated as substituting the values of t_1 and t_2 in Eqs. (7.6) and (7.7):

$$s_1 = (t_1 + a/t_1) \bmod n$$

$$= (4 + 9/4) \bmod 77 = 64$$

$$s_2 = (t_2 - a/t_2) \bmod n$$

$$= (6 - 9/6) \bmod 77 = 43$$

- After this we move on to the decryption step. The first step is to check which condition the private key satisfies. It is of the form $r^2 \equiv a \pmod{n}$.

  - So, to decrypt the message the recipient uses the formula $x = ((s + 2r)/n)$ where $s = s_1$:

$$x = (s_1 + 2r)/n = (64 + 50)/77$$

$$= (114/77) = +1$$

Then this is decoded to obtain the bit 0 as plaintext.

The first example shows how each step works and how the message bit "0" is sent from the sender to the recipient using Cocks IBE.

Now we need to look at another example with same parameters, that is, p and q, but of the condition $r^2 \equiv -a \pmod{n}$.

- The first step is to set up the parameters. Let $p = 7$ and $q = 11$; then we obtain the value of n as 77. We assume that the public key $a = 10$.
- We have to calculate the value of the private key r as mentioned in the extraction phase. It is given by the equation $r = a^{((\phi(n) + 4)/8)} \bmod n$.

$$\emptyset(77) = \emptyset(7) \cdot \emptyset(11).$$

  - $\phi(77) = 6 \cdot 10 = 60$ (using Euler function properties for prime numbers).

- So $r = 10^{((60 + 4)/8)} \, mod \, 77$.

$$r = 10^8 \, mod \, 77.$$

$$r = 23$$

- In this case $r^2 \equiv \, - a(mod \, n)$.
- Let us assume that the bit "1" is to be encrypted with the public key; the bit "1" has to be encoded as -1 and so we pick two values t_1 and t_2 which satisfy the condition mentioned in the encryption phase.
- We choose t_1 and t_2 as 8 and 2 as

  - The Jacobi value $(8/77) = \, -1$.
  - The Jacobi value $(2/77) = \, -1$.

- Now the ciphertext value is calculated as substituting the values of t_1 and t_2 in Eqs. (7.6) and (7.7):

$$s_1 = (t_1 + (a/t_1)) \, mod \, n = (8 + 10/8) \, mod \, 77 = 67$$

$$s_2 = (t_2 + (a/t_2)) \, mod \, n = (2 - 10/2) \, mod \, 77 = 74$$

- After this we move on to the decryption step. The first step is to check which condition the private key satisfies. It is of the form $r^2 \equiv \, - a(mod \, n)$.

  - So, to decrypt the message the recipient uses the formula $x = ((s + 2r)/n)$ where $s = s_2$:

$$x = (s_2 + 2r) \, /n = (74 + 46) \, /77 = 120/77 = -1$$

  - Then this is decoded to obtain the bit 1 as plaintext.

Through the first two examples we have observed two cases and how the Cocks IBE works under the condition where $r^2 \equiv a(mod \, n)$ and $r^2 \equiv \, - a(mod \, n)$. However, decryption does not work every time. There are some cases under which the plaintext cannot be obtained. We will be using the same parameters as the previous example and check what is the condition where an incorrect answer comes and we will also obtain the probability of an incorrect answer.

- The first step is to set up the parameters. Let $p = 7 \, and \, q = 11$; then we obtain the value of n as 77. We assume that the public key $a = 10$.
- We have to calculate the value of the private key r as mentioned in the extraction phase. It is given by the equation $r = a^{((\emptyset(n) + 4)/8)} \, mod \, n$.

$$\emptyset(77) = \emptyset(7) \cdot \emptyset(11).$$

- $\emptyset(77) = 6 \cdot 10 = 60$ (using Euler function properties for prime numbers).
- So $r = 10^{((60 + 4)/8)} \, mod \, 77$.

$$r = 10^8 \, mod \, 77.$$

$$r = 23$$

- In this case $r^2 \equiv \, - a (mod \, n)$.
- Let us assume that the bit "1" is to be encrypted with the public key; the bit "1" has to be encoded as -1 and so we pick two values t_1 and t_2 which satisfy the condition mentioned in the encryption phase.
- We choose t_1 and t_2 as 12 and 5 as

  - The Jacobi value $(12/77) = \, - 1$.
  - The Jacobi value $(5/77) = \, - 1$.

- Now the ciphertext value is calculated as substituting the values of t_1 and t_2 in Eqs. (7.6) and (7.7):

$$s_1 = (t_1 + (a/t_1)) \, mod \, n = (12 + 10/12) \, mod \, 77 = 0$$

$$s_2 = (t_2 + (a/t_2)) \, mod \, n = (5 - 10/5) \, mod \, 77 = 3$$

- After this, we move on to the decryption step. The first step is to check which condition the private key satisfies. It is of the form $r^2 \equiv \, - a(mod \, n)$.

  - So, to decrypt the message the recipient uses the formula $x = ((s + 2r)/n)$ where $s = s_2$:

$$x = ((s_2 + 2r) \, /n) = (3 + 46) \, /77 = 49/77 = 0.$$

So, we see that we are not getting the right answer after decryption; the answer is supposed to be -1 but we have obtained 0. This is because $gcd(s_2 + 2r, n) \neq 1$. This case arises when either p or q is a factor of $s_1 + 2r$ or $s_2 + 2r$ (depending on the value of r). Total number of cases where this condition arises $q - 1 + p - 1 + - 1$ (for the total number multiples of p and q) is equal to $p + q - 3$. So, the probability of decryption failure is

$$P_r = (p + q - 3) \, /n$$

So, in the real-world cases with n having 1024f bits and 512Ä bit values for p and q, the probability is extremely small. So, there is no use for special cases within Cocks IBE in order to handle this situation.

8 Correctness of Cocks IBE

Now that the Cocks IBE algorithm and the examples have been explored, we need
to understand the accuracy of this algorithm and check if we are able to obtain the
plaintext using the algorithm.

Choose the values of p and q such that they satisfy the conditions mentioned in
the setting up of parameters.

Therefore r is the square of either a or $-a$.

We know this by

$$r^2 = a^{((n+5-p-q)/8)^2}$$

$$r^2 = a^{((n+5-p-q-\Phi(n))/8)^2}$$

$$r^2 = a^{((n+5-p-q-(p-1)(q-1))/8)^2}$$

$$r^2 = a^{(0.5)^2}$$

$$r^2 = \pm a$$

Now we check if the decryption step can give the plaintext. Let us assume the
case as

$$r^2 \equiv a \bmod n$$

Therefore $s = t + a/t$

$$(s + 2r)/n = (t \ + (a/t) + 2r)/n$$
$$= t\left(1 + a/t^2 + 2\,r/t\right)/n$$

We know $a = r^2$ so replacing a with r^2 and converting it to a whole square we
get

$$t(1 + r/t)^2/n = (t/n)\left((1 + r/t)/n\right)^2$$
$$= (t/n)(\pm 1)^2 = t/n$$

So, we have obtained the plaintext using the equation mentioned in the decryption
phase so the algorithm works [24].

9   Security

9.1   Quadratic Residuosity Problem

Cocks IBE system can be defeated by a person if he/she is able to figure out the modulus n. This will allow him/her to do the calculation of the private keys, in turn leading the adversary to decrypt any intercepted messages. However, it is not obvious that Cocks IBE system's security is due to the problem of quadratic residuosity. We know that the adversary's ability to handle the decryption of a message encrypted with Cocks IBE system is on deciding upon whether the value of user public key a is square modulo n or not [25]. We know

$$1/n = (t/n) \cdot ((1/t)/n) = +1$$

Such that

$$(t/n) = ((1/t)/n)$$

Thus,

$$((a/t)/n) = (a/n)((1/t)/n)$$
$$= (a/n)(t/n)$$

Now, we have the following solutions:

$$t_1 = t.e_1 + t.e_2$$

$$t_2 = t.e_1 + (a/t).e_2$$

$$t_3 = (a/t).e_1 + t.e_2$$

$$t_4 = (a/t).e_1 + (a/t).e_2$$

where e_1 and e_2 have the following property:

$$e_1 \equiv 1 \pmod{p}$$

or

$$e_2 \equiv 0 \pmod{p}$$

$$e_1 \equiv 0 \pmod{q}$$

or

$$e_2 \equiv 1 \pmod{q}$$

Now, the solutions to the above equations will also be having the following properties:

$$t_1/n = (t/p)\,(t/q)$$

$$t_2/n = (t/p)\,((a/t)\,/q) = (t/p)\,(a/q)\,(t/q)$$

$$t_3/n = ((a/t)\,/p)\,(t/q) = (a/p)\,(t/p)\,(t/q)$$

$$t_4/n = ((a/t)\,/p)\,((a/t)\,/q) = (a/p)\,(t/p)\,(a/q)\,(t/q)$$

If a is a square, then we will be having

$$a/p = (a/q) = +1$$

Such that

$$t_1/n = (t_2/n) = (t_3/n) = (t_4/n)$$

If a is not a square, then we will be having

$$a/p = (a/q) = -1$$

Such that

$$t_1/n = (t_4/n)$$

with

$$t_2/n = (t_3/n)$$

but

$$t_1/n = -\,(t_2/n)$$

We should note that if any of t_1, t_2, t_3, or t_4 is used as the random input in our encryption, the ciphertext created will be the same in all the cases.

For instance, if we use t_1 as the random input,

$$s = (t_1 + a/t_1)$$
$$= (t.e_1 + t.e_2 + a/(t.e_1 + t.e_2))$$
$$= t + a/t$$

However, if we use t_2 as the random input,

$$x = ((s + 2r)/n)$$

$$= (t.e_1 + (a/t).e_2 + a/(t.e_1 + (a/t).e_2))$$
$$= t + a/t$$

Similarly, same value for s will be calculated if we use t_3 and t_4 as the random inputs.

Thus, in the case of a not being a square, the same value of ciphertext can come from different values of plaintext. To distinguish between these different cases, there is only one way, which is the process of determining if a is a square modulo n, and hence the problem of quadratic residuosity.

9.2   Chosen Ciphertext Security

Because of the encryption of a single bit at a point of time by the Cocks IBE scheme, the scheme is prone to a chosen ciphertext attack. For instance, let us say that there is an attacker Rohit having a plaintext $(m_a, m_b, \ldots, m_k)$ and the corresponding ciphertext $(c_a, c_b, \ldots, c_k)$ which has been encrypted to a user Ritvik, who wants the plaintext of the ciphertext $(ca_2, cb_2, \ldots, ck_2)$. Rohit can send a message $(ca_2, cb, \ldots, ck)$ to Ritvik and check his reaction. Rohit can observe if Ritvik uses the ciphertext as the shared secret that he uses to derive a session key. In a similar way, Rohit can repeat this entire process of recovering a single bit each time, to recover all the bits of decryption [25, 26].

9.3   Proof of Security

It can be easily found out using the random oracle model that solving the quadratic residuosity problem is almost equivalent to defeating the Cocks IBE scheme, where an adversary can use his/her Cocks IBE decryption algorithm to solve the problem of quadratic residuosity [24, 27]. Thus, if we can prove the intractability of quadratic residuosity problem, we can prove that Cocks IBE scheme is sufficiently secure.

10 Summary

Here is a summary of all the mentioned algorithms in the Cocks IBE scheme:

1. Cocks IBE Setup
   INPUT: A parameter κ
   OUTPUT: p, q, n and H_1

   (a) Pick a random prime number p where p is congruent to 3 (*mod* 4), and satisfies the security parameter criteria.

   Table 7.2 denotes the size of Cocks IBE ciphertext corresponding to the symmetric key lengths

   (b) Pick a random prime number q where q is congruent to 3 (*mod* 4), and satisfies the security parameter criteria.
   (c) Let $n = p.q$.
   (d) Choose the hash function appropriately $H_1 : \{0,1\}^* \rightarrow Z_n$ so that $((H_1(ID))/n) = +1$ for any $ID \in \{0,1\}^*$.

2. Cocks IBE Public Key Calculation.
   INPUT: n, a character string ID representing the identity, and a hash function H_1

   (a) H_1Calculation of $H_1(ID)$.

3. Cocks IBE Private Key Extraction.
   INPUT: $a, p,$ and q
   OUTPUT: r

   (a) Calculate the value of r:

   $$r = a^{(\phi(n)+4)/8} \bmod n = a^{(pq-p-q+5)/8} \bmod n$$

4. Cocks IBE Encryption
   INPUT: n, and a plaintext bit m

Table 7.2 Output of Ciphertext Size for different size key lengths

Symmetric Key Length	Cocks IBE Ciphertext Size
80 bits	166,710 bits
112 bits	458,752 bits
128 bits	768,432 bits
256 bits	7,864,320 bits

OUTPUT: Ciphertext (s_1, s_2), an integer modulo n for each component

(a) Encode m as $x = (-1)^m$.
(b) Pick t_1 and t_2 randomly using.

$$(t_1/n) = (t_2/n) = x$$

(c) Calculate s_1 using.

$$s_1 = (t_1 + (a/t_1)) \mod n$$

(d) Calculate s_2 using.

$$s_2 = (t_2 - (a/t_2)) \mod n$$

5. Cocks IBE Decryption
   INPUT: Private key r, ciphertext (s_1, s_2), and n
   OUTPUT: Plaintext bit m

(a) If r^2 is equivalent to $a (mod\ n)$, then $s = s_1$; otherwise $s = s_2$.
(b) Calculation of plaintext bit x will be done using.

$$x = ((s + 2r)/n)$$

(c) If x is equal to -1, the value of m will be equal to 0; otherwise the value m will be 1.

References

1. Clifford Cocks, An identity based Encryption scheme based on quadratic residues 2007-02-06 at the Wayback machine, proceedings of the 8th IMA international conference on cryptography and coding, 2001
2. Shamir, A. (1984). Identity-based cryptosystems and signature schemes. In *Proceedings of CRYPTO '84, Santa Barbara, CA, August 19–22* (pp. 47–53).
3. Pengqi Cheng, Yan Gu, Zihong Lv, Jianfei Wang, Wenlei Zhu, Zhen Chen, and Jiwei Huang A. Performance Analysis of Identity-Based Encryption Schemes.
4. Dan Boneh and Xavier Boyen., Secure Identity Based Encryption Without Random Oracles.
5. Joye, M., & Neven, G. *Identity Based Cryptography* (pp. 45–65).
6. Chatterjee, S., & Sarkar, P. *Identity-Based Encryption* (p. 121.e).
7. Michael Clear, Hitesh Tewari, Ciaran, McGoldrick. Anonymous IBE from Quadratic Residuosity with Improved Performance.
8. Zhi-Wei Sunn A congruence for primes Proc. Amer. Math.
9. Hardy, G. H., & Wright, E. M. (2008). *An Introduction to the Theory of Numbers (6th ed.)*. Oxford University Press.
10. Niven, I., & Zuckerman, H. S. (1966). *An introduction to the theory of numbers (2nd ed.)*. John Wiley & Sons.

11. Pettofrezzo, A. J., & Byrkit, D. R. (1970). *Elements of number theory*. Englewood Cliffs: Prentice Hall.
12. Sandifer, C. (2007). *The early mathematics of Leonhard Euler*. MAA.
13. Bach, E., & Shallit, J. (1996). *Algorithmic number theory (Vol I: Efficient algorithms), MIT press series in the foundations of computing*. Cambridge, MA: The MIT Press.
14. Dickson, L. E. (1952). History of the theory of numbers. In *vol 1, chapter 5 "Euler's function, generalizations; Farey Series"*. Chelsea Publishing.
15. Gauss, C. F., & Clarke, A. A. (1986). *Disquisitiones Arithmeticae (Second corrected ed.)*. New York: Springer.
16. Bach, E., & Shallit, J. (1996). *Efficient algorithms, algorithmic number theory, I*. Cambridge: The MIT Press.
17. Gauss, C. F., & Maser, H. (1965). *Untersuchungen über höhere Arithmetik (Disquisitiones Arithmeticae & other papers on number theory) (second edition)*. New York: Chelsea.
18. Gauss, C. F., & Clarke, A. A. (1986). *Disquisitiones Arithmeticae (second, corrected edition)*. New York: Springer.
19. Cohen, H. (1993). *A course in computational algebraic number theory*. Berlin: Springer.
20. Lemmermeyer, F. (2000). *Reciprocity Laws: From Euler to Eisenstein*. Berlin: Springer.
21. Ireland, K., & Rosen, M. (1990). *A classical introduction to modern number theory* (2nd ed.). New York: Springer.
22. Pardo, J. L. G. *Introduction to Cryptography with Maple* (pp. 587–611).
23. Anca-Maria Nica and Ferucio Laurenţiu Ţiplea., On Anonymization of Cocks' Identity-based Encryption Scheme.
24. Bica, I., & Reyhanitabar, R. *Innovative Security Solutions for Information Technology and Communications* (pp. 68–69).
25. Ferucio Laurentiu Tiplea, Sorin Iftene, George Teseleanu, and Anca-Maria Nica., Security of Identity-based Encryption Schemes from Quadratic Residues
26. Ran Canetti, Shai Halevi, Jonathan Katz., Chosen-Ciphertext Security from Identity-Based Encryption
27. Man Ho, A., Barbara Carminati, C.-C., & Kuo, J. (2014). *Network and system security: 8th international conference, NSS 2014, Xi'an, China, October 15-17* (p. 272).

Chapter 8
Boneh-Franklin IBE

Deepak Kumar Sharma, Bhanu Tokas, Venkata Rohit Jakkinapalli, and Ritvik Nagpal

Abstract There has been an unprecedented rise in insecure computer networks in the last few decades. This has led to a rise in the need for large-scale cryptography to improve security. Public key encryption is a relatively new concept that has risen in prominence due to the limitations in the more traditional symmetric cryptographic methods like key management which becomes impractical when used at a large scale.

Identity-based encryption (IBE) systems are defined as a category of public key encryption wherein the user's public key is derived from a distinctive identity of a user (for example, email ID). In this chapter, we will talk about the first IBE scheme with the comprehensive analysis that will include reviewing all the four stages involved: **setup**, **extract**, **encrypt**, and **decrypt**. We will also discuss the different conceptions of security, such as chosen ciphertext security and semantically secure identity-based systems, with detailed examination about the security of this scheme.

Keywords Identity-based encryption IBE · Private key extraction · Boneh-Franklin scheme · Encryption · Decryption · Security

1 Introduction

As some of the readers may already know, Boneh-Franklin was amongst the first secure and practical IBE schemes to be invented in the early 2000s. Now, many readers may wonder if, in our fast-changing world, something decades old is still relevant. Well, the Boneh-Franklin IBE is far from optimized. It is a part of the full-domain hash family of the IBE schemes. This means that the identity used *ID* is

D. K. Sharma (✉)
Department of Information Technology, Netaji Subhas University of Technology, (Formerly Netaji Subhas Institute of Technology), New Delhi, India

B. Tokas · V. R. Jakkinapalli · R. Nagpal
Department of Computer Engineering, Netaji Subhas University of Technology, New Delhi, India

© Springer Nature Switzerland AG 2021
K. A. B. Ahmad et al. (eds.), *Functional Encryption*, EAI/Springer Innovations in Communication and Computing, https://doi.org/10.1007/978-3-030-60890-3_8

mapped onto an elliptic curve [1] at a point Q_{ID} which is subsequently applied in the encrypting and decrypting process. This process of transforming an identity to an elliptic curve requires modular exponentiation, thus making it quite expensive to calculate. Hence, recent works tend to focus more on systems that can translate an ID to an integer, rather than using an expensive full-domain hash function. Another shortcoming of the Boneh-Franklin IBE is its requirement of calculation of pairings, a fairly expensive operation during encryption as well as decryption stage.

The reader might be contemplating if Boneh-Franklin IBE is indeed so expensive to perform and better alternatives exist, then what is the point of studying it. Well, the reason is the same as to why we are still taught Bohr's model of an atom because firstly, we must start from the basics and secondly, older models, though inefficient, are able to help us realize the thought process and reasoning followed in the development of the newer models. Keeping this in mind, let us start with this fascinating IBE scheme.

Boneh-Franklin IBE has adopted some of its attributes from "ElGamal encryption" [2] and "Joux's three-way key exchange" [3]. Similar to the ElGamal encryption, which encrypts a plaintext message using a key from Diffie-Hellman key exchange [4], Boneh-Franklin IBE also utilizes the shared secret key to encrypt the plaintext messages. Just like the exchange, which generalizes the "Diffie-Hellman key exchange" to three parties, Boneh-Franklin also has three secret keys involved. One is a random key generated by the sender, other refers to a master key used in the system, and lastly the final key is a discrete logarithm of the identity of the receiver.

Some readers who might have read or are interested in reading the original paper, "Identity-Based Encryption from the Weil Pairing" [5, 6], should note that there is a slight difference in the notation used here as compared with the original paper. Namely, the values represented by the variables p and q have been replaced. In the original paper, p was used to denote the finite field p's order and q was a prime integer denoting the order of the group $E(F_p)[q]$. Subsequent works interchanged the terms such that now q is used to denote the finite field F_q's order and p now denotes the order of group G_1. Thus, the author advises the readers to clearly understand the meaning of all variables before referring to another text, in order to avoid confusion.

1.1 Identity-Based Encryption (IBE)

Before we further delve into Boneh-Franklin IBE scheme, let us understand the IBE scheme by discussing how it works and who proposed it.

In this category of public key cryptography, some unique parameter is extracted from the user's public addresses (i.e., their name or their address), so as to create the public key. This allows the sender who knows the public parameters of the scheme to encrypt communication to use a key that is based on a public identity of the person who is receiving the message such as their email address or full name. The decryption key is imparted using a private key generator (PKG), which generates secret keys for every user and hence needs to be trusted.

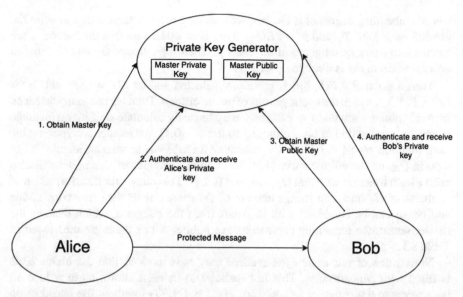

Fig. 8.1 IBE system

The basic functioning of an IBE system is demonstrated in Fig. 8.1.

Adi Shamir [7] first introduced the IBE system in 1984. However, he was able to only provide an instance of ID-based signatures. Thus, IBE systems persisted to be an unsolved problem for a long time until the Boneh-Franklin IBE [5, 8] and Cocks IBE [8, 9] were introduced in 2001.

2 Boneh-Franklin (Basic Scheme)

Before we dive deep into the Boneh-Franklin scheme, we should take a look into the Boneh-Franklin basic scheme. Herein, the sender and receiver shall use a secretly shared key to perform the encryption and decryption of the message data. This scheme is simpler than the full Boneh-Franklin scheme but as you may have guessed it is also much less secure.

2.1 Setting up the Parameters

Like any other scheme, the first step of Boneh-Franklin scheme is setup of parameters. We begin with a security criterion that will help indicate the order of bit strength provided by the scheme. Next, we define groups G_T and G_1, and a pairing $e^\wedge$ such that $e^\wedge : G_1 \times G_T$. This is achieved by using an elliptic curve $E \backslash F_q$ which has an embedding degree of k. Further, we require a prime integer p that satisfies the conditions $p \mid \#E(F_q)$ and $p^2 \nmid \#E(F_q)$. This is to make sure that the identities are hashed into a unique subgroup of the order p. Therefore, groups G_T and G_1 have an order p wherein G_T is a subgroup of $F_{q^k}^*$.

Then a point $P \in E(F_q)[p]$ is randomly selected and let $G_1 = \cdot P \cdot$ and $G_T = \cdot e^\wedge (P, P) \cdot$, which are cyclic groups of prime order p. Then we randomly select an integral value s, such that $s \in F_p^*$. Thus, we can easily calculate sP. Since a message is usually a collection of bits belonging to the set $\{0,1\}$, we require a cryptographic hash function $H_1 : \{ 0,1 \}^* \to G_1$, which will enable us to map an identity ID to a point Q_{ID} on the elliptic curve [10]. To encrypt messages of size n, we will also need a hash function such that $H_2 : G_T \to \{ 0,1 \}^n$. This allows for transformation of elements of G_T into n bit strings that are of the same size as the plaintext message and hence can be combined with it. Apart from the integer s, which denotes the master secret, the remaining parameters are public. All of these are mentioned in Table 8.1.

Now, some of our eagle-eyed readers may have noticed that the above table is filled with dependencies. This is beneficial to us as it allows us to reduce all parameters to a function of $(e^\wedge, n, \ H_1, H_2, sP, G_1, G_T)$ without the introduction of uncertainty.

2.2 (Basic Scheme) Private Key Extraction

After the master key and public parameters are defined one can find the equivalent private key for an *ID* by mapping it to a point P on the elliptic curve E. Thus, we use the formula

Table 8.1 Setting up parameters

Element	Type	Comments
q	Prime power	Order of finite field F_q
E/F_q	Elliptic curve	$E(F_q)$ has embedding degree k
p	Prime	$p \mid \# E(F_q), p^2 \nmid \# E(F_q)$
G_1	Cyclic group	Subgroup of $E(F_q)$, $G_1 = \langle P \rangle$
G_T	Cyclic group	Subgroup of F_q^*, $G_T = \langle e^\wedge (P, P) \rangle$
$e^\wedge$	Pairing	$e^\wedge : G_1 \times G_1 \to G_T$
n	Integer	Length of plaintext (in bits)
P	Point on elliptic curve	$P \in G_1$
sP	Point on elliptic curve	$sP \in G_1$
H_1	Cryptographic hash function	$H_1 : \{0.1\}^* \to G_1$
H_2	Cryptographic hash function	$H_2 : G_T \to \{0.1\}^n$

$$Q_{ID} = H_1(ID)$$

And we deduce the private key by simply multiplying Q_{ID} with the private key s.

2.3   Encryption (Basic Scheme)

Obtaining of the corresponding ciphertext of message M using the receiver's ID is described below:

1. Select a random integer $r \in Z_p^*$ and use it to calculate value rP.
2. Use the recipient's identity ID to find the value of Q_{ID} using the formula

$$Q_{ID} = H_1(ID)$$

3. Use Q_{ID} from the Previous Step, to Calculate K Using the Formula

$$K = H_2\left(e^{\wedge}(rQ_{ID}, sP)\right)$$

4. Set the Equivalent Ciphertext C to $C = (C_1, C_2)$ where C_1 and C_2 Are Defined by

$$C_1 = rP$$

$$C_2 = M \oplus K$$

2.4   Decryption (Basic Scheme)

On obtaining the ciphertext C, the recipient follows the below-mentioned steps to obtain the plaintext message:

1. Uses his/her private key sQ_{ID} and ciphertext component C_1 to calculate K using the formula

$$K = H_2\left(e^{\wedge}(sQ_{ID},\ C_1)\right)$$

2. Calculates $M = C_2 \oplus K$

Herein, plaintext M is recovered as the sender is able to deduce the value of K using the equation

$$K = H_2\left(e^{\wedge}(Q_{ID}, sP)^{rs}\right)$$

Subsequently the receiver deduces the value of K using the formula

$$K = H_2 \left(e^{\wedge} (Q_{ID}, P)^{rs} \right)$$

3 Boneh-Franklin IBE (Full Scheme)

As some of our readers may have already noted, a chosen ciphertext attack may be effective against the Boneh-Franklin basic scheme as the value of K is independent of the message M. Therefore, if someone can decrypt the ciphertext $(C_1, C_2 \oplus \in)$ to obtain plaintext message $M \oplus \in$ and subsequently from the plaintext text they can obtain the original message sent by the user M as $M = (M \oplus \in) \oplus \in$. This vulnerability is eliminated by using the "Fujisaki-Okamoto transform." Using this extra level of hashing, as required by Fujisaki-Okamoto transform, adds more complexity to the IBE scheme. While this transformation secures our scheme against chosen ciphertext attack, it also increases the complexity of the encryption and decryption stages.

3.1 Setup of Parameters

While we retain all the parameters from the basic scheme, mentioned in Table 8.1, we need to introduce two additional hash functions to add more complexity to the IBE scheme:

$$H_3 : \{0, 1\}^n \times \{0, 1\}^n \rightarrow Z_p^*$$
$$\text{and}$$
$$H_4 : \{0, 1\}^n \times \{0, 1\}^n \rightarrow Z_p^*$$

3.2 (Full Scheme) Private Key Extraction

The process for extraction of parameters in the full scheme is indistinguishable from the one followed in the basic scheme.

3.3 Encryption (Full Scheme)

Obtaining of the corresponding ciphertext of message M using the receiver's *ID* is described below:

1. Calculate the corresponding private key for the recipient's identity ID using the formula

$$Q_{ID} = H_1(ID)$$

2. . Select a random value σ, such that $\sigma \in \{0, 1\}^n$. Using this, define r as

$$r = H_3(\sigma, M)$$

3. Define C_1 as

$$C_1 = rP.$$

4. Define C_2 as

$$C_2 = \sigma \oplus H_2\left(e^{\wedge}(rQ_{ID}, sP)\right)$$

5. Define C_3 as

$$C_3 = M \oplus H_4(\sigma)$$

6. Finally, define ciphertext C as

$$C = (C_1, C_2, C_3)$$

3.4 Decryption (Full Scheme)

After obtaining the ciphertext C, the recipient follows the below-mentioned steps to obtain the plaintext message:

1. Calculate σ using the formula

$$\sigma = C_2 \oplus H_2\left(e^{\wedge}(sQ_{ID}, C_1)\right)$$

2. Find the plaintext message M, using the formula

$$M = C_3 \oplus H_4(\sigma)$$

3. Calculate rP where $r = H_3(\sigma, M)$. If $rP \neq C_1$, then ciphertext is invalid and should be rejected.

4 Security

It would be appropriate to state that the difficulty of solving this scheme is directly proportional to the difficulty of the "bilinear Diffie-Hellman Problem" (BDHP).

We can say that for a particular (unknown) value of t, $Q_{ID} = tP$. Thus, we can also say tha

$$e^\wedge (rQ_{ID}, sP) = e^\wedge (rtP, sP) = e^\wedge(P, P)^{rst}$$

This means that the ciphertext C can also be represented as

$$C = \left(rP, M \oplus H_2 \left(e^\wedge(P, P)^{rst} \right) \right)$$

While P and tP can be easily obtained by an adversary from the public parameters and using the recipient's identity, $Q_{ID} = tP$ can be calculated further, and one can observe the ciphertext for rP. Thereafter, if one is able to compute $e^\wedge(P, P)^{rst}$ from the above-deduced parameters, then it would be possible to find the plaintext message M recovered by using the following formula:

$$\left(M \oplus H_2 \left(e^\wedge(P, P)^{rst} \right) \oplus H_2 \left(e^\wedge(P, P)^{rst} = M \right.$$

Alas, to calculate $e^\wedge(P, P)^{rst}$ using the above approach would essentially mean solving the BDHP. So, as long as the assumption that the BDHP is adequately difficult holds true, it is safe to assume that it will be difficult for an external agent to successfully perform the decryption of the encrypted message to obtain the plaintext. This is easily achieved by careful selection of G_1 and G_T.

In the original Boneh-Franklin paper, the random oracle model was used to demonstrate that an attacker who is capable of decrypting a communication encrypted using the Boneh-Franklin IBE will be capable of solving the BDHP using his decryption algorithm, so as long as solving BDHP remains adequately difficult, decrypting the Boneh-Franklin IBE would also remain adequately difficult. Both the basic scheme and the full scheme are almost impervious to adaptive chosen identity attacks and chosen ciphertext attacks.

5 Further Works on Boneh-Franklin IBE

Just like any other result of the human mind, the Boneh-Franklin IBE is not flawless. Given that it was introduced almost two decades ago in 2001, it is obvious that other researchers have introduced new variants or improvements in the original IBE scheme. In this section, we will be looking at some of these new and improved IBE schemes.

Using subgroups of curves:

We can significantly improve the efficiency of the Boneh-Franklin IBE system by using subgroups of the curve instead of the whole curve. To further explore this idea, let us take an example wherein we choose two primes p and q, where q is a 160-bit prime and p is a 1024-bit prime with $p = aq - 1$ and $p = 2 \, mod \, 3$. We then choose point P such that it is of order q. Every ID is mapped to a point Q using a hash function and multiplied by a to get a group point. As long as P generates a group that satisfies the BDH assumption, it can be assumed to be secure. Since the Weil computation is applied to points of a smaller order, it is much faster and hence efficient.

Distributed master key:

Much like the private key in a certificate authority system, the master key in the standard implementation of Boneh-Franklin IBE must be protected. We can achieve this by using threshold cryptography to distribute the master key in numerous sites. This process can be extremely robust and efficient in the Boneh-Franklin IBE system. Thus PKG, by computing $Q_{priv} = sQ_{ID}$, generates the private key. Thereafter, we can give each PKG a share (s_i) of the Shamir's secret sharing of $s \, mod \, q$; this allows us to easily distribute the private key. When we need to generate the private key for x PKGs, we can do so following the simple principle $Q_{priv(i)} = s_i Q_{ID}$. The user can calculate Q_{priv} using the formula $Q_{priv} = \sum \lambda_i Q_{priv(i)}$, where λ_i represents the Lagrange coefficients.

It is worth noting that because in the case of G_1 decisional Diffie-Hellman [11] (DDH) assumption is easy, we can easily make the scheme robust in case of dishonest PKGs. In the setup, for each PKG a unique P_{pub} is defined such that $P_{pub(i)} = s_i P$. In the key generation stage, a user may verify the validity of response for each PKG using the formula $e^\wedge \left(Q_{priv(i)}, P \right) = e^\wedge \left(Q_{ID}, P_{pub(i)} \right)$. Therefore, any corrupted or malicious PKG can be detected instantaneously. Thus, unlike conventional robust threshold Schemes [12], herein there is no requirement of any zero-knowledge proof [13]. Using the processes described by Rabin [14], one can create the master key for the PKGs. Another interesting fact worth noting is that with a distributed master key, it is possible to perform threshold decryption [15] on a per-message basis, such that we would not be required to calculate the correspondent decryption key.

Use of IBE as Signature:

As observed by Moni Naor, we can easily construct a public key signature Scheme [14, 16] that is based on this IBE scheme. Let us have a look at the process required to achieve this. The master key of the IBE scheme acts as the signature scheme's private key. The global system parameters of the IBE scheme act as the signature scheme's public key. The IBE decryption key is used as the signature for message M for $ID = M$. The verification of a signature is carried out through the ensuing process; choose a random message M_k and use the public key $ID = M$ to encrypt it after which we try decrypting the ciphertext by defining the decryption key as the signature for M. In case of chosen message attacks, the signature scheme should be practically impossible to forge as long as it follows the scheme IND-ID-

CCA. An important point worth noting here is that the verification process in this instance is randomized contrary to conventional signature schemes.

Using other pairings:

Boneh-Franklin IBE system, considering the BDH assumption to be valid, can be used with other more computationally efficient bilinear pairings of the form $e^\wedge : G_1 \times G_T \to G_2$. There exist several curves, mostly belonging to the Abelian variety [17] that can be used for this specific application. An example of the abovementioned variety of curves would be Tate pairing [18–20].

6 Examples

1. If we take an elliptic curve E, $E/F_q : y^2 = 1 + x^3$ where q is represented using $q \equiv 11(mod\ 12)$, q is a prime number, and G_1 is a subgroup of order p of $E(F_q)$. The hash function H_1 can be defined as $H_1 : \{0, 1\}^* \to G_1$. We first use the hash function to create a string in order to represent the public address of the user as an integer modulo of q. We can obtain the value of H in two possible ways either by going through the different values for H till we get a value that exists within the required range or by reducing the integral output of H to integer modulo of q. The value of H thus obtained denotes the y coordinate for $Q \in E(F_q)$ and we can deduce the value of the x coordinate from the curve

$$x = \left(y^2 - 1\right)^{1/3}$$
$$a^{q-1} \equiv 1 \ (mod\ q) \ \text{(Euler Theorem)}$$
$$a^{2q-1} \equiv a \ (mod\ q)$$
$$a^{(2q-1)/3} \equiv a^{1/3} \ (mod\ q)$$

We require a value $(2q - 1)$ which divides 3; the condition $q \equiv 11(mod\ 12)$ satisfies the above requirement. We can obtain the x coordinate for the point Q_{ID} using the following formula:

$$Q_{ID} = \left(\frac{\#E\left(F_q\right)}{P}\right) Q$$

With the curve $E/F_q: y^2 = x^3 + 1$ we can write $\#E(F_q) = q + 1$ so the value of Q_{ID} can be written as

$$Q_{ID} = \left(\frac{q+1}{p}\right) Q$$

So this means we need a value $p \mid \#E(F_q)$, and we already have a unique order p, so this results in $Q_{ID} \in G_1$ as needed.

2. If we do want to hash the value of the identity onto an elliptic curve, we can try to hash the value of the identity ID into an integer value, say t. We can then use the point tP as the public key. However, there is a problem with this method. This method will make it easier for the hacker to calculate the shared secret which means that the scheme's security is defeated.

3. All elements of G_T belong to the finite field F_{q^k}. So an element of the group G_T can be expressed to be of the form $\alpha = (\alpha_1, \alpha_2, \ldots\ldots, \alpha_k)$ and for every $\alpha_i \in F_q$. So, we create a hash function $H_2 : \{0, 1\}^n \to G_T$ by concatenating the coordinates α as an input for the hash function and reducing it to a value that is in between the range of 0 to $2^n - 1$, by truncating the value of the hash function to give an output of only n bits.

4. For instance, Ritvik wants to perform encryption of a data to send it to Rohit using Boneh-Franklin IBE. Assuming E to be an elliptic curve such that $E/F_{131} : y^2 = x^3 + 1$, $P = (98, 58) \in E(F_{131})[11]$, $G_1 = \langle P \rangle$, and $G_T = \langle e\,\hat{}\,(P, P) \rangle$, $e\,\hat{}\, : G_1 \times G_1 \to G_T$ refers to the reduced modified Tate pairing such that

$$e\,\hat{}\,(P, Q) = e(P, \phi(Q))^{1560}$$

such that ϕ refers to distortion map $\phi(x, y) = (zx, y)$ for $z = 65 + 112i$. Let us suppose that integer $s = 7$ is the master secret of this system, such that $sP = (33, 100)$, and suppose that Rohit's identity reveals that $H_2(ID_{Rohit}) = Q_{ID} = (128, 57)$. So, Rohit's private key is $sQ_{ID} = (113, 8)$ Table 8.2.

Using the values discussed above, Ritvik can encrypt a message for Rohit. Suppose that he generates the random $r = 5 \in Z_{11}^*$ to do this. Ritvik then calculates $rQ_{ID} = (5)(128, 57) = (98, 73)$ and uses it to calculate

$$rP = 5P = (34, 23)$$

and

$$K = H_2\left(e\,\hat{}\,(rQ_{ID}, sP)\right) = H_2\left(e\,\hat{}\,(98, 73), (33, 100))\right) = H_2\left(49 + 58i\right)$$

Table 8.2 Brief description of example 5 values

Parameters	Type	Value	Comments
P	Elliptic curve point	(98,58)	$P \in E(F_{131})$ [12]
sP	Elliptic curve point	(33,100)	
Q_{ID}	Elliptic curve point	(128,57)	$Q_{ID} \in E(F_{131})$ [12]
sQ_{ID}	Elliptic curve point	(113,8)	Rohit's private key
r	Integer	5	Generated randomly by Ritvik
s	Integer	7	Master secret

which he then uses to create the ciphertext (C_1, C_2) where $C_1 = rP$ and $C_2 = M \oplus K$. When Rohit receives this ciphertext, he then calculates

$$K = H_2 \left(e\hat{}\, (s\, Q_{ID}, C_1) \right) = H_2 \left(e\hat{}\, (113, 8), (34, 23) \right) = H_2 (49 + 58i)$$

which he then uses to recover the plaintext M using the following formula:

$$M = C2 \oplus K = (M \oplus K) \oplus K$$

7 Conclusion

In this chapter, we have explored the Boneh-Franklin IBE scheme. In our exploration, the processes of both the basic scheme and full scheme are divided into four stages: setup, extraction of parameters, encryption, and decryption. We also discussed the security of this scheme and observed that it is dependent on the "bilinear Diffie-Hellman problem" (BDH), such that the complexity in decrypting the message encrypted by Boneh-Franklin IBE is directly proportional to the complexity of solving the BDH problem. Further, we also noted that while the IBE system is a quite efficient and reliable system, it is still far from an optimum state; hence we have explored the possibilities of using different curves (such as Tate), using subgroups of the curve, and using a distributed master key. There still remain several topics that were out of the scope of this book and thus the author would like to suggest the readers explore the references for further increasing their knowledge.

References

1. Lang, S. (1973). *Elliptic functions*. Reading: Addison-Wesley.
2. ElGamal, T. (1985). A public-key cryptosystem and a signature scheme based on discrete logarithms. *IEEE Transactions on Information Theory, IT-31*(4), 469–472.
3. Joux, A. (2000). A one-round protocol for tripartite Diffie-Hellman. In *Proceedings of the 4th international algorithmic number theory symposium, Leider, the Netherlands* (pp. 385–394).
4. Diffie, W., & Hellman, M. (1976). New directions in cryptography. *IEEE Trans Inf Theory, IT-22, 6*, 644–654.
5. Boneh, D., & Franklin, M. Identity based encryption from the Weil pairing. *SIAM Journal of Computing*.
6. Boneh, D., & Franklin, M. (2001). Identity based encryption from the Weil pairing. In *Extended abstract in advances in cryptology—Crypto 2001, Lecture Notes in Computer Science, Vol. 2139* (pp. 231–229). SpringerVerlag. http://eprint.iacr.org/2001/090/.
7. Shamir, A. (1984). Identity-based cryptosystems and signature schemes. In *Proceedings of CRYPTO 84, Santa Barbara, CA* (pp. 47–53).
8. Martin, L. (2008). *Introduction to identity-based encryption*. Norwood, Massachusetts: Artech House.

9. Cocks, C. (2001). An identity based encryption scheme based on quadratic residues. In *Proceedings of the eighth IMA international conference on cryptography and coding, Cirencester, UK* (pp. 360–363).
10. Silverman, J. (1986). *The arithmetic of elliptic curve.* Springer-Verlag.
11. Boneh, D. (1998). The decision Diffie-Hellman problem. In *Proc. third algorithmic number theory symposium, Lecture Notes in Computer Science, Vol. 1423* (pp. 48–63). springer-Verlag.
12. Gennaro, R., Jarecki, S., Krawczyk, H., & Rabin, T. (2000). Robust and efficient sharing of RSA functions. *Journal of Cryptology, 13*(2), 273–300.
13. Feige, U., Fiat, A., & Shamir, A. (1988). Zero-knowledge proofs of identity. *Journal of Cryptology, 1*, 77–94.
14. Gennaro, R., Jarecki, S., Krawczyk, H., & Rabin, T. (1999). Secure distributed key generation for discrete-log based cryptosystems. In *Advances in cryptology—Eurocrypt '99, Lecture Notes in Computer Science, Vol. 1592* (pp. 295–310). Springer-Verlag.
15. Gemmell, P. (1997). An introduction to threshold cryptography. In *CryptoBytes, a technical newsletter of RSA Laboratories, Vol. 2, No. 7.*
16. Boneh, D., & Lynn, B. Short signatures from the Weil pairing. In H. Shacham (Ed.), *Advances in Cryptology—AsiaCrypt 2001, Lecture Notes in Computer Science.*
17. Rubin, K., & Silverberg, A. (2002). Supersingular abelian varieties in cryptography. In *Advances in cryptology—Crypto 2002, Lecture Notes in Computer Science.* Springer-Verlag.
18. Frey, G., Muller, M., & Ruck, H. (1999). The Tate pairing and the discrete logarithm applied to elliptic curve cryptosystems. *IEEE Tran. Info. Th., 45*, 1717–1718.
19. Galbraith, S., Harrison, K., & Soldera, D. Implementing the Tate-pairing. In *Proc. fifth algorithmic number theory symposium, Lecture Notes in Computer Science* (p. 2002). Springer-Verlag.
20. Joux, A. The Weil and Tate pairings as building blocks for public key cryptosystems. In *Proc. fifth algorithmic number theory symposium, Lecture Notes in Computer Science* (p. 2002). Springer-Verlag.
21. Odyurt, U., Application of Fuzzy Logic. In *Identity-based cryptography, dissertation*, 2014.

Chapter 9
Boneh-Boyen IBE

Ankita Karale, Vladimir Poulkov, Milena Lazarova, and Pavlina Koleva

Abstract Identity-based encryption (IBE) is a form of public key infrastructure (PKI) which has become a hot topic of research among the research community as IBE enables senders to encode data without interfering with the public key certificate. IBE computes the public key from the unique recipient's identity and authenticates by a third party so as to overcome the limitations faced by traditional cryptographic techniques. In 2004, D. Boneh and X. Boyen worked towards IBE without using random oracles which is one of the most popular and efficient IBE techniques. This chapter focuses on the Boneh-Boyen IBE technique which is based on selective identity and gives good results in standard model. It also enlightens on the basic IBE model introduction and its working to get a clear idea about IBE. Later on, this document covers the Boneh-Boyen basic scheme that is classified into two categories, additive scheme and multiplicative scheme. These schemes are explained in detail with the help of four main stages: parameter setup, extraction, encryption, and decryption process. Apart from the basic scheme, Boneh-Boyen full scheme is explained here which helps to compute the public key of the receiver. At the end, the chapter emphasizes on the security concept of the Boneh-Boyen IBE.

Keywords Public key infrastructure · Identity-based encryption · Random oracles · Cryptography · Decryption

1 Introduction

An identity-based encryption (IBE) is one of the popular types of public key encryption. This technique mostly addresses the limitation of public key infrastructure and

A. Karale (✉) · V. Poulkov · M. Lazarova
Technical University of Sofia, Sofia, Bulgaria
e-mail: vkp@tu-sofia.bg; milaz@tu-sofia.bg

P. Koleva
SITRC, Sandip Foundation, Nashik, India

© Springer Nature Switzerland AG 2021
K. A. B. Ahmad et al. (eds.), *Functional Encryption*, EAI/Springer Innovations in Communication and Computing, https://doi.org/10.1007/978-3-030-60890-3_9

has evolved as a better solution to cryptography-related issues which makes IBE suitable to work in real time. The IBE system provides featured key evaluation when compared to traditional public key systems and offers an economic solution to the problems like PKI need and communication overhead which are difficult to resolve with the traditional public key systems.

Public key infrastructure (PKI):

If sender and receiver are unknown to each other there is a communication gap to share keys with each other. Hence a successful public key encryption process demands reliable third-party key distribution authority which will ensure authenticity of keys. To overcome these limitations there is the necessity of trusted third parties to distribute public keys. This need emerges from the public key infrastructure (PKI). One of the most popular methods of encryption is public key encryption named by researchers as public key infrastructure [1] also known as PKI. PKI uses certification authorities (CA). User registers himself/herself to the registration authority. Authority will check its authenticity and the verifier will verify his/her identity and after verification CA will issue corresponding keys to the user for encryption purpose. So, this trusted third party will deliver keys to the sender which further works like the public key encryption technique stated above. Figure 9.1 shows the key generation process in the traditional public key system.

1.1 Limitation

PKI is the most powerful secure technique but the main problem with PKI is certificate size and its management. Wherever communication-established certificates get stored in RAM, it leads to large space consumption and communication overhead while distributing the certificate. Also, certificate processing consumes processor capacity which affects computation cost and connectivity with certificate authority (CA).

But to obtain the data privacy it requires public keys of approved users so as to transmit data independently to the target user which ultimately increases the demand for bandwidth [2]. To provide a solution to this drawback broadcast encryption was introduced. This technique addresses the mentioned issue but it is mostly applicable when the data provider has prior knowledge of the target user. It uses a public key for the process of encryption and original data is retrieved by decryption by using only a single secret key. So the more advanced encryption solutions are required.

1.2 Identity-Based Encryption (IBE)

Research is always thrust towards betterment. Can there be a better solution to implement public key encryption? This question is answered by smart innovation of

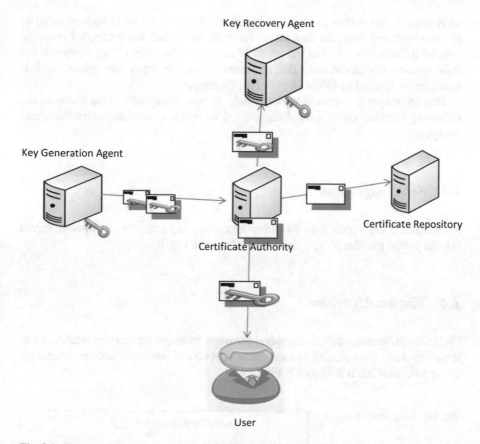

Fig. 9.1 Key generation process in a traditional public key system

identity-based encryption (IBE). A. Shamir [3] introduced an encryption technique based on identity also known as IBE.

Without interfering with the public key certificate here sender can encode data which simplifies the process. Due to this feature of IBE it is appropriate for real-time applications. Here a combination of character is treated as identity. So ultimately A can send a message to B without the help of PKI and it can work smoothly as it reduces communication overhead. The concept is to use the user's identity as a public key, for example email ID of the user. Identity of the user will be considered as a public key and the centralized key server will be responsible for creation of a private key. Here the basic difference between PKI and IBE is that IBE eliminates the need of the certificate lookup process required by PKI.

For example A has the identity of himself as an email address. a@example.com. This identity of A will be used by him to get a private key from a centralized key server. Email address of A will be used by B to encrypt the message. Only A can decrypt the encoded message as email address which acts as an identity belonging to A only. A has access to his identity a@example.com so he is authenticated to get a private key from the key server in order to decode the encrypted message sent by B. Here key server is the center of attraction. Security of key servers is the main concern in order to make IBE a successful mechanism of encryption which is successfully tackled by Boneh-Boyen IBE technique.

IBE technique is executed with the help of four basic algorithms given in the following section. These are usually named as setup, extraction, encryption, and decryption.

1.3 Setup Algorithm

A centralized key server runs the **Setup** algorithm and generates its **master secret** (*s*) and public parameters *<params>*. It is shown in Fig. 9.2.

1.4 Extract Algorithm

PKG also generates a private key (d_{ID}) for users from the parameters introduced in the setup phase that calculates its **master secret** (*s*) and public parameters *<params>* along with user ID. It is shown in Fig. 9.3.

Fig. 9.2 Setup algorithm

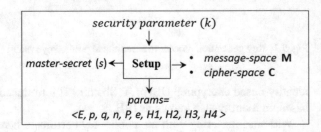

Fig. 9.3 Extract algorithm

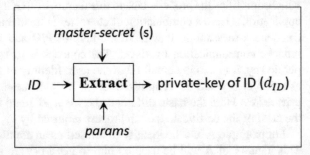

Fig. 9.4 Encryption
algorithm

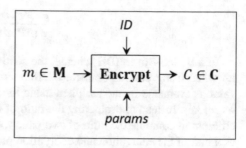

Fig. 9.5 Decryption
algorithm

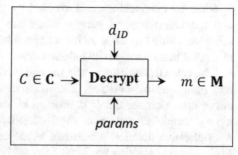

1.5 Encryption Algorithm

Parameters introduced in the setup phase are used to evaluate the IBE public key
which is referred by senders (users) to encrypt the message (M) and the user's
identity and generate ciphertext (C). It is shown in Fig. 9.4.

1.6 Decryption Algorithm

A recipient's identity and the private key of the PKG are used to compute IBE
secret key; along with it the receiver decrypts the message (M) and then generates
the plaintext. It is shown in Fig. 9.5.

1.7 Boneh-Boyen IBE

Boneh and Boyen [4] introduced two effective techniques which are proved secure
against selective-identity attack. They are named as BB1-IBE and BB2-IBE, and
their speciality is that they are designed without using random oracles. The
concept of BB1-IBE is dependent on commutative blinding [5] schema. Here secret
coefficients and blinding factors commute with one another under the pairing:

$$\frac{e\ (aP, bQ)}{e\ (bP, aQ)}$$

In a Boneh-Boyen IBE scheme, the sender encrypts the message and calculates its value using public parameters. After computing a ratio of pairings, the same value is evaluated by the recipient using the resultant encrypted data along with its secret key. Instead of evaluating the ratio of two pairing individuals, it will be more efficient to compute the ratio of two pairs together. Integer value is hashed with ID as a part of the encryption and decryption phases.

Modular exponentiation is avoided by the above method. Due to the modular exponentiation method such systems are more efficient than full-domain hash schemes, similar to the Boneh-Franklin scheme which needs hashing of an identity to a point that resides on an elliptic curve.

The basic Boneh-Boyen scheme can be described in two ways: additive notation and multiplicative notation. The additive notation is generally a part of elliptic curve operation set and it is utilized in various cryptographic standards. This is a very simplified version of the Boneh-Boyen technique. Another scheme is the multiplicative notation which can be described as the additive notation scheme. Multiplicative notation has been commonly utilized in recent literature on pairing-based cryptography.

1.8   Classification of IBE Schemes

This section gives a classification of IBE systems. These classification schemes support fundamental safety reduction in standard model or in random oracle model.

1.8.1   "Quadratic Residuosity" IBE

This is the only scheme of IBE which does not use bilinear pairings because of Cocks [6]. This system depends upon the toughness of the quadratic residuosity problem. This problem is deciding whether $\exists y x = y^2 \ (mod\ N)$, where $x \in Z_N$ and $N = N_1 N_2$ are given. Here x is the modular residue and N is the composite modulus. Speed of this system is fast enough but consumption of bandwidth is very high. Also security is a major issue in this system, as it is not reliable against identity attacks.

1.8.2   "Full Domain Hash" IBE

Boneh and Franklin [7] introduced the first real-time practical IBE system. This system depends on wise complexity hypothesis and it makes use of pairings. This system is efficient and it uses minimal bandwidth when compared with Cocks, which makes it effective and popular. Wide utilization of cryptographic hashes is the main disadvantage of BF systems. It suffers from the shortcoming that it makes an assumption that hash functions are dispersed uniformly with images on elliptic curve like in $h = H(ID) \in G$. Here ID represents a string and G denotes bilinear group. This hashing is expensive for utilization, and choices of curves are limited to users. This drawback degrades the performance of the overall system.

1.8.3 "Exponent Inversion" IBE

Sakai and Kasahara [8] proposed the idea of this scheme. In this perspective various IBE schemes have been suggested that depend on the common idea that it is a tough task to invert exponents. For example it is difficult to calculate $g^{1/x}$ when g and g^x are known. So, by using pairing, exponents can be cancelled even without exploring them (by passing $g^{1/x}$). The objective is to encrypt identities as fragments of this x, so that we can omit the necessity of hashing right on the curve.

To evaluate the performance of the scheme in terms of security is a tough job. Sakai and Kasahara [9] suggested the first technique of this method but there was a lack of security proof. Boneh and Boyen [4] proposed the first relatively protected IBE scheme established on the reversal idea. The main goal of designing this system is to achieve security and that too without random oracles. Later, Chen et al. [10] gave a proof of the novel SK system by initializing random oracles in the BB2 proof. Gentry [11] suggested a variant of the general theme, with improved security than earlier systems.

In all the schemes depending on the "exponent inversion" one common feature could be observed: they needed strong hypotheses. According to standard hypothesis it is difficult to calculate

$g^{1/x}$ (or $e(g, g)^{1/x}$) given g, g^x, and also g^{x^2}, g^{x^3} up to g^{x^q} Here q is a max number of private key holders that the opponent may destroy. Here q should be of large value in order to prove meaningful but simultaneously it will make supposition less reliable.

1.8.4 "Commutative Blinding" IBE

Boneh and Boyen [4] introduced the commutative blinding scheme with BB1 technique. This scheme overcomes disadvantages of Boneh-Franklin and SK/BB2 techniques. Boneh-Boyen and Boneh-Franklin assume an equivalent weak hypothesis. They allow encryption of identities as integers by avoiding hashing on the curve.

Approximately blinding factors are generated by the BB1 concept with two or more secret coefficients which can be able to apply in any order. Many versions of the basic scheme are suggested with diverse modifications. Sahai and Waters [12] suggested a fuzzy IBE system which replaces the receiver identity. In an associated result, Waters [13] proved that strong security can be achieved by a slight modification to BB1. Chatterjee and Sarkar [14] and Naccache [15] have given succeeding enhancements of Waters' proof.

2 Boneh-Boyen IBE (Basic Scheme: Additive Notation)

Basic scheme of Boneh-Boyen IBE makes use of secret parameters. Sender and receiver can calculate this shared secret at their end which is used to encode plaintext data. The public parameters along with receiver's identity arc used by the sender to compute the shared secret of the message, whereas the recipient's private key and the ciphertext are used to calculate the shared secret of the message. Though it is comparatively easy than the full Boneh-Boyen IBE scheme, security is compromised.

2.1 Additive Notation

Here description of additive notations is given; these operations are generally used in the elliptic curve:

$$E\left(F_q\right) = \text{Elliptic curve groups}$$

$$P, Q = \text{Elements in } E\left(F_q\right)$$

$$P + Q = \text{Group operation of } E\left(F_q\right)$$

$$aP = \text{Point } P \text{ multiplication by value } a \text{ where } a \text{ is an integer}$$

2.2 Setup of Parameters (Basic Scheme: Additive Notation)

For the execution of Boneh-Boyen IBE technique, there is a requirement of a security parameter. This parameter will describe the bit strength level which will be provided by the encryption. Then groups G_1 and G_T are needed to define the pairing $\hat{e} : G_1 \times G1 \rightarrow G_T$. To implement elliptic curve $E \mid F_q$ is taken with k as the degree of embedding and prime p so as $p \mid \# E(F_q)$. Scope of the sets G_1 and G_T will be described by the security parameter.

P is the point which is randomly picked up where $P \in E(F_q)[p]$.

Let $G_1 = \langle P \rangle$ and $G_T = \langle \hat{e}\ (P, P)\rangle$. These are cyclical sets of order p.

There is a necessity of a hash function which can perform mapping of strings to symbolize identities to integers. So we use $H_1 : \{0, 1\}^* \rightarrow Z_p$ which is a cryptographic hash function.

For encryption of a message using Boneh-Boyen IBE, there is a requirement of one more hash function defined by $H_2 : G_T \to \{0, 1\}^n$. This function performs hashing on the elements of G_T to merge it with the original data in the form of bit string having length n. $\alpha, \beta, \gamma \in Z_p$ are three integer values which are master secret. By using them public parameters $\alpha P, \beta P, \gamma P$ can be calculated.

$$v = \hat{e}(P_1, P_2) = \hat{e}(\alpha P, \beta P) = \hat{e}(P, P)^{\alpha\beta} \text{ is a constant.}$$

This v can be provided to the user as an element of public parameters. Otherwise constants are pre-calculated by users before performing encryption. It will be assumed that constant v is an element of public parameter; this is the case where βP need not be enumerated in a public parameter. Table 9.1 gives public parameters and Table 9.2 gives master secret.

Attributes of Table 9.1 are interdependent among each other. For example ingroup definition of G_1, p, q, and E values is implicit. Due to this it is able to decrease the count of public parameters required to a smaller list. So it is possible to define public parameters of BB_1 basic scheme: additive notation $= (G_1, G_T, \hat{e}, n, P, P_1, P_3, H_1, H_2, v)$ without having any ambiguity.

Table 9.1 Basic scheme—additive notation parameters

S no.	Attribute	Type	Description
1	q	Prime power	Order of finite field F_q
2	E/F_q	Elliptic curve	$E(F_q)$ has embedding degree k
3	p	Prime	$p \mid \# E(F_q)$
4	G_1	Cyclic group	Subgroup of $E(F_q)$, $G_1 = \langle P \rangle$
5	G_T	Cyclic group	Subgroup of $F^*{q^k}: G_T = \langle \hat{e}(P, P) \rangle$
6	$\hat{e}$	Pairing	$\hat{e} : G_1 \times G1 \to G_T$
7	n	Positive integer	Length of plaintext (in bits)
8	P	Point on elliptic curve	$P \in G_1$
9	P_1	Point on elliptic curve	$P_1 = \alpha P$
10	P_2	Point on elliptic curve	$P_2 = \beta P$
11	P_3	Point on elliptic curve	$P_3 = \gamma P$
12	H_1	Cryptographic hash function	$H_1 : \{0, 1\}^* \to Z_p$
13	H_2	Cryptographic hash function	$H_2 : G_T \to \{0, 1\}^n$
14	v	Element of $F^*{q^k}$	$v = \hat{e}(P_1, P_2) = \hat{e}(\alpha P, \beta P)$ $= \hat{e}(P, P)^{\alpha\beta}$

Table 9.2 Basic scheme—additive notation master secret

S no.	Attribute	Type	Description
1	α, β, γ	Integer	$\alpha, \beta, \gamma \in Z_p$

Table 9.3 Private key for Boneh-Boyen IBE system

S no.	Attribute	Description
1	$D_{ID} = (q_{ID}.rP_1 + \alpha P_2 + rP_3, rP)$ $= (D_0, D_1)$	Private key analogous to identity ID, $q_{ID} = H_1(ID)$

2.3 Extraction of the Private Key (Basic Scheme: Additive Notation)

When all the public parameters (attributes) as mentioned in Table 9.1 and the master secret as mentioned in Table 9.2 are firm, then the private key can be calculated. This secret key is linked with the identity ID. By calculation of $q_{ID} \in Z_p$ per-user random value $r \in Z_p$ is produced. It is used to compute the two private key components $D_{ID} = (q_{ID}.rP_1 + \alpha P_2 + rP_3, rP) = (D_0, D_1)$. Refer Table 9.3 for the same.

2.4 Encryption with Boneh-Boyen IBE (Basic Scheme: Additive Notation)

For the encryption of message the sender will execute the following listed steps to the receiver with identity ID where message $= M \in \{0, 1\}^n$.

1. Evaluate $q_{ID} = H_1(ID)$.
2. Choose random $s \in Z_p$.
3. Evaluate $k = \upsilon^s$.
4. Evaluate $M \oplus H_2(k)$.
5. Evaluate $C_0 = sP$.
6. Evaluate $C_1 = q_{ID}(sP_1) + sP_3$.
7. Set ciphertext to $C = (c, C_0, C_1)$.

2.5 Decrypting with Boneh-Boyen IBE (Basic Scheme: Additive Notation)

After receiving ciphertext $C = (c, C_0, C_1)$ the following phases are executed:

1. Evaluate $k = \frac{\hat{e}(C_0, D_0)}{\hat{e}(C_1, D_1)}$.
2. Evaluate $M = c \oplus H_2(k)$.

Take into consideration that

$$\hat{e}\,(C_0, D_0) = \hat{e}\,(sP, q_{ID}.rP_1 + \alpha P_2 + rP_3)$$
$$= \hat{e}\,(sP, q_{ID}.rP_1)\,\hat{e}\,(sP, \alpha P_2)\,\hat{e}\,(sP, rP_3)$$
$$= \hat{e}\,(sP, \alpha q_{ID}.rP)\,\hat{e}\,(sP, \alpha\beta P)\,\hat{e}\,(sP, \gamma rP)$$
$$= \hat{e}(P, P)^{\alpha q_{ID}.rs}\,\hat{e}(P, P)^{\alpha\beta s}\,\hat{e}(P, P)^{\gamma rs}$$

And also

$$\hat{e}\,(C_1, D_1) = \hat{e}\,(q_{ID}.sP_1 + sP_3, rP)$$
$$= \hat{e}\,(\alpha q_{ID}.sP + \gamma sP, rP)$$
$$= \hat{e}\,(\alpha q_{ID}.sP, rP)\,\hat{e}\,(\gamma sP, rP)$$
$$= \hat{e}\,\big((P, P)^{\alpha q_{ID}.rs}\big)\,\hat{e}(\,P, P)^{\gamma rs}$$

This gives

$$\frac{\hat{e}(C_0, D_0)}{\hat{e}(C_1, D_1)} = \frac{\hat{e}\,(P,P)^{\alpha q_{ID}.rs}\hat{e}(P,P)^{\alpha\beta s}\hat{e}(P,P)^{\gamma rs}}{\hat{e}(P,P)^{\alpha q_{ID}.rs}\hat{e}(P,P)^{\gamma rs}}$$
$$= \hat{e}(P, P)^{\alpha\beta s} = v^s$$

v^s enables the receiver to decode the ciphertext accurately.

3 Boneh-Boyen IBE (Basic Scheme: Multiplicative Notation)

Basic scheme of Boneh-Boyen IBE makes use of secret parameters. Sender and receiver can calculate this shared secret at their end which is used to encode plaintext data. The public parameters along with receiver's identity are used by the sender to compute the shared secret of the message, whereas the recipient's private key and the ciphertext are used to calculate the shared secret of the message. Though it is comparatively easy than the full Boneh-Boyen IBE scheme, security is compromised.

The following part explains the common notations:

$$E\left(F_q\right) = \text{Elliptic curve groups}$$

$$g_1, g_2 = \text{Elements in } E\left(F_q\right)$$

$$g_1 g_2 = \text{Group operation of } E\left(F_q\right)$$

To indicate multiplying of the point g_1 with the integer value u, we can say that $g_1 g_2$ indicates the group operation of $E(F_q)$ applied to the group's elements g_1, g_2 and g^a.

3.1 Setup of Parameters (Basic Scheme: Multiplicative Notation)

For implementing Boneh-Boyen IBE technique, there is a requirement of a security parameter. This parameter will describe the bit strength level which will be provided by the encryption. Then groups G_1 and G_T are needed to define the pairing $\hat{e}$: $G_1 \times G1 \to G_T$. For implementation of elliptic curve $E(F_q)$ is taken with k as the degree of embedding and prime p so as $p \mid \# E(F_q)$. P is the point which is randomly picked up where $P \in E(F_q)[p]$.

Let $G_1 = \langle P \rangle$ and $G_T = \langle \hat{e} \ (P, P) \rangle$. These are cyclical sets of order p.

There is a requirement of a hash function which can perform mapping of strings to symbolize identities to integers. So, we use

$$H_1 : \{0, 1\}^* \to Z_p = \text{cryptographic hash function.}$$

For encryption of a n bit message using Boneh-Boyen IBE, there is a requirement of one more cryptographic hash function that can be defined as $H_2 : G_T \to \{0, 1\}^n$. This function performs hashing on the elements of G_T to merge it with the original data in the form of string of length n.

The Boneh-Boyen scheme needs three integers $\alpha, \beta, \gamma \in Z_p$ which are master secret and these integers are used to compute three extra public parameters αP, βP, γP and extra constant υ which can be elaborated like $\upsilon = \hat{e}(g_1, g_2) = \hat{e}(g^\alpha, g^\beta) = \hat{e}(g, g)^{\alpha\beta}$. This υ can be provided to users as an element of public parameters. Otherwise constants may be pre-calculated by users before performing encryption.

It will be assumed that constant υ is an element of public parameter; this is the case where g_2 need not be enumerated in public parameter because it is only used for calculating υ outside a PKG. Table 9.4 gives public parameters and Table 9.5 gives master secret.

Attributes of Table 9.4 are interdependent among each other. For example in G_1 group definition p, q, E values are implicit. Due to this, it is able to minimize the count of needed public parameters to an extremely short list. It can be stated that public parameter of a Boneh-Boyen IBE multiplicative basic scheme notation can be defined to be BB1 Basic Params Multiplicative $= (G_1, G_T, \hat{e}, n, g, g_1, g_3, H_1, H_2, \upsilon)$.

3.2 Extraction of the Private Key (Basic Scheme: Multiplicative Notation)

When all the public parameters (attributes) as mentioned in Table 9.4 and the master secret as mentioned in Table 9.5 are firm, then the private key can be calculated. This private key is connected with the identity ID. Computing the $q_{ID} = H_1(ID)$, we can calculate the secret key linked with the identity *ID* and it is mapped with an integer,

Table 9.4 (Basic scheme—additive notation) parameters

S no.	Attribute	Type	Description
1	q	Prime power	Order of finite field F_q
2	E/F_q	Elliptic curve	$E(F_q)$ has embedding degree k
3	p	Prime	$p \mid \# E(F_q)$
4	G_1	Cyclic group	Subgroup of $E(F_q)$, $G_1 = \langle g \rangle$
5	G_T	Cyclic group	Subgroup of $F*q^k$, $G_T = \langle \hat{e}\,(g, g) \rangle$
6	$\hat{e}$	Pairing	$\hat{e} : G_1 \times G1 \to G_T$
7	n	Positive integer	Length of plaintext (in bits)
8	g	Point on elliptic curve	$g \in G_1$
9	g_1	Point on elliptic curve	$g_1 = g^\alpha$
10	g_2	Point on elliptic curve	$g_2 = g^\beta$
11	g_3	Point on elliptic curve	$g_3 = g^\gamma$
12	H_1	Cryptographic hash function	$H_1 : \{0, 1\}^* \to Z_p$
13	H_2	Cryptographic hash function	$H_2 : G_T \to \{0, 1\}^n$
14	υ	Element of $F*q^k$	$v = \hat{e}\,(P_1, P_2) = \hat{e}\,(\alpha P, \beta P)$ $= \hat{e}(P, P)^{\alpha\beta}$

Table 9.5 Master secret (basic scheme—multiplicative notation)

S no.	Attribute	Type	Description
1	α, β, γ	Integer	$\alpha, \beta, \gamma \in Z_p$

Table 9.6 Private key (basic scheme—multiplicative notation)

S no.	Attribute	Description
1	$d_{ID} = \left(g_1^{g_{ID}.r} g_2^\alpha g_3^r g^r\right) = (d_0, d_1)$	Private key analogous to identity ID $q_{ID} = H_1(ID)$

and then per-user random value $r \in Z_p$ is produced. It is used to compute the two private key components as mentioned in Table 9.6:

$$d_{ID} = \left(g_1^{q_{ID}.\,r} g_2^\alpha g_3^r g^r\right) = (d_0, d_1)$$

3.3 Encrypting with Boneh-Boyen IBE (Basic Scheme: Multiplicative Notation)

For the encryption of message to the receiver with identity ID, the sender will execute the subsequent listed steps, where message $= M \in \{0, 1\}^n$:

1. Evaluate $q_{ID} = H_1(ID)$.
2. Choose random $s \in Z_p$.
3. Evaluate $k = \upsilon^s$.
4. Evaluate $c = M \oplus H_2(k)$.

5. Evaluate $c_0 = g^s$.
6. Evaluate $c_1 = g_1^{g_{ID} \cdot s} + g_3^r$.
7. Assign ciphertext to $C = (c, c_0, c_1)$.

3.4 Decrypting with Boneh-Boyen IBE (Basic Scheme: Multiplicative Notation)

After receiving ciphertext $C = (c, c_0, c_1)$ the following phases are executed:

1. Evaluate $k = \frac{\hat{e}(c_0, d_0)}{\hat{e}(c_1, d_1)}$.
2. Evaluate $M = c \oplus H_2(k)$.

Consider that

$$
\begin{aligned}
\hat{e}(c_0, d_0) &= \hat{e}\left(g^s, g_1^{q_{ID} \cdot r} g_2^\alpha g_3^r\right) \\
&= \hat{e}\left(g^s, g_1^{q_{ID} \cdot r}\right) \hat{e}\left(g^s, g_2^\alpha\right) \hat{e}\left(g^s, g_3^r\right) \\
&= \hat{e}\left(g^s, g_1^{q_{ID} \cdot r}\right) \hat{e}\left(g^s, g^{\alpha\beta}\right) \hat{e}\left(g^s, g^{\gamma r}\right) \\
&= \hat{e}(g, g)^{\alpha q_{ID} \cdot rs} \hat{e}(g, g)^{\alpha\beta s} \hat{e}(g, g)^{\gamma rs}
\end{aligned}
$$

$$
\begin{aligned}
\hat{e}(c_1, d_1) &= \hat{e}\left(g_1^{q_{ID} \cdot r} g_3^s, g^r\right) \\
&= \hat{e}(g^{\alpha q_{ID} \cdot s} g^{\gamma s}, g^r) \\
&= \hat{e}(g^{\alpha q_{ID} \cdot s}, g^r) \hat{e}(g^{\gamma s}, g^r) \\
&= \hat{e}(g, g)^{\alpha q_{ID} \cdot rs} \hat{e}(g, g)^{\gamma rs}
\end{aligned}
$$

Also consider that

$$
\begin{aligned}
\frac{\hat{e}(c_0, d_0)}{\hat{e}(c_1, d_1)} &= \frac{\hat{e}(g, g)^{\alpha q_{ID} \cdot rs} \hat{e}(g, g)^{\alpha\beta s} \hat{e}(g, g)^{\gamma rs}}{\hat{e}(g, g)^{\alpha q_{ID} \cdot rs} \hat{e}(g, g)^{\gamma rs}} \\
&= \hat{e}(g, g)^{\alpha\beta s} = v^s
\end{aligned}
$$

4 Boneh-Boyen IBE (Full Scheme)

The first introduced basic Boneh-Boyen technique is susceptible to a chosen ciphertext attack. Consider a case where the opponent desires to compute plaintext into ciphertext (c, c_0, c_1). This ciphertext (c, c_0, c_1) relates to the message M which consists of plaintext data. For achieving the goal he/she can decode the ciphertext $(c + \in, c_0, c_1)$ to retrieve the original message $M \oplus \in$ and M can be recovered as $M = (M \oplus \in) \oplus \in$.

This possibility of being attacked can be simply removed by using the Fujisaki-Okamoto transform.

For the description of full Boneh-Boyen scheme one can refer multiplicative notation defined in Sect. 3. The full scheme is strong against various attacks like chosen identity and chosen ciphertext.

4.1   Setup of Parameters (Full Scheme)

Along with the attributes enlisted in Table 9.7, there is a need of an auxiliary hash function to enhance the security of chosen ciphertext. That hash function is

$$H_3 : G_T \to \{0, 1\}^n \times G_1 \times G1 \to Z_p$$

Public parameters for the full scheme are enlisted in Table 9.7. The master secret is the same as stated in the basic scheme, and is given in Table 9.8.

Table 9.7 Parameters of Boneh-Boyen IBE system (full scheme)

S no.	Attribute	Type	Description
1	q	Prime power	Order of finite field F_q
2	$E(F_q)$	Elliptic curve	$E(F_q)$ has embedding degree k
3	p	Prime	$p \mid \# E(F_q)$
4	G_1	Cyclic group	Subgroup of $E(F_q)$, $G_1 = \langle P \rangle$
5	G_T	Cyclic group	Subgroup of $F*_{q^k} : G_T = \langle \hat{e}\ (P, P) \rangle$
6	$\hat{e}$	Pairing	$\hat{e} : G_1 \times G1 \to G_T$
7	n	Positive integer	Length of plaintext (in bits)
8	g	Point on elliptic curve	$g \in G_1$
9	g_1	Point on elliptic curve	$g_1 = g^\alpha$
10	g_2	Point on elliptic curve	$g_2 = g^\beta$
11	g_3	Point on elliptic curve	$g_3 = g^\gamma$
12	H_1	Cryptographic hash function	$H_1 : \{0, 1\}^* \to Z_p$
13	H_2	Cryptographic hash function	$H_2 : G_T \to \{0, 1\}^n$
14	H_3	Cryptographic hash function	$H_3 :\ : G_T \to \{0, 1\}^n \times G_1 \times G1 \to Z_p$
15	υ	Element of $F*_{q^k}$	$v = \hat{e}(P_1, P_2) = \hat{e}(\alpha P, \beta P)$ $= \hat{e}(P, P)^{\alpha\beta}$

Table 9.8 (Boneh-Boyen IBE system full scheme) master secret

S no.	Attribute	Type	Description
1	α, β, γ	Integer	$\alpha, \beta, \gamma \in Z_p$

Table 9.9 Private key for Boneh-Boyen IBE system

S no.	Attribute	Description
1	$d_{ID} = \left(g_1^{q_{ID}.r} g_2^{\alpha} g_3^r, g^r\right) = (d_0, d_1)$	Private key analogous to identity ID, $q_{ID} = H_1(ID)$

Attributes of Table 9.8 are interdependent among each other.

For example ingroup definition of G_1, p, q, E values is implicit. Due to this, it is able to decrease the count of public parameters needed to a smaller list. So it is possible to define Boneh-Boyen IBE system public parameters to be Boneh-Boyen parameters $= \left(G_1, \ G_T, \hat{e}, n, g, g_1, g_3, H_1, H_2, H_3, \upsilon\right)$ without having any ambiguity.

4.2   Extraction of the Private Key (Full Scheme)

When all the public parameters as mentioned in Table 9.7 and the master secret mentioned in Table 9.8 are firm, then the private key can be calculated. This private key is connected with the identity ID. Computing the $q_{ID} = H_1(ID)$.we can calculate the secret key linked with the identity *ID* and it is mapped with an integer, and then per-user random value $r \in Z_p$ is produced. It is used to compute the two private key components mentioned in Table 9.9:

$$d_{ID} = \left(g_1^{q_{ID}.r} g_2^{\alpha} g_3^r, g^r\right) = (d_0, d_1)$$

4.3   Encrypting with Boneh-Boyen IBE (Full Scheme)

For the encryption of message M to the receiver with identity ID, the sender will execute the subsequent listed steps:

1. Evaluate $q_{ID} = H_1(ID)$.
2. Choose random $s \in Z_p$.
3. Evaluate $k = \upsilon^s$.
4. Evaluate $c = M \oplus H_2(k)$.
5. Evaluate $c_0 = g^s$.
6. Evaluate $c_1 = g_1^{q_{ID}.s} + g_3^s$.
7. Evaluate $t = s + H_3(k, c, c_0, c_1)$.
8. Assign ciphertext to $C = (c, c_0, c_1, t)$.

4.4 Decrypting with Boneh-Boyen IBE (Full Scheme)

After receiving ciphertext $C = (c, c_0, c_1, t)$ the following phases are executed by the receiver:

1. Evaluate $k = \frac{\hat{e}(C_0, d)}{\hat{e}(C_1, d_1)}$.
2. Evaluate $s = t - H_3(k, c, c_0, c_1)$.
3. Verify that $k = v^s$ and $c_0 = g^s$. If any of the conditions fails, raise an error condition and exit.
4. Compute $M = c \oplus H_2(k)$.

5 Security of the Boneh-Boyen IBE Scheme

The security concept of the Boneh-Boyen IBE can be implemented under selective-ID model [1]. As this is the selective-ID model, the intruder will declare the duplicate ID^* which is the ID of the user whom the intruder wants to attack.

As explained earlier, IBE works by using four algorithms; setup is the first and most important algorithm to generate key pairs of the user.

Here, we are using a few different notation for the brief description of the setup phase to elaborate the security concept in IBE.

- Randomly, select three integers g, h, u; G will be a group that must be in prime order p such as $p : g, h, u \in G$.
- Then choose arbitrary $a, b \in Z_P$.
- MSK would be the master secret key such that $MSK = g^{ab}$.
- Pp is a public parameter such that $Pp = (g, h, u, e(g, g)^{ab}$.
- The hash function can be represented as $f(ID) = u^{ID}h$ where $ID \in Z_p$.

The Boneh-Boyen challenger will have few parameters like g, $A = g^a$, $B = g^b$, and $C = g^c$ and T is a random element from G_T or $T \in G_T$ whereas $T = e(g, g)^{abc}$. We must check whether an intruder on the Boneh-Boyen system is able to resolve the decisional bidirectional Diffie-Hellman problem. To find out the solution to DBDH, the intruder has to generate half plus more than minor probability appropriately through $T = e(g, g)^{abc}$ or it may be generated randomly.

Early setup for the simulator will have the following terms:

- Select a random $y \in Z_P$.
- Pp is a public parameter; it can be evaluated as

$$Pp = \left(g, e(g, g)^{ab} = e(A, B), u = A, h = A^{-ID^*} \cdot g^y \right).$$

- Remember that the attacker will attack with the identity ID^* earlier.
- The hash function $f(ID) = u^{ID}h = A^{ID-ID^*}g^y$.

However the previous declarations become very different for the simulator. Initially, $f(ID^*) = g^y$; this is a significant parameter. We use the random parameter g^y in the hash function; as a result the intruder does not acquire $f(ID^*) = 1$ and stops.

Furthermore, from the simulator's point of view, ciphertext can be as follows:

$$CT = (C_0, C_1, C_2) = \left(Me(g,g)^{abs}, g^s, g^{a.\Delta ID \, s} g^{ys} \right)$$

where $\Delta ID = ID - ID$.

Note that as we attempt to achieve the blinding factor $e(g,g)^{abs}$, we cannot decrypt CT, if $\Delta ID = 0$ as a simulator. However, if $\Delta ID \neq 0$, it yields

$$\left(\frac{C_2}{C_1^y} \right)^{\frac{1}{\Delta ID}} = \left(g^{a.\Delta ID.s} \right)^{\frac{1}{\Delta ID}} = g^{as}$$

We can compute the blinding factor $e(g^b, g^{abs})$ by taking $e(g^b, g^{as})$. We cannot do it, if $\Delta ID = 0$.

During the key generation phase, key must be randomly added. The intruder assumes that keys are disseminated randomly. Initially we can generate a key for a specific ID; further keys can be randomly generated and can be derived from public parameters as well as the previous key.

Let us provide the next key along with random parameter $r : K_1 = g^{ab} f(ID)^r$, $K_2 = g^r$. So that we can select a random $t \in Zp$ and randomly generate a key $rr' = t + r$. It can be likely as below:

$$K_1' = g^{ab} f(ID)^{r+t}$$
$$= g^{ab} f(ID)^r f(ID)^t$$
$$= K_1 f(ID)^t$$

$$K_2' = g^{r+t} = g^r g^t = K_2 \, g^t$$

"Decisional BDH is hard $\Rightarrow$ Boneh-Boyen IBE is CPA selective-ID secure," as stated in Boneh-Boyen IBE.

Few parameters are provided by the simulator (the DBDH intruder); those parameters are $A = g^a$, $B = g^b$, $C = g^c$, and T as $T = e(g,g)^{abs}$ or it may be random. Let us consider that the intruder has worked on setup and key generation phases, as the concept explained earlier. The next step will be evaluating the challenge phase.

The intruder sends two messages M_0 and M_1 to the simulator for encryption. Υ will be selected randomly as $\Upsilon \in \{0, 1\}$. Later, authenticated encryption M_v and the intruder will be presented by the simulator. To pursue the same, the simulator must compute the following:

$C_0 = M_v T$.

$C_1 = g^c = C$: here, c plays the role of s | simulator.

$C_2 = f(ID^*)^c = g^{yc} = C^y$: y is known because we select it randomly.

The intruder receives $CT = = (C_0, C_1, C_2)$ from the simulator and then the intruder predicts Υ and replies with Υ'. Actually, before the simulator acts the intruder may ask for various set of secret keys for many different identities apart from ID^* and then the simulator can reply with a random secret key.

If the DBDH combination were followed by the (A, B, C, T) values then the intruder will answer with $\Upsilon' = \Upsilon$; also probably, the simulator will agree, hence solving DBDH. In other situations, if T were chosen randomly, then both intruder and simulator will answer but the intruder replies with 0.5 probability whereas the simulator will respond appropriately in half of the situations. Remember, unlike Boneh-Franklin proof, reply with probability 0 will be aborted because whatever ID the intruder wishes to attack, beforehand we will be aware of that ID^*. This is the odd thing identified in the selective-ID model.

At the end of security proof we can conclude that the success ratio of the Boneh-Boyen intruder is $\frac{1}{2} + \in$, whereas $\in$ is not negligible.

After evaluation of DBDH, the probability is

$$P_r \text{ (success)} = P_r \left(\text{success} | T = e(g, g)^{ab} \right)$$
$$P_r \left(T = e(g, g)^{ab} \right) + P_r \left(\text{success} | T = R \right) P_r \left(T = R \right)$$
$$= \left(\tfrac{1}{2} + \in \right) \tfrac{1}{2} + \tfrac{1}{2} \cdot \tfrac{1}{2} = \tfrac{1}{2} + \tfrac{\in}{2}$$

6 Conclusion and Future Scope

This chapter has covered IBE techniques that are proved secure with respect to selective-identity attacks in the standard model without using random oracle. Here, the BB1 scheme of the Boneh-Boyen IBE has been concentrated that is based on the decisional bilinear Diffie-Hellman (BDH) assumption. The Boneh-Boyen IBE scheme and other commutative blinding schemes in which the user's ID is hashed with randomly selected integer are used in the encryption and decryption operations. Basically, two basic notation schemes of Boneh-Boyen, additive scheme and multiplicative scheme, are explained in detail with its standard algorithms. Additive notation scheme is utilized for elliptic curve and multiplicative notation scheme is used in pairing-based cryptography which is followed by a fully secure version of the scheme with its algorithm. At the end, by considering the intruder attack, we have examined whether an intruder on the Boneh-Boyen system is capable to resolve the decisional bidirectional Diffie-Hellman problem and the security concept of the Boneh-Boyen IBE is implemented under selective-ID model.

It is observed that security parameters should be raised to prevent inefficiencies of security reduction. There is a need for constructive research on developing a fully secure IBE scheme which is safe against adaptive ID attacks without random oracles. Though Boneh-Boyen scheme is proven secured in the standard model the more practical and real-time solution to this problem could be an important research direction. To provide strong security, the performance of the system must be excelled and attaining security with inheriting complexity by preserving the performance is an open research topic.

References

1. Rivest, R. L., Shamir, A., & Adleman, L. (1978). A method for obtaining digital signatures and public-key cryptosystems. *Communications of the Association for Computing Machinery, 21*(2), 120–126.
2. Pang, L., Li, H., & Wang, Y. (2013). NMIBAS: A novel multi-receiver ID-based anonymous signcryption with decryption fairness. *Computing and Informatics, 32*(3), 441–460.
3. Pang, L., Li, H., & Pei, Q. (2012). Improved multicast key management of Chinese wireless local area network security standard. *IET Communications, 6*(9), 1126–1130.
4. Boneh, D., & Boyen, X. (2004). Efficient selective-ID secure identity-based encryption without random oracles. In C. Cachin & J. L. Camenisch (Eds.), *EUROCRYPT 2004. LNCS* (Vol. 3027, pp. 223–238). Heidelberg: Springer.
5. Boyen, X. (2007). General ad hoc encryption from exponent inversion IBE. In M. Naor (Ed.), *EUROCRYPT 2007. LNCS* (Vol. 4515, pp. 394–411). Heidelberg: Springer.
6. Cocks, C. (2001). An identity based encryption scheme based on quadratic residues. In *Proceedings of the 8th IMA International Conference on Cryptography and Coding*.
7. Boneh, D., & Franklin, M. (2001). Identity-based encryption from the Weil pairing. In *Advances in cryptology—CRYPTO 2001, volume 2139 of lecture notes in computer science* (pp. 213–229). Springer-Verlag.
8. Mitsunari, S., Sakai, R., & Kasahara, M. (2002). A new traitor tracing. *IEICE Transactions on Fundamentals, E85-A*(2), 481–484.
9. Ryuichi Sakai and Masao Kasahara. ID based cryptosystems with pairing over elliptic curve. Cryptology ePrint archive, Report 2003/054, 2003. http://eprint.iacr.org/2003/054/.
10. Liqun Chen, Zhaohui Cheng, John Malone-Lee, and Nigel P. Smart. An efficient ID-KEM based on the Sakai-Kasahara key construction. Cryptology ePrint Archive, Report 2005/224, 2005. http://eprint.iacr.org/2005/224/.
11. Gentry, C. (2006). Practical identity-based encryption without random oracles. In *Advances in cryptology—EUROCRYPT 2006, lecture notes in computer science*. Springer-Verlag.
12. Sahai, A., & Waters, B. (2005). Fuzzy identity-based encryption. In *Advances in cryptology—EUROCRYPT 2005, volume 3494 of lecture notes in computer science*. Springer-Verlag.
13. Waters, B. (2005). Efficient identity-based encryption without random oracles. In *Advances in cryptology—EUROCRYPT 2005, volume 3494 of lecture notes in computer science*. Springer-Verlag.
14. Sanjit Chatterjee and Palash Sarkar. Trading time for space: Towards an efficient IBE scheme with short(er) public parameters in the standard model. In Proceedings of ICISC 2005, 2005.
15. Naccache, D. (2005). Secure and practical identity-based encryption. In *Cryptology ePrint Archive, Report 2005/369*. http://eprint.iacr.org/2005/369/.

Chapter 10
Sakai-Kasahara IBE

Hamza Mutaher and Mahmoud E. Hodeish

Abstract Public key cryptography (PKC) provides a very robust encryption in networking and electronic communication. The strength of PKC comes from the idea of paired keys that are independent (but mathematically dependent). The encryption-decryption process of PKC requires both parties of communication, i.e., sender and receiver, to provide each other with its public key and the digital certificate of authority, and each party has to keep a directory to store all parties' public keys so these requirements are considered as drawbacks of PKC. To overcome these drawbacks, the identity-based encryption (IBE) came to existence. IBE is a form of PKC which uses a third-party server to distribute the public parameters to all the parties and extract the private key from the arbitrary public key. To encrypt the message, the sender will use the receiver public key, and to decrypt the message, the receiver will use the extracted private key. In this chapter, we discuss the Sakai-Kasahara IBE and how it differs from other IBE schemes. The additive, multiplicative, and full schemes of IBE are explained along with the encryption and decryption process. The security of this scheme is also discussed and proved.

Keywords Cryptography · IBE · Encryption · Decryption · Additive · Multiplicative · Security

H. Mutaher (✉)
Department of Computer Science and Information Technology, Maulana Azad National Urdu University, Hyderabad, India

M. E. Hodeish
Faculty of Computer & Information Technology, Al-Razi University, Sana'a, Yemen

Department of Computer, Faculty of Education-Zabid, Hodeidah University, Hodeidah, Yemen

© Springer Nature Switzerland AG 2021
K. A. B. Ahmad et al. (eds.), *Functional Encryption*, EAI/Springer Innovations in Communication and Computing, https://doi.org/10.1007/978-3-030-60890-3_10

1 Introduction

Identity-based encryption (IBE) or the so-called ID-based encryption is a scheme that uses public key encryption in which any bit string can be represented as a public key in which the public key of a user can be some unique meaningful identity like name and email address.

The motivation of introducing the IBE scheme is to solve such problems of traditional public key systems like the necessity for directories and digital certificates to manage public keys and the expensive computations of generating public-private key pairs. However, Shamir [1] was the first who introduced the concept of IBE that eliminates the use of directories and digital certificates. He considered the receiver identity as the representation of the public key. Despite solving some of the related problems of identity-based signature, IBE proved much more challenging.

The Cocks IBE scheme [2] is one of the encryption algorithms. This type of encryption algorithm encrypts the plaintext into ciphertext. The assurance of this algorithm depends on the durability of the quadratic residuosity problem and the computational difficulty of integer factorization as well. The system authority of this algorithm generates a modulus m which is universally available. To create this modulus, system authority calculates two primes p and q; thus the modulus m will be the product of this calculation, where both primes p and q must be congruent to 3 mod 4. This system ensures the use of a universally available hash function to the text that needs to be encrypted to represent it as a value to a modulo m. Therefore, when the user A wants to get encrypted data, he/she needs to send any of his/her identities such as (username or email address) to the system authority. Mutually, user A will receive a private key from the system authority. The user B who seeks to send an encrypted data to user A will be able to deliver it by knowing only the public identity of user A and universal public parameters where there is no need to know the public key.

On the other hand, there is another type of IBE algorithm called Boneh-Franklin IBE [3]. It is an algorithm-based identity that encrypts the data for security. It is considered as the first secure and practical scheme of IBE and it is an example of an IBE full domain hash scheme family. This scheme maps the identity to the elliptic curve to accomplish the process of encryption and decryption. Modular exponentiation is required to start the process of mapping between the identity and the point in the elliptic curve. The expensive calculation is considered as a drawback for the performance of the full domain hash scheme.

The Boneh-Boyen IBE scheme [4] is also used to encrypt the identity of the users. In this scheme, the sender and receiver have to use the same value to encrypt and decrypt the identity where the receiver also uses its private key at the decryption side. In this scheme, the user identity is hashed to an integer to be used in the process of encryption and decryption. The hashed integer avoids the calculation of modular exponentiation and it is considered more rapid than the full domain hash scheme.

This chapter aims to discuss in detail the Sakai-Kasahara IBE scheme [5, 6]. This scheme depends on the bilinear pairing and elliptic curve to provide security solutions. The private key is the system element that is responsible to decrypt the ciphertext. It is one of the security solutions that belong to the exponent inversion scheme family. The encryption and decryption procedures are applied through a hashed integer on an identity in a form of string. Such algorithms like Boneh-Franklin IBE use the full domain hashed scheme which is considered slower than Sakai-Kasahara IBE. Due to the avoidance of modular exponentiation, Sakai-Kasahara IBE is quite faster than Boneh-Franklin IBE [7] which is going to be discussed in this chapter. Before the discussion of Sakai-Kasahara IBE, we have to explain the procedure that occurs in the IBE system to encrypt and decrypt the message.

When the sender A wants to send an encrypted message to receiver B, he/she simply encrypts the message using B's public identity, for example email address, and there is no need for A to check the B's public key certificate. When B receives the encrypted message, he/she will communicate the private key generator (PKG), also called system authority, which will send him/her a private key to allow him/her to decrypt the message. Note that A and B have to authenticate themselves to the PKG before starting the message exchange procedure; see Fig. 10.1.

IBE has four major operations explained as follows:

1. Setup of parameters: PKG will generate public parameters θ and master secret S and distribute public parameters to both A and B.
2. Extraction of the private key: PKG will use the master secret S to extract the private key S_{ID_B} which corresponds to an arbitrary public identity of string ID_B.
3. Encryption: The sender A will encrypt the message using the receiver B's public identity ID_B.
4. Decryption: The receiver B will decrypt the message using the corresponding private key S_{ID_B} that has been sent by the PKG.

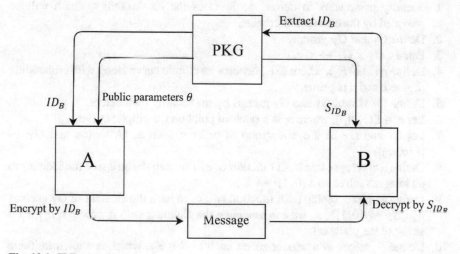

Fig. 10.1 IBE operations

This chapter is divided into two main parts; the first one discusses the encryption and decryption of Sakai-Kasahara IBE basic scheme with its additive notation and multiplicative notation. The second part discusses encryption and decryption of the Sakai-Kasahara IBE full scheme with its security proof. Both parts will explain the setup of the parameters to accomplish the encryption and decryption process.

2 Sakai-Kasahara IBE (Basic Scheme: Additive Notation)

The S-K IBE basic scheme is less secure than the S-K IBE full scheme but easier to understand. In the basic scheme, two parties need to exchange encrypted messages safely. Both parties must agree on a unique shared secret to encrypt the message in plaintext form. The first party (sender) calculates the shared secret from its public parameters and identity of the second party (receiver). The receiver gets the shared secret by calculating the ciphertext and its private key.

2.1 Setup of Parameters

This scheme deals with additive notation; thus $E(F_q)$ is an elliptic curve group, σ_1, σ_2 are two elements of the elliptic curve, and $\sigma_1 + \sigma_2$ indicates the $E(F_q)$ group operation to be applied to the group elements σ_1, σ_2 and multiply σ by integer s which is indicated as $s\sigma$.

To implement this scheme, we need to define some basic essential elements shown in Table 10.1 and explained as follows:

1. Security parameters to define the level of the bit durability which will be provided by the encryption process.
2. Define G_1 and G_T groups.
3. Pair $\hat{e} : G_1 \times G_1 \rightarrow G_T$.
4. Define $p \mid \# E(F_q)$, where E/F_q denotes an elliptic curve along with embedding degree k and p is prime.
5. Define the size of G_1 and G_T groups by the security parameters.
6. Let $\sigma \in E(F_q)[p]$, where σ is a random point on the elliptic curve.
7. Let G_1 and G_T be a cyclic group of order σ such as $G_1 = \langle \sigma \rangle$ and $G_T = \langle \hat{e}(\sigma, \sigma) \rangle$.
8. Define a cryptographic hash function one h_1 to map the string of the identity to an integer such as $h_1 : \{0, 1\}^* \rightarrow \mathbb{Z}_p$.
9. Define a cryptographic hash function two h_2 to hash the element of G_T such as $h_2 : G_T \rightarrow \{0, 1\}^n$, so we can associate the plaintext with it, where n is the bit string of the plaintext.
10. Define S integer as a master secret such as $S \in \mathbb{Z}_p$, which is shown in Table 10.1.

The public parameters of this scheme are $(G_1, G_T, \hat{e}, \sigma, s\sigma, h_1, h_2, v)$.

Table 10.1 Parameters of Sakai-Kasahara IBE scheme

Element	Description
p	Prime
q	Prime power
E/F_q	Elliptic curve
G_1	Cyclic group
G_T	Cyclic group
$\hat{e}$	Pairing
n	Positive integer
σ	A point on elliptic curve
$s\sigma$	A point on elliptic curve
h_1	A cryptographic one-way hash function
h_2	A cryptographic one-way hash function
h_3	A cryptographic one-way hash function
h_4	A cryptographic one-way hash function
v	Element of $F_{q^k}^*$
S	Master secret $S \in \mathbb{Z}_p$
$Priv_{ID} = \frac{1}{S+q_{ID}}\sigma$	A private key for additive notation
$Priv_{ID} = \sigma^{1/(S+q_{ID})}$	A private key for multiplicative notation and full scheme

Algorithm 11.1: Parameters _Setup ()

Input: A security parameter, an elliptic curve E, and a plaintext length n

Output: Public parameters $\theta_1 = \left(G_1, G_T, \hat{e}, \sigma, s\sigma, h_1, h_2, v\right)$ and a master secret S.

Procedure:

Begin

1. Select a prime p and a prime power q with $p \mid \# E(F_q)$ which meets the security parameter requirement.
2. Pick up a random $\sigma \in E(F_q)[p]$ and let $G_1 = \langle \sigma \rangle$.
3. Embed the degree k to $F_{q^k}^*$ and pair $\hat{e} : G_1 \times G_1 \to F_{q^k}^*$.
4. Let $G_T = \langle \hat{e}(\sigma, \sigma) \rangle$.
5. Pick up a random $S \in \mathbb{Z}_p$.
6. Use cryptographic hash functions:

    (a) $h_1 : \{0, 1\}^* \to \mathbb{Z}_p$
    (b) $h_2 : G_T \to \{0, 1\}^n$

End

2.2   Extraction of the Private Key

The extraction of the private key is the responsibility of the receiver party. After listing out the security parameters, elements, and master key, the receiver party extracts the private keys as follows:

1. Map the *ID* to the integer $q_{ID} \in \mathbb{Z}_p$ by calculating $q_{ID} = h_1(ID)$.
2. Use master secret S to calculate the private key such as $Priv_{ID} = \frac{1}{S+q_{ID}}\sigma$, where the calculation of $\frac{1}{S+q_{ID}}$ occurs in $\mathbb{Z}_p^*$.

Algorithm 11.2: Private_Key_Extraction ()
   Input: An identity *ID*, public parameters $\theta_1 = (G_1, G_T, \hat{e}, \sigma, s\sigma, h_1, h_2, v)$, and a master secret S
   Output: A private ID $Priv_{ID}$
   Procedure:
   Begin

        1. Calculate $Priv_{ID} = 1/S + q_{ID}$.

End

2.3   Sakai-Kasahara IBE Encryption

In this section, the sender needs to encrypt the message $M \in \{0, 1\}^n$ and send it to the receiver along with identity *ID*, so the sender will perform some steps to encrypt the message as follows:

1. Map the identity to an integer and hash it using hash function one as $q_{ID} = h_1(ID)$.
2. Pick up a random number $R \in \mathbb{Z}_p$.
3. Calculate $L = R(s\sigma + q_{ID^\sigma}) = R(S + q_{ID^\sigma})$.
4. Calculate $\lambda = h_2(v)^R$.
5. Calculate $\omega = M \oplus \lambda$.
6. Define $C = (L, \omega)$ as a ciphertext.

Algorithm 11.3: Encryption ()
   Input: A plaintext message $M \in \{0, 1\}^n$, a string *ID*, public parameters $\theta_1 = (G_1, G_T, \hat{e}, \sigma, s\sigma, h_1, h_2, v)$, and a master secret S
   Output: A ciphertext $C = (L, \omega)$
   Procedure:
   Begin

        1. Calculate $q_{ID} = h_1(ID)$.
        2. Pick up a random number: $R \in \mathbb{Z}_p$.
        3. Calculate $L = R(s\sigma + q_{ID^\sigma}) = R(S + q_{ID^\sigma})$.

   4. Calculate $\lambda = h_2(v)^R$.
   5. Calculate $\omega = M \oplus \lambda$.
   6. Calculate $C = (L, \omega)$.

End

2.4   Sakai-Kasahara IBE Decryption

In the section, the receiver needs to decrypt the message that has been sent by the sender to get the plaintext by performing the following steps:

1. Calculate $\lambda = h_2\left(\hat{e}\left(L, Priv_{ID}\right)\right)$.
2. Calculate $M = (\omega \oplus \lambda)$.

   Note that

$$\hat{e}\left(L, Priv_{ID}\right) = \hat{e}\left(R\left(S + q_{ID}\right)\sigma, \frac{1}{S + q_{ID}}\sigma\right) = \hat{e}(\sigma, \sigma)^R$$

So, step 5 of the encryption section and step 2 of the decryption section are calculating the same λ that permits the receiver to decrypt the ciphertext correctly.

Algorithm 11.4   Decryption ()
   Input: A ciphertext $C = (L, \omega)$, public parameters $\theta_1 = \left(G_1, G_T, \hat{e}, \sigma, s\sigma, h_1, h_2, v\right)$, and a private key $Priv_{ID}$
   Output: A plaintext message M
   Procedure:
   Begin

   1. Calculate $\lambda = h_2\left(\hat{e}\left(L, Priv_{ID}\right)\right)$.
   2. Calculate $M = (\omega \oplus \lambda)$.

End

3   Sakai-Kasahara IBE (Basic Scheme: Multiplicative Notation)

The S-K IBE basic scheme is less secure than the S-K IBE full scheme but easier to understand. In the basic scheme, two parties need to exchange encrypted messages safely. Both parties must agree on a unique shared secret to encrypt the message in plaintext form. The first party (sender) calculates the shared secret from its public parameters and identity of the second party (receiver). The receiver gets the shared secret by calculating the ciphertext and its private key.

3.1   Setup of Parameters

This scheme deals with multiplicative notations; thus $E(F_q)$ is an elliptic curve group and σ_1, σ_2 are two elements of the elliptic curve; then we consider $\sigma_1\sigma_2$ to point out $E(F_q)$ group operation to be applied to the group elements σ_1, σ_2 and multiply σ by integer s indicated as σ^s.

To implement this scheme, we need to define some basic essential elements shown in Table 10.1 and explained as follows:

1. Security parameters to define the level of the bit durability which will be provided by the encryption process.
2. Define G_1 and G_T groups.
3. Pair $\hat{e} : G_1 \times G_1 \to G_T$.
4. Define $p \mid \# E(F_q)$, where E/F_q denotes an elliptic curve along with the embedding degree k and p is prime.
5. Define the size of G_1 and G_T groups by the security parameters.
6. Let $\sigma \in E(F_q)[p]$ where σ is a random point on the elliptic curve.
7. Let G_1 and G_T be a cyclic group of order σ such as $G_1 = \langle \sigma \rangle$ and $G_T = \langle \hat{e}(\sigma, \sigma) \rangle$.
8. Define a cryptographic hash function one h_1 to map the string of the identity to an integer such as $h_1 : \{0, 1\}^* \to \mathbb{Z}_p$.
9. Define a cryptographic hash function two h_2 to hash the element of G_T such as $h_2 : G_T \to \{0, 1\}^n$, so we can associate the plaintext with it, where n is the bit string of the plaintext.
10. Define S integer as a master secret such as $S \in \mathbb{Z}_p$, which is shown in Table 10.1.

The public parameters of this scheme are $(G_1, G_T, \hat{e}, n, \sigma, \sigma^s, h_1, h_2, v)$.

Algorithm 11.5 Parameters_ Setup ()

Input: A security parameter, an elliptic curve E, and a plaintext length n

Output: Public parameters $\theta_2 = (G_1, G_T, \hat{e}, n, \sigma, \sigma^s, h_1, h_2, v)$ and a master secret S

Procedure:

Begin

1. Select a prime p and a prime power q with $p \mid \# E(F_q)$ which meets the security parameter requirement.
2. Pick up a random $\sigma \in E(F_q)[p]$ and let $G_1 = \langle \sigma \rangle$.
3. Embed the degree k to $F_{q^k}^*$ and pair $\hat{e} : G_1 \times G_1 \to F_{q^k}^*$.
4. Let $G_T = \langle \hat{e}(\sigma, \sigma) \rangle$.
5. Pick up a random $S \in \mathbb{Z}_p$.
6. Use cryptographic hash functions:

    (a) $h_1 : \{0, 1\}^* \to \mathbb{Z}_p$
    (b) $h_2 : G_T \to \{0, 1\}^n$

End

3.2   Extraction of the Private Key

The extraction of the private key is the responsibility of the receiver party. After listing out the security parameters, elements, and master key, the receiver party extracts the private keys as follows:

1. Map the ID to the integer $q_{ID} \in \mathbb{Z}_p$ by calculating $q_{ID} = h_1(ID)$.
2. Use master secret S to calculate the private key such as $Priv_{ID} = \sigma^{1/(S+q_{ID})}$.

Algorithm 11.6 Private_Key_Extraction ()
   Input: An identity ID, public parameters $\theta_2 = (G_1, G_T, \hat{e}, n, \sigma, \sigma^s, h_1, h_2, v)$, and a master secret S
   Output: A private ID $Priv_{ID}$
   Procedure
   Begin

   1. Calculate $Priv_{ID} = \sigma^{1/(S+q_{ID})}$.

End

3.3   Sakai-Kasahara IBE Encryption

In this section, the sender needs to encrypt the message $M \in \{0, 1\}^n$ and send it to the receiver along with identity ID, so the sender will perform some steps to encrypt the message as follows:

1. Map the identity to an integer and hash it using hash function one as $q_{ID} = h_1(ID)$.
2. Pick up a random number $R \in \mathbb{Z}_p$.
3. Calculate $L = R(\sigma^S \sigma^{q_{ID}})^R = \sigma^{R(S+q_{ID})}$.
4. Calculate $\lambda = h_2(v)^R$.
5. Calculate $\omega = M \oplus \lambda$.
6. Define $C = (L, \omega)$ as a ciphertext.

Algorithm 11.7 Encryption ()
   Input: A plaintext message $M \in \{0, 1\}^n$, a string ID, public parameters $\theta_2 = (G_1, G_T, \hat{e}, n, \sigma, \sigma^s, h_1, h_2, v)$, and a master secret S
   Output: A ciphertext $C = (L, \omega)$
   Procedure
   Begin

   1. Calculate $q_{ID} = h_1(ID)$.
   2. Pick up a random number: $R \in \mathbb{Z}_p$.
   3. Calculate $L = R(\sigma^S \sigma^{q_{ID}})^R = \sigma^{R(S+q_{ID})}$.

   4. Calculate $\lambda = h_2(v)^R$.
   5. Calculate $\omega = M \oplus \lambda$.
   6. Calculate $C = (L, \omega)$.

End

3.4 Sakai-Kasahara IBE Decryption

In the section, the receiver needs to decrypt the message that has been sent by the sender to get the plaintext by performing the following steps:

1. Calculate $\lambda = h_2 \left(\hat{e} (L, Priv_{ID}) \right)$.
2. Calculate $M = (\omega \oplus \lambda)$.

   Note that

$$\hat{e} (L, Priv_{ID}) = \hat{e} \left(\sigma^{R(S+q_{ID})}, \sigma^{1/(S+q_{ID})} \right) = \hat{e}(\sigma, \sigma)^R$$

So, step 5 of the encryption section and step 2 of the decryption section are calculating the same λ that permits the receiver to decrypt the ciphertext correctly.

Algorithm 11.8 Decryption ()
   Input: A ciphertext $C = (L, \omega)$, public parameters
$\theta_2 = \left(G_1, G_T, \hat{e}, n, \sigma, \sigma^s, h_1, h_2, v \right)$, and a private key $Priv_{ID}$
   Output: A plaintext message M
   Procedure
   Begin

   1. Calculate $\lambda = h_2 \left(\hat{e} (L, Priv_{ID}) \right)$.
   2. Calculate $M = (\omega \oplus \lambda)$.

   End

4 Sakai-Kasahara IBE (Full Scheme)

The basic scheme is insecure to chosen ciphertext attack: if an attacker wants to get the plaintext back, the attacker will decrypt the ciphertext $C(L, \omega \oplus \varepsilon)$ to get the plaintext $M \oplus \varepsilon$, and the attacker reconstructs M as $M = (M \oplus \varepsilon)$. The full scheme is intended to overcome this insecurity by adding the Fujisaki-Okamoto transformation technique [8] to the basic scheme [7].

4.1 Setup of Parameters

The setup of parameters in the full scheme is similar to the basic scheme along with some extra parameters. We need extra hash function parameters to impose the security against chosen ciphertext attack. Principally, we need two hash functions $h_3 : \{0,1\}^n \to \mathbb{Z}_p$ and $h_4 : \{0,1\}^n \to \{0,1\}^n$ and need to add these two hash functions into the list of public parameters of the full scheme. The master secret remains the same as in the basic scheme. The public parameters of this scheme are $\left(G_1, G_T, \hat{e}, n, \sigma, \sigma^s, h_1, h_2, h_3, h_4, v\right)$.

Algorithm 11.9 Parameters _Setup ()
   Input: A security parameter, an elliptic curve E, and a plaintext length n
   Output: Public parameters $\theta_3 = \left(G_1, G_T, \hat{e}, n, \sigma, \sigma^s, h_1, h_2, h_3, h_4, v\right)$ and a master secret S
   Procedure
   Begin

1. Select a prime p and a prime power q with $p \mid \# E(F_q)$ which meets the security parameter requirement.
2. Pick up a random $\sigma \in E(F_q)[p]$ and let $G_1 = \langle \sigma \rangle$.
3. Embed the degree $k$to $F_{q^k}^*$ and pair $\hat{e} : G_1 \times G_1 \to F_{q^k}^*$.
4. Let $G_T = \langle \hat{e}\,(\sigma, \sigma) \rangle$.
5. Pick up a random $S \in \mathbb{Z}_p^*$.
6. Use cryptographic hash functions:

   (a) $h_1 : \{0,1\}^* \to \mathbb{Z}_p$
   (b) $h_2 : G_T \to \{0,1\}^n$
   (c) $h_3 : \{0,1\}^n \to \mathbb{Z}_p$
   (d) $h_4 : \{0,1\}^n \to \{0,1\}^n$

   End

4.2 Extraction of the Private Key

The extraction of the private key in the full scheme occurs as follows:

1. Map the ID to the integer $q_{ID} \in \mathbb{Z}_p$ by calculating $q_{ID} = h_1(ID)$.
2. Use master secret S to calculate the private key such as $Priv_{ID} = \sigma^{1/(S+q_{ID})}$.

Note that the extraction of the private key in the full scheme is similar to the extraction of the private key in the basic scheme.

Algorithm 11.10 Private_Key_Extraction ()
   Input: An identity ID, public parameters $\theta_3 = \left(G_1, G_T, \hat{e}, n, \sigma, \sigma^s, h_1, h_2, h_3, h_4, v\right)$, and a master secret S

Output: A private ID $Priv_{ID}$
Procedure
Begin

      1. Calculate $Priv_{ID} = \sigma^{1/(S+q_{ID})}$.

End

4.3 Sakai-Kasahara IBE Encryption

In this section, the sender needs to encrypt the message $M \in \{0, 1\}^n$ and send it to the receiver along with identity ID, so the sender will perform some steps to encrypt the message as follows:

1. Map the identity to an integer and hash it using hash function one as $q_{ID} = h_1(ID)$.
2. Pick up a random number $\tau \in \mathbb{Z}_p$.
3. Calculate $R = h_3(\tau, M)$.
4. Calculate $L = \left(\sigma^S \sigma^{q_{ID}}\right)^R = \sigma^{R(S+q_{ID})}$.
5. Calculate $\lambda = \tau \oplus h_2(v)^R$.
6. Calculate $\omega = M \oplus h_4(\tau)$.
7. Calculate $\delta = h_4(M)$.
8. Define $C = (L, \omega, \delta)$ as a ciphertext.

Algorithm 11.11 Encryption ()
    Input: A plaintext message $M \in \{0, 1\}^n$, a string ID, public parameters $\theta_3 = \left(G_1, G_T, \hat{e}, n, \sigma, \sigma^s, h_1, h_2, h_3, h_4, v\right)$, and a master secret S
    Output: A ciphertext $C = (L, \omega, \delta)$
Procedure
Begin

      1. Calculate $q_{ID} = h_1(ID)$.
      2. Pick up a random number: $\tau \in \mathbb{Z}_p$.
      3. Calculate $R = h_3(\tau, M)$.
      4. Calculate $L = \left(\sigma^S \sigma^{q_{ID}}\right)^R = \sigma^{R(S+q_{ID})}$.
      5. Calculate $\lambda = \tau \oplus h_2(v)^R$.
      6. Calculate $\omega = M \oplus h_4(\tau)$.
      7. Calculate $\delta = h_4(M)$.
      8. Calculate $C = (L, \omega, \delta)$.

End

4.4   Sakai-Kasahara IBE Decryption

In this section, the receiver needs to decrypt the message that has been sent by the sender to get the plaintext by performing the following steps:

1. Calculate $q_{ID} = h_1(ID)$.
2. Calculate $N = \hat{e}(L, Priv_{ID})$.
3. Calculate $\tau = \lambda \oplus h_2(N)$.
4. Calculate $M = \delta \oplus h_4(\tau)$.
5. Calculate $R = h_3(\tau, M)$.
6. If $L \neq \left(\sigma^{q_{ID}}\sigma^S\right)^R$, then an error has occurred, so exit.
7. Else assign the plaintext to M.

Algorithm 11.12   Decryption ()
  Input: A ciphertext $C = (L, \omega, \delta)$, public parameters $\theta_3 = \left(G_1, G_T, \hat{e}, n, \sigma, \sigma^s,\right.$
$\left. h_1, h_2, h_3, h_4, v\right)$, and a private key $Priv_{ID}$
  Output: A plaintext message M
  Procedure
  Begin

       1. Calculate $q_{ID} = h_1(ID)$.
       2. Calculate $N = \hat{e}(L, Priv_{ID})$.
       3. Calculate $\tau = \lambda \oplus h_2(N)$.
       4. Calculate $M = \delta \oplus h_4(\tau)$.
       5. Calculate $R = h_3(\tau, M)$.
       6. If $L \neq \left(\sigma^{q_{ID}}\sigma^S\right)^R$ exit.
       7. Else plaintext $= M$.

  End

5   Security of the Sakai-Kasahara IBE Scheme

In this section, we are going to prove that the S-K IBE scheme is secure against the adversary $\bar{E}$ using the random oracle model (ROM); therefore we have to define the one-way hash function (OWH) before we start the analysis.

  Definition 1.1: The OWH function $f : \{0, 1\}^* \rightarrow \{0, 1\}^n$ that is considered to be infeasible to invert is that which can take any input $x \in \{0, 1\}^*$ of arbitrary length and give an arbitrary-length output value $y = f(x) \in \{0, 1\}^n$ which is called digest or hash value. While using the hash function, we have to consider the following properties:

1. $y = f(x) \in \{0, 1\}^n$ is irreversible.

$$y = h(x) \neq h\left(x'\right).$$

2. It is impossible to get $h(x^{'})$ if $x \neq x^{'}$.

Theorem 1.1: We assume that the OWH function closely operates as a random oracle. According to our assumption, the Sakai-Kasahara IBE scheme is provably secure against an adversary $\bar{E}$ to derive the message M.

Proof 1.1: We assume that the adversary $\bar{E}$ can derive the message M that has been sent from the sender to the receiver. To find out the message M, an adversary $\bar{E}$ has to use the experimental algorithm

$$EXPT^{h3(\tau,M)}_{HASH,\phi}$$

The probability of success of the experimental algorithm is defined as

$$\left| SUCCESS^{h3(\tau,M)}_{HASH,\phi} = \Pr\left[EXPT^{h3(\tau,M)}_{HASH,\phi} = 1 \right] - 1 \right|$$

where Pr denotes the probability of success of $EXPT^{h3(\tau,M)}_{HASH,\phi}$. The experimental algorithm is dependent on the advantage function that is defined as

$$ADVAT^{h3(\tau,M)}_{HASH,\phi}(et,qR) = \max_{\phi}\left\{ SUCCESS^{h3(\tau,M)}_{HASH,\phi} \right\}$$

where *max* is specified by three factors:

1. Overall $\bar{A}$
2. The number of queries (qR) obtained from the execution time (et)
3. Reveal oracle

We can say that the S-K IBE scheme is vulnerable to the adversary $\bar{E}$ to derive the message M if

$$ADVAT^{h3(\tau,M)}_{HASH,\phi}(et) \leq \varepsilon, \forall \varepsilon > 0.$$

Contemplating Algorithm 11.1, the adversary $\bar{E}$ can derive the message M if and only if it can invert the OWH function. According to Definition 11.1, the OWH function is infeasible to be inverted by cause of

$$ADVAT^{h3(\tau,M)}_{HASH,\phi}(et,qR) \leq \varepsilon$$

Since $ADVAT^{h3(\tau,M)}_{HASH,\phi}(et,qR)$ depends on $ADVAT^{h3(\tau,M)}_{HASH,\phi}(et)$, the S-K IBE scheme is provably secure against the adversary $\bar{E}s$ to derive the message M.

References

1. Shamir, A. (1984). Identity-based cryptosystems and signature schemes. In *Workshop on the theory and application of cryptographic techniques* (pp. 47–53). Berlin, Heidelberg: Springer.
2. Cocks, C. C. (1973). *A note on non-secret encryption* (p. 20). CESG Memo.
3. Boneh, D., & Franklin, M. (2003). Identity-based encryption from the Weil pairing. *SIAM J Comp, 32*(3), 586–615.
4. Boneh, D., & Boyen, X. (2004). Efficient selective-ID secure identity-based encryption without random oracles. In *International conference on the theory and applications of cryptographic techniques* (pp. 223–238). Berlin, Heidelberg: Springer.
5. Sakai, R., Ohgishi, K., & Kasahara, M. (2000). *Cryptosystems based on pairing. The 2000 Symposium on Cryptography and Information Security, Japan* (Vol. 45, pp. 26–28).
6. Chen, L., Cheng, Z., Malone-Lee, J., & Smart, N. P. (2005). *An efficient ID-KEM based on the Sakai-Kasahara key construction* (p. 224). IACR Cryptology ePrint Archive.
7. Martin, L. (2008). *Introduction to identity-based encryption*. Artech house.
8. Fujisaki, E., & Okamoto, T. (1999). Secure integration of asymmetric and symmetric encryption schemes. In *Annual international cryptology conference* (pp. 537–554). Berlin, Heidelberg: Springer.

Chapter 11
HIBE: Hierarchical Identity-Based Encryption

Tawseef Ahmed Teli, Faheem Syeed Masoodi, and Alwi M. Bahmdi

Abstract Cryptosystems fundamentally deal with the issue of securing data communication. Public key infrastructure (PKI) model requires an authenticated public key for encryption that has to be obtained prior to initiating communication. The identity-based encryption (IBE) essentially removes the public key distribution by using an arbitrary string, e.g., an email address or a phone number, as the public key. A private key generator (PKG) delegates private keys to corresponding users for decryption. In trivial identity-based encryption schemes, a single private key generator for all the users makes key generation computationally inefficient. The hierarchical identity-based encryption (HIBE) eliminates this bottleneck of verifying the proofs of identity, generating private keys and establishing secure paths to transfer these keys. In HIBE, a root PKG with a secret master key distributes the workload by mirroring an organization hierarchy, delegating the process of key generation, and authenticating identities of users to lower level PKGs.

Keywords Cryptosystems · Public key infrastructure · Identity-based encryption · Private key generator · Hierarchical identity-based encryption · Public key · Master secret

T. A. Teli (✉)
Department of Computer Sciences, Amar-Singh College, Srinagar, J&K, India
e-mail: tawseef@kashmiruniversity.net

F. S. Masoodi
Department of Computer Sciences, University of Kashmir, Srinagar, J&K, India
e-mail: masoodifahim@uok.ac.in

A. M. Bahmdi
Computing College AlQufudah, Umm Al-Qura University, Mecca, Kingdom of Saudi Arabia
e-mail: saambamhdi@uqu.edu.sa

1 Introduction

1.1 Public Key Cryptography

Cryptosystems, based on the visibility of key, are generally classified into two types, viz. symmetric (private key) systems and asymmetric (public key) systems [1]. The public key cryptosystems use the concepts of digital certificates to work with public keys. The key pair (public key, private key) is generated randomly by a user or a third-party entity. In a traditional public key cryptosystem scenario, if Alice wants to send a message to Bob, Alice uses Bob's public key, which can be any randomly generated string. The public key and the owner's identity are used as certificate, digitally signed by a certificate authority (CA), for key management. Alice obtains the authenticated key from a certificate repository that is designated to store public keys and the identity of the owners of the public keys (digital certificates) in a traditional public key encryption scheme as shown in Fig. 11.1. The private keys may also be stored by a recovery system to cater for the lost private keys or keys that may be unavailable due to some unforeseeable reasons. Like public keys, if private keys are generated by a centrally designated authority/agent, they are also archived and a copy of the private key is sent to the owner by the certificate authority.

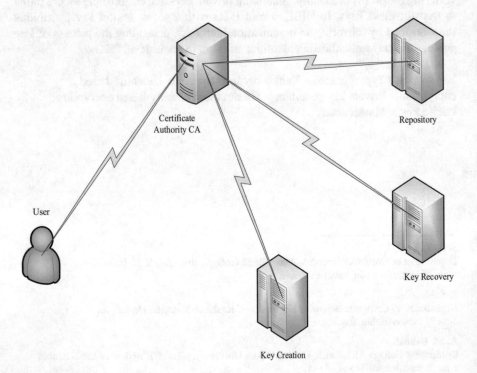

Fig. 11.1 Traditional public key encryption (key generation)

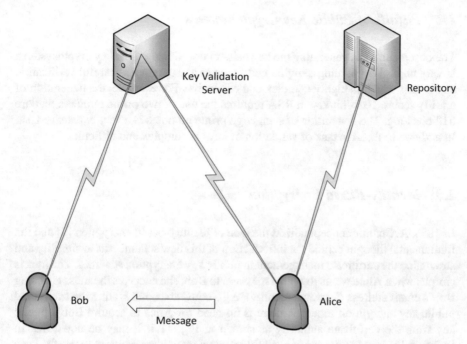

Fig. 11.2 Validation of keys

The digital certificates come with expiry dates that are usually kept as long periods of time due to the high computational and verification costs. The expiration in digital certificates asks for a validation check on keys before using it. If Alice desires to transmit a message to Bob, she receives the certificate from the CA and before she can use it to encrypt the message, she needs to verify the key validity from a validation authority as shown in Fig. 11.2.

The traditional public key encryption can thus be defined in terms of three main algorithms: key generation, encryption, and decryption. The key generation involves the creation of key pair (public-private). In this step a key is generated randomly by the user or a third-party entity while the other key is derived from the already generated key. Before the encryption step is implemented, a certificate authority (CA) digitally signs the public key and the owner's identity to create a digital certificate. The public key in the digital certificate is used for encryption after the keys are successfully validated in the validation step. Finally, the decryption is done using the private key.

1.2   Pitfalls of Public Key Cryptosystems

The cost incurred in generating the key pairs in traditional public key cryptosystems is very high. Also, issuing a digital certificate to a user requires careful verification of the user which is highly complex and expensive. For example, the generation of a fairly secure 1024-bit key in RSA requires the use of two prime numbers that are 512 bits long. The generation of such large prime numbers is a very expensive task. In addition to this, the task of validation of keys is complex and difficult.

1.3   Identity-Based Encryption

In 1854, A. Shamir conceptualized the idea of identity-based encryption [2] and the fundamental thought behind the introduction of this new scheme was to simplify and streamline the certificate management in public key encryption schemes. The idea is simple: when Alice wants to send a message to Bob, she encrypts the message using Bob's email address (a known identity), e.g., bob@abc.com. Unlike the traditional public key encryption schemes, there is no need for Alice to acquire Bob's public key from the certificate authority as shown in Fig. 11.3. It may be noted that in traditional public key encryption, all the parameters that are required to use the keys are contained in the public keys while in case of IBE, a trusted third party is used to get a set of parameters.

After receiving the message from Alice, Bob obtains his private key from a neutral entity called as private key generator (PKG) but not before authenticating himself to the private key generator (PKG). A master secret key in addition to the user's identity is used by PKG to calculate the private key which is then transmitted to the authorized user (Bob). Now Bob uses the private key to decipher the text shown in Fig. 11.4.

In this new public key encryption system, there are four algorithms that define the whole scheme:

1. Setup
2. Extraction
3. Encrypt
4. Decrypt

In the setup algorithm, an initialization of many necessary global parameters is performed that are used for the calculations of IBE; for example, the master secret key is initialized which is used to calculate the private keys by the PKG. The extraction algorithm mainly deals with the generation of the private key corresponding to a particular user, using the parameters initialized in the setup algorithm that includes the identification of the user and master secret key. The encryption algorithm uses the user's identity and the public key calculated in the setup algorithm to encrypt the message. The decryption algorithm uses the private

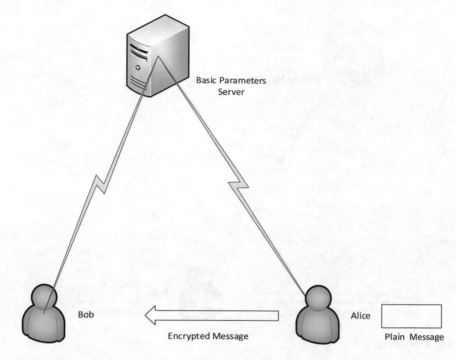

Bob

Alice

Plain Message

Encrypted Message

Fig. 11.3 Encryption

key calculated in the IBE extraction algorithm and the user's identity from the PKG to decrypt the message.

Even though Shamir introduced IBE in 1984 it was not until after the contributions of Boneh and Franklin [3, 4] and Cocks [5] in 2001 that we saw a practical identity-based encryption scheme. Cocks uses quadratic residuosity problem as the basis of his scheme, achieving fast speeds for encryption and decryption, but the message length becomes profoundly large. Boneh-Franklin scheme is essentially based on the bilinear Diffie-Hellman problem [6–8], achieving fast speeds and reasonable ciphertext lengths using Weil or Tate pairings. It is pertinent to note here that IBE replaced the problem of acquiring the public keys with the problem of obtaining public parameters of PGSs. With just a fewer number of PKGs, the latter problem should not be as cumbersome as the former one.

A common issue in traditional public key encryption schemes is key validation. IBE consists of keys that are short lived, which means that the keys are valid for just 12 hours or a day and cannot be revoked or suspended during this period of time. Although this scheme takes away the ability to revoke or suspend a key immediately it makes the implementation and key validation fairly simple.

Key recovery is yet another issue in traditional public key encryption schemes that needs to be resolved when the keys are lost or unavailable due to any unforeseeable situations [9]. The issue is catered by archiving the copies of keys

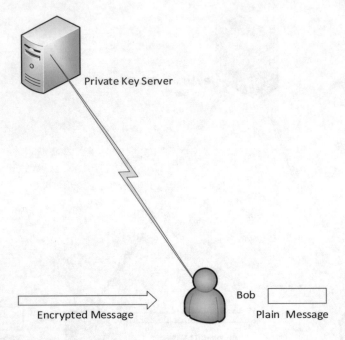

Fig. 11.4 Decryption

that can be recovered as required. It is a difficult task to securely archive keys and provide controlled access to these keys. Identity-based encryption solves this problem as the keys are generated as needed. From the master key and the identity of the owner, PKG can easily calculate the private keys, making the whole task much easier than the traditional PKI schemes. IBE is thus equipped with the capabilities of cost reduction in terms of computations and maintenance than the traditional public encryption schemes.

1.4 Hierarchical Identity-Based Encryption

Private key generator is the single most significant entity in identity-based encryption that is burdened with most of the work. PKG calculates private keys for the users from the master secret key and the users' identities, which then transmit the extracted private keys to the authorized users after verifying the identities of the users successfully. The task of calculating thus becomes very expensive and if a single PKG exists for a large network with tens of thousands of users, it would be a bottleneck to the whole network. In hierarchical identity-based encryption (HIBE) [10], a root PKG has the capability to distribute its workload to lower level PKGs. The distribution of workload is done by adding a hierarchy of PKGs. The root PKG calculates private keys for PKGs in its domain and the PKGs in turn generate private

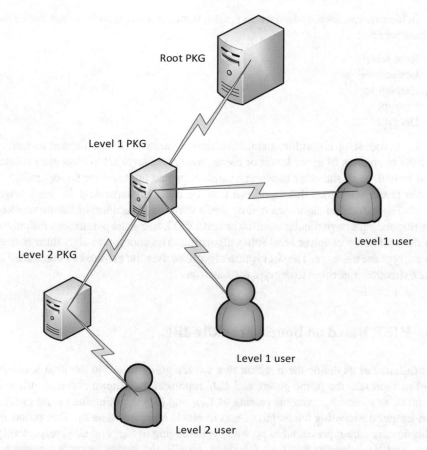

Root PKG

Level 1 PKG

Level 1 user

Level 2 PKG

Level 1 user

Level 2 user

Fig. 11.5 Two-level hierarchy

keys for their respective users. Hence the verification of users and the transmission of keys are done locally. The master secret key of more than one PKG is used to generate all private keys.

Some of the important contributions in hierarchical identity-based encryption schemes were introduced by Hanaoka et al. [11, 12]. A two-level scheme with a total and partial collision resistance at the first and second levels, respectively, was introduced by Horwitz and Lynn [13] as shown in Fig. 11.5.

It can be seen that the work done by a particular PKG at a specific level depends on the work done by the PKGs that are above in its hierarchy. With different security policies at different levels of hierarchy, a security policy at a higher level in hierarchy can enforce it on all its subordinate levels. Also, a breach/compromise In that happened in one level will not necessarily affect other parts of the hierarchy. Hence, the recovery from such a breach is easy as only a part of the system needs to be recreated rather than the whole system.

In hierarchical identity-based encryption, there are five algorithms that define the whole scheme:

1. Root setup
2. Lower level setup
3. Extraction
4. Encrypt
5. Decrypt

In the root setup algorithm, an initialization of parameters is performed necessary for the operations of upper levels of hierarchy. Lower level setup algorithm creates and initializes all the other necessary parameters that are used for the operation of lower levels and hence this algorithm may be executed separately for each lower level. The extraction algorithm mainly deals with the generation of the private key corresponding to a particular user/lower level PKG, using the parameters initialized in the root setup or lower level setup algorithm. The encryption algorithm is used to encrypt the message. The decryption algorithm uses the private key calculated in the extraction algorithm to decrypt the message.

2   HIBE Based on Boneh-Franklin IBE

Notation: Let us define the notation that we are going to use in the next sections. Let p_p represent the prime power and E_c/f represent the elliptic curve. A prime is denoted as p_r and e represents pairing of C_{G1} and C_{G2} denoting the cyclic groups. An integer designating the length of text in bits is represented as b_1. Two points on elliptic curve are represented as p_{t1} and p_{t2} belonging to C_{G1} and C_{G2}, respectively. H_{S1} and H_{S2} represent two hash functions. Finally, the master secret is represented as MS.

2.1   HIBE (Based on Boneh-Franklin IBE) Root Setup

In the root setup algorithm of HIBE based on Boneh-Franklin [14] IBE, the following parameters are defined:

Basic Parameters

- C_{G1} | Cyclic group #1
- C_{G2} | Cyclic group #2
- e | Pairing,
- b_l | Plaintext length
- p_{t1} | Elliptic curve point #1
- p_{t2} | Elliptic curve point #2
- H_{S1} | Hash function #1

- H_{S2} | Hash function #2

In the root setup step, the master key is also defined as *MS*.

2.2   HIBE (Based on Boneh-Franklin IBE) Lower Level Setup

Since there are multiple levels in the hierarchy and at each level there shall be PKGs, a lower level setup needs to be performed. All the PKGs in the lower levels shall have the same parameters as the root level with the additional parameter of master secret defined for its level. For example, at level i, the master secret parameter is MS_1 which belongs to $Z^*{}_p$.

2.3   HIBE (Based on Boneh-Franklin IBE) Extraction of the Private Key

For a hierarchy with l levels, the identity of a user is l-tuple; for example, for a single user u, the identity can be defined as $U = (I_1, I_2, \ldots, I_n)$ where $n \leq l$. After applying H_{S1} on the identity components of U, D_{IK} is calculated as

$$D_{IK} = H_{S1} \ (I_1, I_2, \ldots, I_n)$$

Hence the private key for U shall comprise n components from the n identities of the user which is defined as follows:

$$P_K = (K_0, K_1, \ldots, K_{n-1})$$

The first element of P_K is calculated as

$$K_0 = \sum_{i=1}^{l} MS_{i-1} D_{Ii}$$

The rest of the elements are calculated as

$$K_i = MS_i \ p_{t1} \text{ where } 1 \leq i \leq n-1$$

2.4   HIBE (Based on Boneh-Franklin IBE) Encryption

The encryption in a HIBE scheme with l levels can be achieved as follows:

If the plaintext *MSG* needs to be encrypted using the identity of a user U the identity is defined as

$$U = (I_1, I_2, \ldots, I_n) \quad \text{where } n \leq 1$$

Let $r = \text{e}\,(MS_0\, p_{t1}, D_{U1})$.

A number x is picked randomly from Z_p and the $k + 1(M, N_0, N_1, N_2, \ldots N_K)$ elements of the ciphertext C_T are calculated as follows:

$$M = MSG_e \oplus HS_2\left(r^x\right)$$

$$N_0 = x\ p_{t1}$$

$$N_i = x\ D_{ui} \ where\ 2 \leq i \leq k$$

2.5   HIBE (Based on Boneh-Franklin IBE) Decryption

Being private at the receiving end, the ciphertext $C_T = (M, N_0, N_1, N_2, \ldots N_K)$ is received by the receiver and he/she has to decipher it back to the plaintext.

The process of deciphering is done making the following calculations:

$$M \oplus HS_2\left(\frac{e\,(N_0, K_0)}{\prod_{i=2}^{l} e\,(K_{i-1}, N_i)}\right) = M \oplus HS_2\left(\frac{e\left(x p t1, \sum_{i=1}^{l} MS_{i-1}D_{ui}\right)}{\prod_{i=2}^{l} e\,(MS_{i-1}p t1, x D_{ui})}\right)$$

$$= M \oplus HS_2\left(\frac{\prod_{i=1}^{l} e\,(x p t1, MS_{i-1}D_{ui})}{\prod_{i=2}^{l} e\,(MS_{i-1}p t1, x D_{ui})}\right)$$

$$= M \oplus HS_2\left(\frac{\prod_{i=1}^{l} e(p t1, D_{ui})^{x MS_{i-1}}}{\prod_{i=2}^{l} e(p t1, D_{ui})^{x MS_{i-1}}}\right)$$

$$= M \oplus HS_2\left(e(p t1, D_{ui})^{x MS_0}\right)$$

$$= M \oplus HS_2\left(r^x\right)$$

$$= MSG \oplus HS_2\left(r^x\right) \oplus HS_2\left(r^x\right)$$

$$= MSG$$

3 HIBE Based on Boneh-Boyen IBE

Notation: Let us define the notation that we are going to use in the next section for BB-IBE (additive notation).

Let p_p represent the prime power, and E_c/f represent the elliptic curve. A prime is denoted as p_r and e represents pairing of C_{G1} and C_{G2} denoting the cyclic groups. An integer designating the length of the text in bits is represented as b_l. Four points on elliptic curve are represented as p_a, p_b, p_c, and p_d where p_a belongs to C_{G1}, $p_b = \alpha\ p_a$, $p_c = \beta p_a$, and $p_d = \gamma\ p_a$. H_{S1} and H_{S2} represent two hash functions: $v = e(p_b, p_c) = e(\alpha\ p_a, \beta\ p_a) = e(p_a, p_a)^{\alpha\beta}$. Finally, the master secret is represented by α, β, and γ which are the elements of Z_p.

3.1 HIBE (Based on Boneh-Boyen IBE) Setup

In the setup algorithm for Boneh-Boyen IBE [15], the following parameters (additive notation) are defined:

Basic Parameters

- C_{G1} | Cyclic group #1
- C_{G2} | Cyclic group #2
- □ | Pairing,
- b_l | Plaintext length
- p_a | Elliptic curve point #1
- p_b | Elliptic curve point #2
- p_d | Elliptic curve point #4
- H_{S1} | Hash function #1
- H_{S2} | Hash function #2
- v | Element of $F^*{}_q$.

In order to extend this scheme to HIBE, the following randomly generated parameters need to be added to the list of parameters defined above:

$$S_1, S_2, \ldots, S_l$$

It is pertinent to mention that all the new parameters belong to the cyclic group C_{G1} and the final list of parameter setup for the HIBE is as follows:

Basic Parameters Boney-Boyen $= (C_{G1}, C_{G2}, \ \square \ , b_l, p_a, p_b, p_d, S_1, S_2, \ldots,$
$S_l, H_{S1}, H_{S2}, v \ )$

3.2 HIBE (Based on Boneh-Boyen IBE) Extraction of the Private Key

The algorithm for generating private key for a user is executed in this step of HIBE. Suppose the identity of a user is defined as $U = (I_1, I_2, \ldots, I_l)$ for l levels of hierarchy, and the private key for the user u is defined as $K_u = (K_1, K_2)$. A random number x is generated by the root PKG that belongs to Z_p and the following calculations are made by the root PKG:

$$p_p I_i = H_{S1}(I_1) \text{ where } 1 \leq i \leq k$$
$$K_1 = x p_a \text{ and}$$
$$K_2 = a p_a + x \sum_{(i=1)}^{k} p_p I_i S_k$$

3.3 HIBE (Based on Boneh-Boyen IBE) Encryption

The encryption of plaintext that needs to be sent to the receiver is performed in this step. Alice has to send a message to Bob with identity $U = (I_1, I_2, \ldots, I_k)$, where $k \leq l$. A random element m is picked up from Z_p to calculate the ciphertext C_T from the plaintext MSG:

$$C_T = (M, N_1, N_2)$$

And the elements are calculated as follows:

$$M = MSG \oplus H S_2 (v^m)$$

$$N_1 = m p_a$$

$$N_2 = m \sum_{i=1}^{k} p_p I_k S_k$$

3.4 HIBE (Based on Boneh-Boyen IBE) Decryption

At the receiving end the ciphertext $C_T = (M, N_1, N_2)$ is received by Bob and he has to decipher it back to the plaintext.

The process of deciphering is done making the following calculations:

$$M \oplus HS_2 \left(\frac{e(N_1, K_2)}{e(K_1, N_2)}\right) = M \oplus HS_2 \left(\frac{e\left(mp_a, \alpha p_c + x\left(\sum_{i=1}^{k} p_p I_i S_i + p_d\right)\right)}{e\left(xp_a, m\left(\sum_{i=1}^{k} p_p I_i S_i + p_d\right)\right)}\right)$$

$$= M \oplus HS_2 \left(e\left(mp_a, \alpha p_c\right)\right)$$

$$= M \oplus HS_2 \left(e(p_a, p_c)^{\alpha m}\right)$$

$$= M \oplus HS_2 \left(e\left(\alpha p_a, p_c\right)^m\right)$$

$$= M \oplus HS_2 \left(e\left(\alpha p_a, p_c\right)^m\right)$$

$$= M \oplus HS_2 \left(v^m\right)$$

$$= MSG \oplus HS_2 \left(v^m\right) \oplus HS_2 \left(v^m\right)$$

$$= MSG$$

4 Master Secret Sharing

In hierarchical identity-based encryption scheme, a user at a specific level in the hierarchy can have multiple identities pertaining to each separate level. For a hierarchy with n levels, the identity of a user is n-tuple; for example, for a single user U with n levels, the identity can be defined as $U = (I_1, I_2, \ldots, I_n)$, where each entry in the tuple may be distinct.

In HIBE, the master secret key that is used to calculate private keys of the users is shared and distributed among PKGs. A system with m PKGs will share master secret and each of the PKG will contribute to generating a part of the private key for the user known as a share. Whenever a user requests a private key from PKGs, each

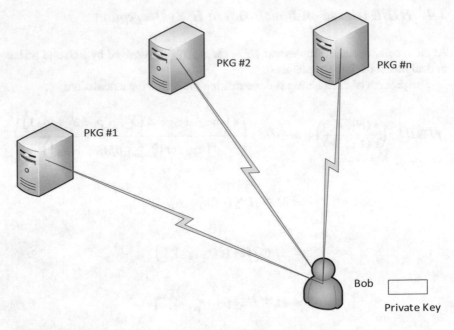

Fig. 11.6 Master key sharing

PKG sends a share and the user combines all the shares to calculate the private key. If a user receives p of the m possible shares and generates the private key from these p shares, the same private key cannot be calculated from any $p - 1$ shares which significantly means not a single share from p shares can be dropped out to calculate the same private key. It may also be noted that for any $p - 1$ PKGs, it is impossible to recreate the master secret as shown in Fig. 11.6.

Sharing of the secret master key removes the single point of failure or security breach point. For example if one of many m PKGs sharing the master secret key is compromised, it would be far less worse a situation than if the master key was stored on a single PKG and that PKG is compromised. Hence, one can easily grasp the advantages of sharing the master secret key. For an attacker to access the master secret key it will require him/her to gain access to all the p of the m possible PKGs which would be fairly easy if only one PKG is used.

Mathematically, it was Boneh et al. [15] and Shamir [16] who showed how a master secret can be shared among many PKGs. Suppose there are m PKGs, and a user is required to get p, $p < = m$, and shares to generate his/her private key. Suppose there is a polynomial of degree $p - 1$; then the master secret can be represented as the coefficients of the polynomial. Each PKG in the hierarchy is assigned a point (x_i, y_i) that solves the polynomial and a user containing p of these points can calculate all the coefficients of the polynomial and hence generate his/her private key. It is important to note that any PKGs with $p - 1$ share cannot calculate the same private key.

Consider the polynomial in which the master secret is put as the constant coefficient of the polynomial:

$$f(x) = MS + c_1 x + c_2 x^2 + c_3 x^3 + \cdots + c_{p-1} x^{p-1}$$

A master secret MS is thus divided and shared among the PKGs as follows:

A total of m numbers are randomly generated and the pair (x_i, y_i), where $1 \leq i \leq m$ and $y_i = f(x_i)$, is assigned to the ith PKG. A user with user identity D_I can request the ith PKG for his/her share and the PKG replies with $(x_i, y_i D_I)$. After receiving the similar pairs p from such PKGs, the user can calculate his/her private key by using Lagrange's interpolation as $MSD_I = (0)D_I$.

Now,

$$f(x) = \sum_i g_i(x) y_i$$

Multiplying both sides by D_I, we get

$$f(x) D_I = \sum_i g_i(x) y_i D_I$$

For $x = 0$, we get

$$MSD_I = f(0) D_I = \sum_i g_i(0) y_i D_I$$

And hence MSD_I is the private key for the user with identity D_I that was generated after receiving p pairs of $(x_i, y_i D_I)$ from the PKGs.

5 Security of the Hierarchical IBE

Identity-based encryption (IBE) has been discussed as the cryptosystem [2] in which the encryption key (public key) is any arbitrary string. In this cryptosystem, there is a central authority that issues decryption keys (private keys) to the users. Another variation of this scheme is the hierarchical identity-based encryption in which there can be many levels of hierarchy. In the initial implementations of hierarchical identity-based encryption schemes, the security of the system was attributed to the bilinear Diffie-Hellman (BDH) assumption in the random oracle model [10]. In one of the other constructions of HIBE also known as selective-ID secure HIBE, the security is due to the bilinear Diffie-Hellman (BDH) assumption without random oracles.

In identity-based encryption, it is assumed that an adversary can request for private keys and in the process, the adversary can acquire the private key corresponding to a particular identity. Similarly, in hierarchical identity-based encryption, an adversary is assumed to be able to execute queries for the extraction of private keys and it can choose a specific identity for which the adversary can be challenged [4]. There are two ways in which an adversary can pick its target identity; one is adaptive and the second is nonadaptive. An adversary with no particular designated target starts finding its target by executing hash and extraction queries in order to choose a target adaptively based on the results of these hash and extraction queries. The chosen target may not even be associated to an entity and the adversary can still be successful in acquiring some identity. As far as nonadaptive adversary is concerned, the target is selected beforehand. The adversary has prior knowledge on who its target is and there is no need to execute hash and extraction queries first. The motive behind such an attack can be personal. It is pertinent to mention here that the security is considered stronger for an adaptively chosen target as compared to an adversary that has chosen the target nonadaptively.

In CHK [17], a user is required to update the private keys at regular intervals of time such that a message that has been encrypted at time t cannot be read using any private key from time $t_1 > t$. Now for $Q = 2^n$ periods, n is the depth and the identities are represented as binary vectors with a length of at most n. Encryption at time t takes place using the identity associated with the tth node of depth in a binary tree. Dodis and Fazio [18] cover NNL [19] into a public key broadcast system.

References

1. Masoodi, F. S., & Bokhari, M. U. (2019). Symmetric Algorithms I. In *Emerging Security Algorithms and Techniques* (p. 79). CRC Press.
2. Shamir, A. (1984). Identity-based cryptosystems and signature schemes. In *Advances in Cryptology – Crypto '84, Lecture Notes in Computer Science 196* (pp. 47–53). Springer.
3. Boneh, D., & Franklin, M. (2001). Identity based encryption from the Weil pairing. In *Advances in Cryptology – Crypto 2001, Lecture Notes in Computer Science 2139* (pp. 213–229). Springer.
4. D. Boneh and M. Franklin. Identity based encryption from the Weil pairing. http://www.cs.stanford.edu/~dabo/papers/ibe.pdf.
5. Cocks, C. (2001). An identity based encryption scheme based on quadratic residues. In *Cryptography and Coding, Lecture Notes in Computer Science* (pp. 360–363). Springer.
6. Diffie, W., & Hellman, M. (1976). New directions in cryptography. *IEEE Transactions on Information Theory, IT-22, 6*, 644–654.
7. Joux, A., & Nguyen, K. (2003). Separating decision Diffie-Hellman from Diffie-Hellman in cryptographic groups. *Journal of Cryptology, 16*(4), 239–247.
8. Boneh, D. (1998). The decision Diffie-Hellman problem. In *Algorithmic Number Theory Third International Symposium, Portland, OR, June 21–25* (pp. 48–63).
9. Nielsen, R. (2005). Observations from the deployment of a large scale PKI. In *Proceedings of the 4th Annual PKI R&D Workshop, Gaithersburg, MD, August 19–21* (pp. 159–165).
10. Gentry, C., & Silverberg, A. (2002). Hierarchical ID-based cryptography. In *Proceedings of ASIACRYPT 2002, Queenstown, New Zealand, December 1–5* (pp. 548–566).
11. Hanaoka, G., Nishioka, T., Zheng, Y., & Imai, H. (1999). An efficient hierarchical identity based key-sharing method resistant against collusion-attacks. In *Advances in Cryptology – Asiacrypt 1999, Lecture Notes in Computer Science* (pp. 348–362). Springer.

12. G. Hanaoka, T. Nishioka, Y. Zheng, and H. Imai, A hierarchical non-interactive key-sharing scheme with low memory size and high resistance against collusion attacks, to appear in The Computer Journal
13. Horwitz, J., & Lynn, B. (2002). Toward Hierarchical Identity-Based Encryption. In *Advances in Cryptology – Eurocrypt 2002, Lecture Notes in Computer Science 2332* (pp. 466–481). Springer.
14. Boneh, D., & Franklin, M. Identity based encryption from the weil pairing. *SIAM Journal of Computing, 32*(3), 586–615.
15. Boneh, D., Boyen, X., & Goh, E. Hierarchical identity-based encryption with constant size ciphertext. In *Proceedings of EUROCRYPT 2005, Aarhus, Denmark, May 22–26* (pp. 440–456).
16. Shamir, A. (1979). How to share a secret. *Communications of the ACM, 22*(1), 612–613.
17. Canetti, R., Halevi, S., & Katz, J. (2003). A forward-secure public-key encryption scheme. In E. Biham (Ed.), *Proceedings of Eurocrypt 2003, volume 2656 of LNCS*. Springer.
18. Dodis, Y., & Fazio, N. (2002). Public key broadcast encryption for stateless receivers. In J. Feigenbaum (Ed.), *Proceedings of the digital rights management workshop 2002, volume 2696 of LNCS* (pp. 61–80). Springer.
19. Naor, D., Naor, M., & Lotspiech, J. (2001). Revocation and tracing schemes for stateless receivers. In J. Kilian (Ed.), *Proceedings of crypto 2001, volume 2139 of LNCS* (pp. 41–62). Springer.

Chapter 12
Extensions of IBE and Related Primitives

Syed Taqi Ali

Abstract Identity-based encryption can be extended to various other primitives, such as identity-based key agreement protocol, fuzzy identity-based encryption, keyword search enabled public key encryption, threshold keyword search enabled public key encryption, wildcard identity-based encryption, identity-based conditional proxy re-encryption, and so on. Every extension has its own application in real world. In this chapter, we discussed all these extended primitives with their formal definitions and basic security requirements.

Keywords Extensions of IBE · Fuzzy IBE · Keyword search · Wildcards key derivation

1 Introduction

Identity-based encryption (IBE) can be extended to the various other primitives. Few of them are discussed in this chapter. Usually, the key agreement protocols are used for establishing secret session key in public environment between two or more parties, to enable secure communications among them. The basic property that any key agreement protocol needs to be achieved is the inability of the passive eavesdropper to compute secret session key. To be more secure, the key agreement protocol should not allow an adversary to distinguish between the given, session key and a random string, with non-negligible probability. Normally, the parties established the secret session key with the help of their long-term keys. If the long-term key gets compromised, then the future session keys can be compromised. To avoid this, a property called perfect forward secrecy is incorporated on the long-term keys, which allows us to update the long-term keys. In this chapter, we discussed

S. Taqi Ali (✉)
Visvesvaraya National Institute of Technology, Nagpur, Maharashtra, India
http://cse.vnit.ac.in/people/stali/

© Springer Nature Switzerland AG 2021
K. A. B. Ahmad et al. (eds.), *Functional Encryption*, EAI/Springer Innovations in
Communication and Computing, https://doi.org/10.1007/978-3-030-60890-3_12

various identity-based key agreement protocols [10–12, 20] with their short comings and strengths.

Fuzzy identity-based encryption (FIBE) is a type of IBE where the identities are replaced by the set of attributes, each associated with some description. Fuzzy IBE allows user to decrypt the ciphertext even his/her identity partially matched till some limit with the secret key, which we called as error-tolerance property. Fuzzy IBE facilitates us to apply it in bio-metric authentication applications and attribute-based cryptographic schemes. The security requirement for the FIBE is that given the ciphertext of one of the two plain texts, the adversary should unable to map the challenge ciphertext with its exact plain text with non-negligible probability. In this chapter, we will discuss few constructions of FIBE [6, 19, 21, 22] with its security properties.

Nowadays, data owners transfer their huge data on the cloud in encrypted form. To enable searching on the encrypted data, a Public Key Encryption with Keyword Search (PEKS) was introduced, in which keywords are also encrypted. Then, to search the data, user sends the corresponding keyword's trapdoor to the cloud server, which enables server to perform search operation. Broadcast Encryption with Keyword Search is a PEKS where the keyword is encrypted for the set of users, so that any user from the set can generate the trapdoor for search operation. To reduce the ability of generating search keyword trapdoor from a single user, a threshold policy was added, called "Threshold Public Key Encryption with Keyword Search" (TPEKS) [13, 25, 26]. In this, the set of users and threshold value is fixed. In "Threshold Broadcast Encryption with Keyword Search" (TBEKS), the data owner for each document chooses the set of users and the threshold value "l" to encrypt a keyword. Then, to search a certain keyword, at least "l" users of that set need to pool their share to enable the cloud server to perform search operation of that keyword. The security requirement for TBEKS is the inability of the adversary to guess that the encrypted keyword belongs to which keyword among two chosen keywords, with non-negligible probability. In this chapter, we see all the related constructions of this.

The asymmetric encryption schemes, such as public key encryption (PKE), identity-based encryption (IBE), attribute-based encryption (ABE), and other similar schemes, whose decryption algorithm has linearity property, i.e., $D(SK, C1, \ldots)t = D(SKt, C1, \ldots)$, can be converted to Linear Encryption with Keyword Search (LEKS) scheme [27]. These features enable us to convert existing asymmetric encryption scheme to the corresponding searchable encryption schemes. Similar to any searchable encryption scheme, this scheme also obeys the basic security properties, such as chosen keyword attack. In this chapter, we look few of these conversions.

"Public Key Encryption with Keyword Search" (PEKS) [7, 9] enables server to search the presence of encrypted keyword in the cloud data without learning anything about the data or keyword. In this, three parties are involved, sender, receiver, and server. Sender creates and uploads encrypted keywords with ciphertexts. Server stores ciphertexts with encrypted keywords and performs search operation upon receiving trapdoors from the receiver. Receiver creates trapdoors and sends to the server. In the literature, PEKS schemes avoid static trapdoors, and this is to prevent server from learning any partial information from the trapdoor, such as frequency of searching same keyword or which other document contains same keyword, etc. There are PEKS schemes that allow to search keywords in range and with relational operators too. In this chapter, we see various constructions of such schemes.

"Wildcard Identity-Based Encryption" (WIBE) [4, 5, 8] allows sender to encrypt a message with the pattern such that the range of receivers whose identity matches the pattern they can only decrypt it. The pattern includes strings with wildcards. Normally, the WIBE is the generalization of Hierarchical IBE scheme where the ancestor of the matching identity is able to derive the secret key of their descendant identities in the same hierarchy.

In "Identity-Based Encryption with Wildcards Key Derivation" (WKD-IBE) [1, 2], the user possessing the secret key for the pattern P can derive the secret key for other patterns P', if P' contains in P. There introduces one key derivation algorithm that enables this feature, as compare to WIBE.

In "Identity-Based Conditional Proxy Re-encryption" (IBCPRE) [15, 23], proxy can convert the ciphertext of one user to the other user if the prescribed condition, set by the delegator, satisfied. This will add extra security over identity-based proxy re-encryption (IBPRE), so that proxy may not able to transform the ciphertext to any unauthorized user. In this chapter, we see all the constructions of above schemes along with their security proofs.

2 Identity-Based Key Agreement (IBKA)

Key agreement protocols are used to establish a secure communication between two or more parties in insecure environment. Using this, a shared secret session key is generated between the parties. Then, all parties use this shared key to secure their communication using a well private key encryption scheme. In identity-based key agreement (IBKA), the public key is the user identity (an arbitrary string such as email or phone number), which is well known to all, and the corresponding private key is generated under the unique setup of identity-based key generation with a fixed master key. This way it extends the concept of identity-based encryption to key agreement protocols. There are many identity-based key agreement protocols in the literature [10, 11, 14, 16, 20]. The identity-based key agreement protocols have applications in vehicular ad hoc networks [12], where two or more vehicles interested in establishing a secure communication under common setup.

Intuition In IBKA, two or more users can establish a shared secret key using their public identity and corresponding secret key, generated under common setup or with common master key. It is similar to any other key agreement protocol, but the public key is not generated, and it is his/her identity.

Next, we give the formal model of IBKA followed by its basic security properties.

2.1 Formal Model

Any identity-based key agreement protocol contains 3 stages [20]: setup, private key extraction, and key agreement.

- Setup: In this phase, under decided security parameter λ, the trusted authority generates the fixed public values PP and master secret key MK.
- Private key extraction: Here, a user with his/her unique ID (an arbitrary string) approaches a trusted authority, and the trusted authority using the master key MK generates the user's secret key for his/her ID, say SK_{ID}, and we call it as long-term secret key.
- Key agreement: This is the interactive protocol phase between the users. Each user with his/her secret key associated with his/her ID under the common trusted authority setup will participate in this protocol. Users exchange the messages and finally agree upon some secret shared key, say $SK_{ID_1, ID_2, \ldots}$.

The above three phases vary in their constructions with respect to the various research papers.

2.2 Security Requirements

For any identity-based key agreement protocol to be used in real-time application should satisfy few security properties, as per the requirement of the application environment. The security properties in identity-based key agreement are

1. **Known-key security (K-KS)**. It says that the revealing of one session key should not affect the security of the other session key. That is, no partial information should be revealed from the session key of the other session key.
2. **Forward secrecy (FC)**. It says that the revelation of long-term secret key should not compromise the previous session keys. This has 3 flavors,

   - Partial forward secrecy. Revelation of few users' long-term secret keys should not reveal any partial information of the previous session keys.
   - Perfect forward secrecy. Revelation of all users' long-term secret keys should not reveal any partial information of the previous session keys.
   - Master key forward secrecy. Revelation of master secret key should not reveal any partial information of the previous session keys. This also implies perfect forward secrecy.

3. **Key-compromise impersonation resilience (K-CI)**. If any user's secret key gets compromised, then it should not allow adversary to impersonate any other user other than the secret key user. That is, no user should be able to impersonate other user.
4. **Unknown key-share (UK-S) resilience.** Each user should truly identify each other while sharing the secret key.

5. **No key control**. This disallows any user to precompute the session key.
6. **Ephemeral secrets reveal resistance (ESRR).** Ephemeral secret keys are the short-lived keys. The key agreement protocol should be resistant to the disclosure of any user's ephemeral secret key. That is, session secret keys should not get compromised of the sessions where that disclosed ephemeral secret key not used.

The key agreement protocol is said to be secure which obeys the maximum security properties. It is difficult for any protocol to achieve all the above security requirements.

2.2.1 Oracles

In the adversary modeling of these security properties, they involved few oracles. Let $\prod_{A,B}^{i}$ denote the behavior of the user B at i-th time when A is communicated to him/her. Then, the common oracles are

- **Extract**(ID). Upon adversary query with ID, it generates the secret key SK_{ID} for the ID and gives it to the adversary.
- **Send**($\prod_{A,B}^{i}$,M). It replies with the response of user A on the message M when user B sends to A at i-th time during the session.
- **Reveal**($\prod_{A,B}^{i}$). It reveals the session key of that session, where $\prod_{A,B}^{i}$ belongs to the session.
- **Corrupt**(ID). It reveals the secret key of the existing user with identity ID.

3 Fuzzy Identity-Based Encryption (FIBE)

"Fuzzy identity-based encryption" was first introduced by Sahai et al. [22], where identity is treated as a set of descriptive attributes. The attributes can be any feature of a user, like age, gender, designation, city, etc., through which we can uniquely or approximately identify the user. In this FIBE scheme, it allows multiple users to decrypt the ciphertext which encrypted using certain attributes if they possess some prescribed similarity with those attributes. Later, many other FIBE schemes were developed [6, 19, 21]. This has applications in encryption schemes where biometric inputs are used as identities, since it can tolerate some error. Later, attribute-based encryptions were derived from it.

Intuition In FIBE, a user can decrypt the data if his/her identity matched till at least the allowed level or percentage, using his/her secret key. Here, data user identity may not be exactly matched but till some percentage with the identity associated with ciphertexts.

The formal definition and security properties of it are discussed below.

3.1 Formal Model

The FIBE model is similar to the IBE model, and it consists of the following algorithms:

- **Setup()**: It takes security parameter as input and generates master key MK and public parameters PP. Note that the tolerance parameter d was included in PP.
- **Extract(MK, ID)**: Upon input MK and ID, it generates secret key associated with ID, say SK_{ID}.
- **Encrypt(PP, ID, M)**: It generates a ciphertext C for the users with similar identities to ID (similarity up to the tolerance parameter d).
- **Decrypt($PP, SK_{ID'}, C$)**: The user with identity ID' is able to decrypt C if $|ID' \cap ID| \geq d$, else the algorithm outputs $\perp$, invalid.

Correctness It says that for all properly generated secret keys of the users, all properly generated ciphertext, the user should able to correctly decrypt the ciphertext if the similarity, between the user identity and the identity associated with ciphertext, is up to the prescribed tolerance parameter. More formally, it is as given below:

$$(MK, PP = \{\ldots, d\}) \overset{\$}{\leftarrow} \text{Setup}() \wedge (SK_{ID'}) \overset{\$}{\leftarrow} \text{Extract}(MK, ID') \wedge$$

$$C \overset{\$}{\leftarrow} \text{Encrypt}(PP, ID, M) \text{ then } \text{Decrypt}(PP, SK_{ID}, C) \rightarrow M$$

3.2 Security Properties

Primary security requirement of FIBE is that given a ciphertext encrypted under the one identity of any two of the messages, it should be difficult to map ciphertext with the correct message. This we may called as "IND-FSID-CPA—indistinguishability of encryptions under fuzzy selective-ID, chosen plain text attack." More formally,

[IND-FSID-CPA]: It is the game between the challenger and the adversary.

- **Phase 1/Selection Phase:** Adversary selects a challenged identity ID^* of her/his choice and gives it to challenger.
- **Phase 2/Setup Phase:** Here, challenger sets all the parameters PP, MK, and PP is given to the adversary.
- **Phase 3/Query Phase:** Adversary makes some queries related to private key extraction of some identities ID' with the restriction that $|ID' \cap ID^*| \not\geq d$, and finally adversary outputs two messages of her/his choice M_0 and M_1.
- **Phase 4/Challenge Phase:** Challenger encrypts message M_b under the identity ID^*, where b is random bit from $\{0, 1\}$ and outputs a challenge ciphertext $C*$.
- **Phase 5/Output Phase:** In this phase, adversary again can query the private key extraction queries with the same restrictions and finally output a bit b', claiming that $M_{b'}$ is encrypted in C^*.

Adversary is said to be successful if $b' = b$. As a security requirement of FIBE, we want the advantage of adversary $Adv_{\mathcal{A}}^{IND-FSID-CPA}(1^\lambda) = Pr[b' = b] - 1/2$ to be negligible. Similarly, the above security property can be defined for CCA adversary, where we also allowed decrypt queries to adversary, and it is said to be stronger notion than above.

4 Threshold Broadcast Encryption with Keyword Search

In "threshold broadcast encryption with keyword search" (TBEKS) [26], the data owner of each document chooses the set of users and also the threshold value l_{k_i} while encrypting a keyword k_i. Then, when there is a search query for the keyword k_j, at least l_{k_j} number of users need to pool their secret share to enable cloud server to perform this search. This restriction of involvement of threshold number of authorized users in searching some keyword is because to reduce the trust from any single user for avoiding disclosure of some sensitive data. It has many applications, for example, research team of a company, which do not want any single user to access the sensitive data but in a group of users. The similar schemes to TBEKS are [24, 25], but not completely same. This is also similar to threshold broadcast encryption scheme [13], where we encrypt the message, and here in TBEKS we encrypt the keyword.

Intuition In TBEKS, encrypted data can be combinedly decrypted by the minimum defined number (threshold value) of authorized users. Here, single user cannot decrypt the data. Data owner decides minimum how many and which set of users need to pool their secret shares to decrypt the data.

The formal definition and desirable security properties are discussed below.

4.1 Formal Model

TBEKS involved the data owner, the data users U_k, and the cloud server. It consists of following probabilistic polynomial time algorithms:

- $PP \xleftarrow{\$} \mathsf{Setup}(1^\lambda)$: It generates the public parameters PP upon input security parameter λ.
- $(PK_i, SK_i) \xleftarrow{\$} \mathsf{KeyGen}(PP)$: Each user runs this algorithm and generates his/her secret key and public key.
- $C_k \xleftarrow{\$} TBEKS(\{PK_{i_j}\}_{j=1}^n, l_k, W_k)$: Here, k-th keyword W_k is encrypted under n user's public key so that at least l_k users among them are required to decrypt it.
- $T_{i,j} \xleftarrow{\$} Trapdoor(SK_i, W_j)$: It generates the trapdoor for the i-th user and j-keyword. It runs by an individual user.

- $1/0 \leftarrow Test(\{T_{i_j,k}\}_{j=1}^{\geq l_k}, C_k)$: It returns 1 if sufficient trapdoors are matched with the k-th keyword ciphertext.

Correctness We say that the above scheme is correct if the following statement is true:

$$\forall PP \xleftarrow{\$} \mathsf{Setup}(1^\lambda) \wedge \forall (PK_i, SK_i) \xleftarrow{\$} \mathsf{KeyGen}(PP) \wedge \forall n, l_k \in \mathbb{Z}^+ \wedge l_k \leq n$$

$$\wedge \forall W_k \in \{0,1\}^* \text{ then } Test(\{T_{i_j,k}\}_{j=1}^{\geq l_k} | T_{i_j,k} \xleftarrow{\$} Trapdoor(SK_{i_j}, W_k),$$

$$C_k | C_k \xleftarrow{\$} TBEKS(\{PK_{i_j}\}_{j=1}^n, l_k, W_k)) = 1$$

4.2 Security Requirements

The basic security requirement is, adversary having inability to map the ciphertext to the correct keyword among the two chosen keywords. Here, we assume server is honest but curious, always interested in knowing the keyword that is encrypted into ciphertext, which is being currently searched by the set of users. We give formal definition of this security model below, and it requires few oracles viz.,

- $\mathcal{O}_{KeyGen}(i)$: It registered and generates key pair for the new honest user i. It gives only public key to the adversary.
- $\mathcal{O}_{Corrupt}(i)$: It returns secret key of the registered user i. Now, this user is no more in honest list.
- $\mathcal{O}_{Trapdoor}(i, k)$: It returns honest user's i trapdoor for the k-th keyword to the adversary.

"Indistinguishability in the Threshold setting against Chosen Keyword Attack" (IND-T-CKA) It is a game between the challenger and the adversary,

- **Phase 1/Setup Phase**: Challenger runs Setup algorithm and generates system parameters PP. It is given to the adversary.
- **Phase 2/Query Phase**: Adversary is given the following oracle accesses— $\mathcal{O}_{KeyGen}(i), \mathcal{O}_{Corrupt}(i)$, and $\mathcal{O}_{Trapdoor}(i, k)$. At the end, adversary outputs users set S^*, two keywords W_0^* and W_1^*, and a threshold value $l_{0,1}^* : l_{0,1}^* \leq |S^*|$.
- **Phase 3/Challenge Phase**: Challenger randomly selects a bit b and executes $C_b^* \xleftarrow{\$} TBEKS(\{PK_{i_j}\}_{U_{i_j} \in S^*}, l_{0,1}^*, W_b^*)$. Challenger gives C_b^* to adversary.
- **Phase 4/Output Phase**: Adversary outputs a bit b'.

We say that the advantage of adversary in winning the above game is $\mathbf{Adv}_{\mathcal{A}}^{IND-T-CKA}(1^\lambda) = |Pr[b' = b] - 1/2|$.

We say that the TBKES scheme is secure against $\mathbf{IND - T - CKA}$ adversary if $\mathbf{Adv}_{\mathcal{A}}^{IND-T-CKA}(1^\lambda)$ is negligible.

5 Linear Encryption with Keyword Search

"Linear encryption with keyword search" (LEKS) scheme is more or less a searchable encryption scheme, which is indeed a PEKS. In [27], the authors have given a general template that converts few types of encryption schemes to a searchable encryption schemes by encrypting keywords without re-encrypting whole data again.

Intuition In LEKS, data owner encrypts the message and also encrypts the keyword with desirable list of identities, so that the user with matching identity can generate a trapdoor for that keyword and enable the search operation on encrypted keywords. Here, data owner decides the list of data user who can search their keyword.

We give formal definition and security requirement below.

5.1 Formal Model

It involved data owner, server, and user. It consists of the following algorithms:

- $(MK, PP) \xleftarrow{\$} \mathsf{Setup}(1^\lambda)$: It generates master secret key MK and public parameters PP, by taking security parameter λ as input.
- $SK_i \xleftarrow{\$} \mathsf{KeyGen}(MK, ID_i)$: It generates secret key for the user with identity ID_i.
- $C_{j,k} \xleftarrow{\$} \mathsf{LEKS}(PP, S_j, W_k)$: It encrypts the keyword W_k for the set of users (in S_j).
- $T_{i,k} \xleftarrow{\$} \mathsf{Trapdoor}(SK_i, W_k)$: This generates the trapdoor value for the i-th user on the k-th keyword.
- $1/0 \leftarrow \mathsf{Test}(C_{j,k}, T_{i,k})$: It returns 1 if $ID_i \in S_j$, else returns 0.

Correctness We say that the above scheme is correct if the following statement is true:

$$\forall (MK, PP) \xleftarrow{\$} \mathsf{Setup}(1^\lambda) \land \forall W_k \in \{0, 1\}^* \land \forall ID_i \in S_j \ then$$

$$\mathsf{Test}(\mathsf{LEKS}(PP, S_j, W_k), \mathsf{Trapdoor}(\mathsf{KeyGen}(MK, ID_i), W_k)) = 1$$

5.2 Security Requirements

Any LEKS can be secured against the "Indistinguishable Adaptive Chosen Keyword Attack" (IND-CKA) adversary or "Indistinguishable Selective-ID Adaptive Chosen Keyword Attack" (IND-sCKA) adversary. When compared to former, later adversary chooses the challenged ID prior to the security game setup phase. Here, we formally define IND-CKA adversary model. To model this adversary, we require few oracles as follows:

- $\mathcal{O}_{Corrupt}(i)$: It returns secret key of the registered user i. Now, this user is no more in honest list.
- $\mathcal{O}_{Trapdoor}(i, k)$: It returns honest user's i trapdoor for the k-th keyword to the adversary.

"Indistinguishable Adaptive Chosen Keyword Attack" (IND-CKA) adversary It is a game between the challenger and the adversary,

- **Phase 1/Setup Phase**: Challenger runs Setup algorithm and generates master key MK and system parameters PP. PP is given to the adversary.
- **Phase 2/Query Phase**: Adversary is given the following oracle accesses— $\mathcal{O}_{Corrupt}(i)$ and $\mathcal{O}_{Trapdoor}(i, k)$. At the end, adversary outputs users set S^* and two keywords W_0^* and W_1^*.
- **Phase 3/Challenge Phase**: Challenger randomly selects a bit b and executes $C_b^* \xleftarrow{\$} \text{LEKS}(PP, S^*, , W_b^*)$. Challenger gives C_b^* to adversary.
- **Phase 4/Output Phase**: Adversary can make similar queries as in Query Phase with the restriction that he/she cannot query W_0, W_1 to $\mathcal{O}_{Trapdoor}(., .)$ oracle and also cannot make query on user i secret key to $\mathcal{O}_{Corrupt}(i)$ oracle if $i \in S^*$. At the end, adversary outputs a bit b'.

We say that the advantage of adversary in winning the above game is $\text{Adv}_{\mathcal{A}}^{IND-CKA}(1^\lambda) = |Pr[b' = b] - 1/2|$. We say that the LEKS scheme is secure against **IND** − **CKA** adversary if $\text{Adv}_{\mathcal{A}}^{IND-CKA}(1^\lambda)$ is negligible.

6   Public Key Encryption with Keyword Search

"Public key encryption with keyword search" (PEKS) enables the user to search encrypted keywords without decrypting the data [7, 9]. It has many applications such as delegation of monitoring urgent emails. Suppose, *Ali* wants to read only important emails with "urgent" keyword during his holiday duration, he does not want to read all emails. Then, using PEKS, *Ali* can send "urgent" keyword trapdoor to the server, so that the server can search the emails with this keyword and forward it to *Ali's* mobile or intimate him by any other mean. Therefore, in this whenever sender is sending any email, he/she needs to encrypt the appropriate keyword along with the encryption of email under the receiver's public key. In [9], PEKS was first introduced, and in [7], the authors have pointed out three issues. First one is what happens if server stores the trapdoor and uses it in future to learn some partial information of the messages. Second, *Ali* and server need to communicate using secure channel that is quite expensive and not suitable in many applications. Third, the PEKS scheme is not given for multiple keywords that are connected through some OR or AND gate. These issues have been solved in [7].

Intuition In PEKS, data owner will upload his/her encrypted data with encrypted keywords so that the data user with suitable trapdoor related the keyword can search that keyword with the help of cloud server. Data user gives keyword trapdoor to the server for search operation.

We give formal model of PEKS and its security requirements below.

6.1 Formal Model

It consists of the following algorithms:

- $PP \xleftarrow{\$}$ Setup(1^λ): It generates common public parameters upon input security parameter λ.
- $(SK_S, PK_S) \xleftarrow{\$}$ KeyGen$_{Server}(PP)$: It generates key pair for the server, secret key and public key.
- $(SK_{R_i}, PK_{R_i}) \xleftarrow{\$}$ KeyGen$_{Receiver}(PP)$: It generates key pair for the receivers, secret key and public key.
- $C_{i,k} \xleftarrow{\$}$ PEKS(PP, PK_{R_i}, PK_S, W_k): It generates searchable ciphertext of k-th keyword for i-th receiver.
- $T_{i,k} \xleftarrow{\$}$ Trapdoor(PP, SK_{R_i}, W_k): This takes input as receiver secret key, keyword W_k and outputs the trapdoor value.
- $1/0 \leftarrow$ Test($PP, T_{i,k}, SK_S, C_{i,k}$): It returns 1 if keyword present in ciphertext matches with the keyword trapdoor.

Correctness We say that the above scheme is correct if the following statement is true:

$$\forall PP \xleftarrow{\$} \text{Setup}(1^\lambda) \wedge \forall W_k \in \{0, 1\}^* \wedge$$

$$\forall (SK_S, PK_S) \xleftarrow{\$} \text{KeyGen}_{Server}(PP) \wedge \forall (SK_{R_i}, PK_{R_i}) \xleftarrow{\$} \text{KeyGen}_{Receiver}(PP)$$

then Test(PP, Trapdoor(PP, SK_{R_i}, W_k), SK_S, PEKS(PP, PK_{R_i}, PK_S, W_k))=1

6.2 Security Requirements

The PEKS scheme is said to be secure against indistinguishable—chosen keyword attack if the advantage of adversary winning the following game is negligible.

IND-CKA Game
- **Phase 1/Setup Phase**: Challenger sets up the PEKS scheme by running the algorithm Setup(1^λ). It also generates the receiver's key pairs, say (SK_{R_i}, PK_{R_i}) and gives only public keys to the adversary.

- **Phase 2/Query Phase**: In this phase, adversary can make trapdoor queries for the keyword W_k on the receiver i. At the end of this phase, adversary outputs two keywords (W_0^*, W_1^*) and receiver id i^*.
- **Phase 3/Challenge Phase**: Challenger generates ciphertext by randomly picking one keyword under receiver i^*, i.e., $C_{i^*,b} \xleftarrow{\$} \mathsf{PEKS}(PP, PK_{R_i^*}, PK_S, W_b^*)$ and gives it to the adversary.
- **Phase 4/Output Phase**: Here, adversary can make similar queries as in Phase 2 with obvious restrictions, and finally adversary outputs a bit b'.

We say that adversary wins the game with advantage $\mathbf{Adv}_{\mathcal{A}}^{IND-CKA}(1^\lambda) = |Pr[b' = b] - 1/2|$. The above scheme is said to be secure if there does not exist any probabilistic polynomial time adversary who can distinguish two keywords given a ciphertext of one of it.

7 Identity-Based Encryption with Wildcards (WIBE)

Abdalla et al. [4] have proposed a "wildcard identity-based encryption" (WIBE), which allows sender to encrypt a message with the pattern such that the range of receivers whose identity matches the pattern can decrypt it. Here, the pattern includes the strings with wildcards. Later, many other such schemes were developed [5, 8]. It has many applications, such as professor *Ali* can send an encrypted email to entire computer engineering department by encrypting it under the identity *@cse.vnit.ac.in, and so on.

Intuition In WIBE, sender can encrypt a message with some pattern so that the receivers with their identities matching to that pattern can decrypt it with their decryption keys.

Next, we give formal definition of WIBE scheme.

7.1 Formal Model

Pattern is denoted by a vector $P = (P_1, \ldots, P_l) \in (\{0, 1\}^* \cup \{*\})^l$, where $*$ is used as a wildcard character. The identity $ID = (ID_1, \ldots, ID_{l'})$ is said to be matched with the above P, denoted by $ID \in_* P$, if $l' \leq l$ and for every $i = 1$ to l' in ID that $ID_i = P_i$ or $P_i = *$. Here we say user with identity $ID' = (ID_1, \ldots, ID_{l'+1})$ is a child of the user with identity $ID = (ID_1, \ldots, ID_{l'})$. WIBE scheme consists of the following four algorithms:

- $(MSK, MPK) \xleftarrow{\$} \mathsf{Setup}(1^\lambda)$: It generates the master key pairs (master secret key and master public key) upon input security parameters.

- $d_{ID'} \xleftarrow{\$} \text{KeyDer}(d_{ID}, ID_{l+1})$: It generates decryption key for the $ID' = ID_{l+1} \cup ID$, where $ID = (ID_1, \ldots, ID_l)$ and d_{ID} is being the decryption key for ID. Thus, $ID' = (ID_1, \ldots, ID_{l+1})$ and $d_\varepsilon = MSK$.
- $C_P \xleftarrow{\$} \text{Encrypt}(MPK, P, M)$: It encrypts the message M under the pattern P.
- $M/\perp \leftarrow \text{Decrypt}(d_{ID}, C_P)$: It decrypts the ciphertext using the decryption key d_{ID} such that $ID \in_* P$ and returns M. If encryption is invalid, then returns $\perp$.

Correctness We say that the above scheme is correct if the following statement is true:

$$\forall (MSK, MPK) \xleftarrow{\$} \text{Setup}(1^\lambda) \wedge \forall 0 \le l \le L, \forall P \in (\{0.1\}^* \cup \{*\})^l,$$

$$\forall ID : ID \in_* P \wedge \forall M \in \{0, 1\}^*, \ we \ have$$

$$\text{Decrypt}(\text{KeyDer}(MSK, ID), \text{Encrypt}(MPK, P, M)) = M$$

7.2 Security Requirements

Here we define formal security model called IND-WID-CPA security model, through the following game between the challenger and the adversary,

- **Phase 1/Setup Phase**: Challenger generates the master key pairs by running the algorithm, $(MSK, MPK) \xleftarrow{\$} \text{Setup}(1^\lambda)$. Challenger gives MPK to adversary.
- **Phase 2/Query Phase**: Adversary is given access to key derivation oracle, through which adversary can get decryption key d_{ID} of any identity $ID = (ID_1, \ldots, ID_l)$. At the end of this phase, adversary outputs two messages (M_0, M_1) and a challenge pattern P^*.
- **Phase 3/Challenge Phase**: Challenger randomly selects a bit $b \in_R \{0, 1\}$ and generates challenge ciphertext, $C_{P*}^* \xleftarrow{\$} \text{Encrypt}(MPK, P^*, M_b)$.
- **Phase 4/Output Phase**: Adversary can make queries similar to Phase 2 with the obvious restrictions. Finally, adversary outputs a bit b'.

We say that the adversary wins the game if $b' = b$, and the adversary never queries the key derivation oracle for the keys of identities that match the target pattern (i.e., $ID \in_* P^*$). The advantage of adversary in winning the above game is defined as $\text{Adv}_{\mathcal{A}}^{IND-WID-CPA}(1^\lambda) = |Pr[b' = b] - 1/2|$. We say that the WIBE scheme is secure against $IND - WID - CPA$ adversary if the $\text{Adv}_{\mathcal{A}}^{IND-WID-CPA}(1^\lambda)$ is negligible. The above game can be converted to IND-WID-CCA security model by giving access to the adversary the decryption oracle.

8 Identity-Based Encryption with Wildcard Key Derivation (WKD-IBE)

"Identity-based encryption with wildcard key derivation" (WKD-IBE) was introduced by Abdalla et al. in [1, 2], which allows key delegation pattern in more general way in HIBE. Here, the secret keys are attached with patterns consisting of identity strings and wildcards (∗). Then, the owner of the key can derive keys for other identity that matches the pattern attached to his/her secret key. This is an extension of HIBE and has many applications, such as, suppose that one wants to allow the university head to derive secret keys for all the departments head email addresses of the form, $head@ * .vnit.ac.in$, where ∗ is a wildcard character that can be replaced with any string (i.e., it can be head@cse.vnit.ac.in or head@ece.vnit.ac.in or etc.). Later on, more efficient schemes were developed [3].

Intuition In WKD-IBE, a user A can derive secret key for the user B, if A is an ancestor of B. Thus, user A can also decrypt the ciphertexts intended for his/her descendant B. The ancestor and descendant relation is defined with the help of pattern with wildcard character.

The formal model of WKD-IBE is given below followed by security requirements.

8.1 Formal Model

The main idea in WKD-IBE is that a user with secret key of pattern P can generate a secret key for any pattern P' that matches P. That is, if $P' = (P'_1, \ldots, P'_{l'})$ and $P = (P_1, \ldots, P_l) \in (\{0, 1\}^* \cup \{*\})^l$, then we say that P' matches P, denoted by $P' \in_* P$, if $l' \leq l$ and $\forall i \in [1, l'], P'_i = P_i$ or $P_i = *$; and $\forall i \in [l' + 1, l], P_i = *$. Formally, WKD-IBE consists of the following algorithms:

- $(MSK, MPK) \xleftarrow{\$} \mathsf{Setup}(1^\lambda)$: It generates the master key pairs (master secret key and master public key) upon input security parameters.
- $sk_{P'} \xleftarrow{\$} \mathsf{KeyDer}(sk_P, P')$: It generates secret key for the $P' : P' \in_* P$. The secret key for the root identity is $MSK = sk_{(*,\ldots,*)}$.
- $C_{ID} \xleftarrow{\$} \mathsf{Encrypt}(MPK, ID, M)$: It encrypts the message M intended for the identity $ID = (ID_1, \ldots, ID_l)$, so that any user with secret key associated with the pattern $P : ID \in_* P$ can decrypt the ciphertext.
- $M/\bot \leftarrow \mathsf{Decrypt}(sk_P, C_{ID}, ID)$: It decrypts the ciphertext using the secret key sk_P such that $ID \in_* P$ and returns M. If encryption is invalid, then returns $\bot$.

Correctness We say that the WKD-IBE scheme is correct if the following statement is true:

$$\forall (MSK, MPK) \xleftarrow{\$} \mathsf{Setup}(1^\lambda) \land \forall 0 \le l \le L, \forall P \in (\{0.1\}^* \cup \{*\})^l,$$

$$\forall ID : ID \in_* P \land \forall M \in \{0, 1\}^*, \ we \ have$$

$$\mathsf{Decrypt}(\mathsf{KeyDer}(MSK, P), \mathsf{Encrypt}(MPK, ID, M)) = M$$

8.2 Security Requirements

Here we define formal security model called IND-WKID-CPA security model, through the following game between the challenger and the adversary,

- **Phase 1/Setup Phase**: Challenger generates the master key pairs by running the algorithm, $(MSK, MPK) \xleftarrow{\$} \mathsf{Setup}(1^\lambda)$. Challenger gives MPK to adversary.
- **Phase 2/Query Phase**: Adversary is given access to key derivation oracle, through which adversary can get secret key sk_P of any pattern $P = (P_1, \ldots, P_l)$. At the end of this phase, adversary outputs two messages (M_0, M_1) and a challenge identity ID^*.
- **Phase 3/Challenge Phase**: Challenger randomly selects a bit $b \in_R \{0, 1\}$ and generates challenge ciphertext, $C^*_{ID^*} \xleftarrow{\$} \mathsf{Encrypt}(MPK, ID^*, M_b)$.
- **Phase 4/Output Phase**: Adversary can make queries similar to Phase 2 with the obvious restrictions. Finally, adversary outputs a bit b'.

We say that the adversary wins the game if $b' = b$, and the adversary never queries the key derivation oracle for the secret keys of patterns that match the target identity (i.e., $ID^* \in_* P$). The advantage of adversary in winning the above game is defined as $\mathbf{Adv}_{\mathcal{A}}^{IND-WKID-CPA}(1^\lambda) = |Pr[b' = b] - 1/2|$. We say that the WIBE scheme is secure against $IND-WKID-CPA$ adversary if the $\mathbf{Adv}_{\mathcal{A}}^{IND-WKID-CPA}(1^\lambda)$ is negligible. The above game can be converted to IND-WKID-CCA security model by giving access to the adversary the decryption oracle.

9 Identity-Based Conditional Proxy Re-Encryption (IBCPRE)

"Identity-based conditional proxy re-encryption" (IBCPRE) was introduced by Shao et al. in [23], which allows proxy to convert the ciphertexts of one user to another user if the prescribed condition met (set by the delegator). This will add extra security over identity-based proxy re-encryption (IBPRE). It has applications related to, suppose the sender wants to encrypt an email content with $Ali's$ identity and condition c to get the ciphertext C_A. Then, the proxy with re-encryption key related to condition c can transform C_A intended for Ali to another ciphertext C_B, which can be opened by Bob. Later on, many other efficient IBCPRE schemes were proposed [15, 17, 18], either with enhanced security or with extra feature related to underling condition.

Intuition In IBCPRE, sender will encrypt the message with some condition c to produce conditional ciphertext such that if proxy with suitable re-encryption key related to the condition c wants to transform that conditional ciphertext to some other targeted receiver's public key as a regular ciphertext can do so. Then, that targeted receiver can decrypt the regular or transformed ciphertext with his/her secret key.

Next, we give the formal definition of the scheme followed by security requirements.

9.1 Formal Model

It consists of the following algorithms:

- $(MPK, MSK) \xleftarrow{\$} \mathsf{KeyGen}(1^\lambda)$: This algorithm generates the master key pairs—master public key and master secret key, upon input security parameter λ.
- $d_{ID} \xleftarrow{\$} \mathsf{Extract}(MSK, ID)$: It generates the private key d_{ID} for the identity ID.
- $rk_{ID_i \xrightarrow{c} ID_j} \xleftarrow{\$} \mathsf{ReKeyGen}(ID_i, d_{ID_i}, ID_j, c)$: It generates re-encryption key for the delegator ID_j from the delegator ID_i under the condition c.
- $\hat{C}_{ID_i} \xleftarrow{\$} \mathsf{Encrypt}_c(ID_i, c, M)$: It encrypts the message intended for the user with identity ID_i under the condition c and outputs the conditional ciphertext.
- $C_{ID_i} \xleftarrow{\$} \mathsf{Encrypt}(ID_i, M)$: It does the regular encryption of the message M to the user with identity ID_i and outputs regular ciphertext.
- $C_{ID_j} \xleftarrow{\$} \mathsf{ReEncrypt}(rk_{ID_i \xrightarrow{c} ID_j}, c, \hat{C}_{ID_i})$: It re-encrypts the ciphertext and outputs a regular ciphertext intended for the ID_j.
- $M \leftarrow \mathsf{Decrypt}_c(d_{ID_i}, c, \hat{C}_{ID_i})$: It decrypts the conditional ciphertext intended for the identity ID_i.
- $M \leftarrow \mathsf{Decrypt}(d_{ID_j}, C_{ID_j})$: It decrypts the regular ciphertext intended for the identity ID_j.

Correctness The correctness of this scheme requires the following 3 statements to be met:

$$\forall (MPK, MSK) \xleftarrow{\$} \mathsf{KeyGen}(1^\lambda)$$

$$\mathsf{Decrypt}(\mathsf{Extract}(MSK, ID_j), \mathsf{Encrypt}(ID_j, M)) = M$$

$$\mathsf{Decrypt}_c(\mathsf{Extract}(MSK, ID_i), c, \mathsf{Encrypt}_c(ID_i, c, M)) = M$$

$$\text{Decrypt}(d_{ID_j}, \text{ReEncrypt}(\text{ReKeyGen}(ID_i, d_{ID_i}, ID_j, c), c,$$
$$\text{Encrypt}_c(ID_i, c, M))) = M$$

9.2   Security Requirements

Here we give formal security model for the *Chosen Ciphertext and Identity security for IBCPRE (IBCPRE-CCIA)* [23] security. It involved 5 oracles given as below:

- $\mathcal{O}_{Extract}$: Private key extract oracle, on input ID, it returns $\text{Extract}(MSK, ID)$.
- $\mathcal{O}_{RK}$: Re-encryption key generation oracle, on input (ID_i, ID_j, c), it returns $\text{ReKeyGen}(ID_i, \text{Extract}(MSK, ID_i), ID_j, c)$.
- $\mathcal{O}_{RE}$: Re-encryption oracle, on input $(ID_i, ID_j, c, \hat{C}_{ID_i})$, it returns $\text{ReEncrypt}(\text{ReKeyGen}(ID_i, \text{Extract}(MSK, ID_i), ID_j, c), c, \hat{C}_{ID_i})$.
- $\mathcal{O}_{CDec}$: Conditional decryption oracle, on input $(ID_i, c, \hat{C}_{ID_i})$, it returns $\text{Decrypt}_c(\text{Extract}(MSK, ID_i), c, \hat{C}_{ID_i})$.
- $\mathcal{O}_{Dec}$: Regular decryption oracle, on input (ID_i, C_{ID_i}), it returns $\text{Decrypt}(\text{Extract}(MSK, ID_i), C_{ID_i})$.

[IBCPRE-CCIA Game]: It is a game between the challenger and the adversary,

- **Phase 1/Setup Phase**: In this phase, challenger sets up the IBCPRE scheme by generating master key pair through $\text{KeyGen}(1^\lambda)$. Challenger gives master public key to the adversary.
- **Phase 2/Query Phase**: Here adversary is given access to the above 5 oracles— $\{\mathcal{O}_{Extract}, \mathcal{O}_{RK}, \mathcal{O}_{RE}, \mathcal{O}_{CDec}, \mathcal{O}_{Dec}\}$, through which adversary can make his/her desirable queries. At the end of this phase, adversary outputs 2 messages (M_0, M_1), a condition c^*, and an identity ID^*, such that ID^* is uncorrupted, and no query of the form $(ID^*, \star, c^*)$ is made to $\mathcal{O}_{RK}$.
- **Phase 3/Challenge Phase**: Challenger picks a random bit $b \in \{0, 1\}$, and for *conditional challenge ciphertext*, it outputs $\text{Encrypt}_c(ID^*, c^*, M_b)$, and for *regular challenged ciphertext*, it outputs $\text{Encrypt}(ID^*, M_b)$.
- **Phase 4/Output Phase**: Adversary again can make queries as did in phase 2 with the obvious restrictions related to challenge ciphertext. At the end, adversary outputs a bit b'.

We say that the adversary wins the game if $b' = b$, and the advantage of adversary in winning this game is defined as $\text{Adv}_{\mathcal{A}}^{IBCPRE-CCIA}(1^\lambda) = |Pr[b' = b] - 1/2|$. To say that the above scheme is secure against IBCPRE-CCIA adversary, $\text{Adv}_{\mathcal{A}}^{IBCPRE-CCIA}(1^\lambda)$ needs to be negligible.

10 Summary

The chapter discusses the various cryptographic primitives derived from IBE. The summary of these primitives is discussed below.

In IBKA, two or more users can establish a shared secret key using their public identity and corresponding secret key, generated under common setup or with common master key. It is similar to any other key agreement protocol, but the public key is not generated, and it is his/her identity.

In FIBE, a user can decrypt the data if his/her identity matched till at least the allowed level or percentage, using his secret key. Here, data user identity may not be exactly matched but till some percentage with the identity associated with ciphertexts.

In TBEKS, encrypted data can be combinedly decrypted by the minimum defined number (threshold value) of authorized users. Here, single user cannot decrypt the data. Data owner decides minimum how many and which set of users need to pool their secret shares to decrypt the data.

In LEKS, data owner encrypts the message and also encrypts the keyword with desirable list of identities, so that the user with matching identity can generate a trapdoor for that keyword and enable the search operation on encrypted keywords. Here, data owner decides the list of data user who can search their keyword.

In PEKS, data owner will upload his/her encrypted data with encrypted keywords so that the data user with suitable trapdoor related the keyword can search that keyword with the help of cloud server. Data user gives keyword trapdoor to the server for search operation.

In WIBE, sender can encrypt a message with some pattern so that the receivers with their identities matching to that pattern can decrypt it with their decryption keys.

In WKD-IBE, a user A can derive secret key for the user B, if A is an ancestor of B. Thus, user A can also decrypt the ciphertexts intended for his/her descendant B. The ancestor and descendant relation is defined with the help of pattern with wildcard character.

In IBCPRE, sender will encrypt the message with some condition c to produced conditional ciphertext such that if proxy with suitable re-encryption key related to the condition c wants to transform that conditional ciphertext to some other targeted receiver's public key as a regular ciphertext can do so. Then, that targeted receiver can decrypt the regular or transformed ciphertext with his/her secret key.

For each of these primitives, we defined suitable security model in formal way.

References

1. Abdalla, M., Kiltz, E., & Neven, G. (2007). Generalized key delegation for hierarchical identity-based encryption. In *European Symposium on Research in Computer Security* (pp. 139–154). Berlin: Springer.
2. Abdalla, M., Kiltz, E., Neven, G. (2008). Generalised key delegation for hierarchical identity-based encryption. *IET Information Security, 2*(3), 67–78.
3. Abdalla, M., De Caro, A., & Phan, D. H. (2012). Generalized key delegation for wildcarded identity-based and inner-product encryption. *IEEE Transactions on Information Forensics and Security, 7*(6), 1695–1706.
4. Abdalla, M., Catalano, D., Dent, A. W., Malone-Lee, J., Neven, G., Smart, & N. P. (2006). Identity-based encryption gone wild. In *International Colloquium on Automata, Languages, and Programming* (pp. 300–311). Berlin: Springer.
5. Abdalla, M., Birkett, J., Catalano, D., Dent, A. W., Malone-Lee, J., Neven, G., et al. (2011). Wildcarded identity-based encryption. *Journal of Cryptology, 24*(1), 42–82.
6. Baek, J., Susilo, W., & Zhou, J. (2007). New constructions of fuzzy identity-based encryption. In *Proceedings of the Second ACM Symposium on Information, Computer and Communications Security* (pp. 368–370).
7. Baek, J., Safavi-Naini, R., & Susilo, W. (2008). Public key encryption with keyword search revisited. In *International Conference on Computational Science and Its Applications* (pp. 1249–1259). Berlin: Springer (2008).
8. Birkett, J., Dent, A. W., Neven, G., & Schuldt, J. C. (2007). Efficient chosen-ciphertext secure identity-based encryption with wildcards. In *Australasian Conference on Information Security and Privacy* (pp. 274–292). Berlin: Springer.
9. Boneh, D., Di Crescenzo, G., Ostrovsky, R., & Persiano, G. (2004). Public key encryption with keyword search. In *International Conference on the Theory and Applications of Cryptographic Techniques* (pp. 506–522). Berlin: Springer.
10. Cao, X., Kou, W., Yu, Y., & Sun, R. (2008). Identity-based authenticated key agreement protocols without bilinear pairings. *IEICE Transactions on Fundamentals of Electronics, Communications and Computer Sciences, E91*.A(12), 3833–3836. https://doi.org/10.1093/ietfec/e91-a.12.3833
11. Chen, L., Cheng, Z., & Smart, N. P. (2007). Identity-based key agreement protocols from pairings. *International Journal of Information Security, 6*(4), 213–241.
12. Dang, L., Xu, J., Cao, X., Li, H., Chen, J., & Zhang, Y. (2018). Efficient identity-based authenticated key agreement protocol with provable security for vehicular ad hoc networks. *International Journal of Distributed Sensor Networks, 14*(4), 1550147718772545.
13. Daza, V., Herranz, J., Morillo, P., & Rafols, C. (2007). CCA2-secure threshold broadcast encryption with shorter ciphertexts. In *International Conference on Provable Security* (pp. 35–50). Berlin: Springer.
14. Fujioka, A. (2017). Adaptive security in identity-based authenticated key agreement with multiple private key generators. In *International Workshop on Security* (pp. 192–211). Berlin: Springer.
15. Ge, C., Susilo, W., Wang, J., & Fang, L. (2017). Identity-based conditional proxy re-encryption with fine grain policy. *Computer Standards & Interfaces, 52*, 1–9.
16. Khatoon, S., & Thakur, T. (2017). Provable secure identity based key agreement protocol with perfect forward secrecy. *International Journal of Computational Intelligence Research, 13*(8), 1917–1930.
17. Liang, K., Liu, Z., Tan, X., Wong, D.S., & Tang, C. (2012). A CCA-secure identity-based conditional proxy re-encryption without random oracles In *International Conference on Information Security and Cryptology* (pp. 231–246). Berlin: Springer.
18. Liang, K., Chu, C.K., Tan, X., Wong, D.S., Tang, C., & Zhou, J. (2014). Chosen-ciphertext secure multi-hop identity-based conditional proxy re-encryption with constant-size ciphertexts. *Theoretical Computer Science, 539*, 87–105.

19. Mao, Y., Li, J., Chen, M.R., Liu, J., Xie, C., & Zhan, Y. (2016). Fully secure fuzzy identity-based encryption for secure IoT communications. *Computer Standards & Interfaces, 44*, 117–121.
20. Ni, L., Chen, G., Li, J., & Hao, Y. (2016). Strongly secure identity-based authenticated key agreement protocols without bilinear pairings. *Information Sciences, 367*, 176–193.
21. Ren, Y., Gu, D., Wang, S., & Zhang, X. (2010). New fuzzy identity-based encryption in the standard model. *Informatica, 21*(3), 393–407.
22. Sahai, A., & Waters, B. (2005). Fuzzy identity-based encryption. In *Annual International Conference on the Theory and Applications of Cryptographic Techniques* (pp. 457–473). Berlin: Springer.
23. Shao, J., Wei, G., Ling, Y., & Xie, M. (2011). Identity-based conditional proxy re-encryption. In *2011 IEEE International Conference on Communications (ICC)* (pp. 1–5). Piscataway, NJ: IEEE.
24. Siad, A. (2012). Anonymous identity-based encryption with distributed private-key generator and searchable encryption. In *2012 5th International Conference on New Technologies, Mobility and Security (NTMS)* (pp. 1–8). Piscataway, NJ: IEEE.
25. Wang, P., Wang, H., & Pieprzyk, J. (2008). Threshold privacy preserving keyword searches. In *International Conference on Current Trends in Theory and Practice of Computer Science* (pp. 646–658). Berlin: Springer.
26. Zhang, S., Mu, Y., & Yang, G. (2015). Threshold broadcast encryption with keyword search. In *International Conference on Information Security and Cryptology* (pp. 322–337). Berlin: Springer.
27. Zhang, S., Yang, G., & Mu, Y. (2016). Linear encryption with keyword search. In *Australasian Conference on Information Security and Privacy* (pp. 187–203). Berlin: Springer.

Chapter 13
Attribute-Based Encryption

Ankita Karale, Vladimir Poulkov, and Milena Lazarova

Abstract In today's world data is growing enormously. With the advancement in data, there is need of effective data security mechanism to handle this stored or transmitted information. This necessity introduced science of cryptography. Being an effective cryptographic system attribute-based encryption (ABE) became hot topic of research among the researchers. ABE is a kind of public key cryptography system. It uses ciphertext and private key of user for the encryption purpose. This scheme hooked up ciphertext with attributes (e.g., e-mail id, country name) which ensures high level of protection. This ABS scheme ensures data confidentiality and exclusive access policy. This chapter studies basic ABS system followed by brief introduction of few popular algorithms like Public Key Encryption, Public Key Infrastructure, and Identity Based Encryption. It also enlightened two major types of ABE, i.e., Key Policy-ABE (KP-ABE) and Ciphertext Policy-ABE (CP-ABE). Finally, it is concluded with security model of ABE with its comparative popular schemes.

Keywords Attributes · Attribute-based encryption · Cryptography · Public key cryptography · Key policy-ABE · Ciphertext policy-ABE

1 Introduction

With the evolution of technology data is increasing tremendously. Security of growing data which is to be transmitted or stored is the most important issue nowadays. Data Owner desires to address various security concerns in order to provide confidentiality of data. Protection of confidential data can be achieved using various mechanisms. To provide security to the data, information is encrypted so that it must be read by a specific party in order to control who can access the data.

A. Karale (✉) · V. Poulkov · M. Lazarova
Technical University of Sofia, Sofia, Bulgaria
e-mail: vkp@tu-sofia.bg; milaz@tu-sofia.bg

© Springer Nature Switzerland AG 2021
K. A. B. Ahmad et al. (eds.), *Functional Encryption*, EAI/Springer Innovations in Communication and Computing, https://doi.org/10.1007/978-3-030-60890-3_13

225

Encryption is conversion of original data or information to the coded form that is not readable by human beings. It is also known as encoding too. The decoding of encoded information into original form is known as decryption. Only authenticated person will have the right to access or translate the information by using encryption. Therefore it gives data privacy. There are many advanced versions of it available until now. Every version has different applications and benefits. But ABE is most popular and reliable among them.

ABE is attribute-based encryption. It uses one to many key encryptions. It will have specific policies or attributes related with every key of decryption. To retrieve data well, every attribute should fulfill access policy and if these attributes of key are identical with attributes of coded text, then only information can be decoded.

Let us have a quick review of a few popular techniques related to encryption before going in deep about ABE.

1.1 Encryption

The translation of data from simple plaintext form to unreadable coded form is known as encryption of data. After encryption of data, only an authenticated person having the private decode key can access the data. Usually this secret key is known as decryption key, and the coded text is known as cipher text. This is the most widely used and effective method of data security. There are basically two types of encryption, Symmetric Encryption and Asymmetric Encryption.

Asymmetric encryption is popular by name of public key encryption.

1.2 Public Key Encryption

This technique uses a pair of keys generated by a cryptographic algorithm. Messages can be encrypted using public key or private key, but it will be only decrypted with receiver's private key which is generated randomly by mathematical cryptographic algorithms. This will provide security to the data as the private key is available only with the intended receiver.

In Fig. 13.1, we can see all public keys are available with each of the users. When B intends to send any secret information to C, B will encode it with the public key of C. When information is delivered to C, he can decrypt it with the help of his own private key. Only C can decode the message as C is the only one who has knowledge of C's secret key.

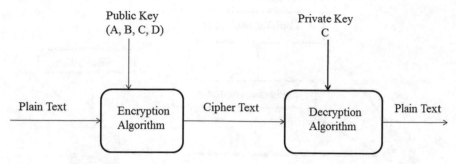

Fig. 13.1 Public key encryption

1.2.1   Limitation

We can observe that this technique uses two different keys: public and private. When *B* sends a message to *C*, he uses the public key of *C*. When *C* wants to decode the message, he uses his own private key. When *B* knows the public key of *C* in advance, then only this method can work efficiently. In short, if *A* and *B* are known to each other, then only they can have communication to share public keys by some authenticated media. But there may be a situation when both parties do not know each other at all, then this method will fail.

1.3   *Public Key Infrastructure (PKI)*

To overcome abovementioned limitations, there is the necessity of trusted third parties to distribute public keys. This need emerges from the public key infrastructure (PKI) (Fig. 13.2). The most popular method of encryption is public key encryption named by researchers as public key infrastructure [1] also known as PKI. PKI uses certification authorities (CA). User registered himself to the registration authority. Authority will check its authenticity, and the verifier will verify his identity, and after verification CA authority will issue corresponding keys to the user for encryption purpose. So this trusted third party will deliver keys to the sender which further works like the public key encryption technique stated above. But to obtain the data privacy, it requires public keys of approved users so as to transmit data independently to the target user which ultimately increases the demand for bandwidth [2]. To provide a solution to this drawback, broadcast encryption [3] was introduced. This technique addresses the mentioned issue, but it is mostly applicable when the data provider has prior knowledge of the target user. It uses a public key for the process of encryption, and original data is retrieved by decryption by using only a single secret key. So the more advanced encryption solutions are required.

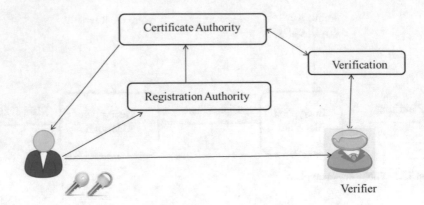

Fig. 13.2 Public key infrastructure

1.4 Identity-Based Encryption (IBE)

Research is always thrust toward betterment. Can there be a better solution to implement public key encryption? This question is answered by smart innovation of identity-based encryption (IBE). A. Shamir [4] introduced an encryption technique based on identity also known as IBE. Without interfering with the public key certificate here, sender can encode data which simplifies the process. Due to this feature of IBE, it is appropriate for real-time applications. Here combination of character is treated as Identity. So ultimately A can send a message to B without the PKI help, and it can work smoothly as it reduces communication overhead.

The concept is to use the user's identity as a public key, for example, e-mail Id of the user. Identity of the user will be considered as a public key, and the centralized key server will be responsible for creation of a private key. Here the basic difference between PKI and IBE is that IBE eliminates the need of the certificate look up process required by PKI.

For example, A have the identity of him as an e-mail address: a@example.com. This identity of A will be used by him to get a private key from a centralized key server. E-mail address of A will be used by B to encrypt the message. Only A can decrypt the encoded message as e-mail address which is acting as an identity belonging to A only. A has access to its identity a@example.com, so he is authenticated to get a private key from the key server in order to decode the encrypted message sent by B. Here key server is the center of attraction. Security of key server is the main concern in order to make IBE a successful mechanism of encryption.

2   Attribute-Based Encryption (ABE)

The concept of IBE is further improved by ABE. This technique was proposed by scientist Sahai and Waters [5]. They introduced an advanced version of IBE known as FIBE. It is the first idea of ABE. Instead of using a single string as identity, it uses a set of attributes. Most of the features of ABE match with IBE like the way of handling identities as sequences of characters. It is a comprehensive form of IBE, but more expressive when compared with it. To encrypt the plaintext, it uses a collection of different attributes. Key servers having master keys only distribute private keys after authentication of attributes possessed by the user. So every private key will have its own set of attributes associated with it. To decrypt the message sent by sender, attributes should be matched then only cipher text could be decoded.

Let us take an example to understand it.

Arya has the attributes "*Role = Manager*" and "Age > 25." Anjali uses these attributes ("*Role = Manager*" AND "Age > 25") to encrypt the message. Arya is able to decrypt the information because she can satisfy both the specified attributes. Anjali sends another message with the help of encryption by using attributes ($Role$ = Director OR Role = Chairman). This time Arya cannot retrieve the message as she does not fulfill requirements of attributes.

So we can observe the key role that is played by attributes here.

What are the working expectations that should be fulfilled by the ideal ABE method, what type of functioning should be executed by ideal ABE is concluded by [6]. They are as follows:

1. Confidentiality of Data:

   - Cipher text should be confidential to provide security from unauthorized access.

2. Fine-grained access control:

   - Though users belong to the same type of group, their right to access information should not be the same. So that it will provide good access control.

3. Scalable:

   - Technique should be scalable in terms of performance. When the number of users using the system increases suddenly or dynamically, it should be able to handle load efficiently. Number of increasingly approved users should not degrade the performance of the scheme.

4. Consumer/Attribute Elimination:

   - If any user leaves the system, it should be able to cancel the user's access right. Same in case of attribute also.

5. Security in forward and Backward mode:

- Any existing user eliminated from the system should not be permitted to access any data going to be published in future, also new users should be prevented from accessing old data published before they joined the system.

6. Accountability:

- Fraudulent users can leak data by distributing their keys or part of keys. So there should be a track to observe who had distributed keys. This problem of key abuse also called as Boye problem. There should be solution to this problem using accountability.

7. System should be safe from fraud users who can decrypt the ciphertext.
8. Computational cost of the whole system should be minimal.

2.1 Basic Model of ABE

In this section let us discuss the basic model of ABE. The first model introduced by Sahai and Waters [5] known as FIBE is considered as the basic model of ABE. Here collection of attributes is treated as identity. Here we will see complexity rules stated in the basic model, and then we will go for algorithms on it.

2.1.1 Complexity Assumptions

Definition 1 DBDH supposition

It is also known as Diffie-Hellman decisional bilinear supposition. Suppose competitor selects p, q, r, z $\in \mathbb{Z}_q$ randomly. The supposition stated that in polynomial time, no opposition is able to differentiate the tuple.

$(P = g^p, Q = g^q, R = g^r, \mathbb{Z} = e(g, g)^{pqr}$ from the tuple $(P = g^p, Q = g^q, R = g^r, \mathbb{Z} = e(g, g)^z$.

With a minor advantage.

Definition 2 MBDH supposition

It is also known as modified decisional bilinear supposition of Diffie-Hellman.

Suppose competitor selects p, q, r, z $\in \mathbb{Z}_q$ randomly. The supposition stated that in polynomial time, no opposition is able to differentiate the tuple.

$P = g^p, Q = g^q, R = g^r, \mathbb{Z} = e(g, g)^{pq/r}$. From the tuple $P = g^p, Q = g^q, R = g^r, \mathbb{Z} = e(g, g)^Z$ with minor advantage.

2.2 Access Policy

Policy means rules or strategy. The access policy [7] here outlines which kind of users can have rights to access or read the information. Let us understand it by taking

Fig. 13.3 Access policy

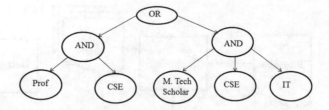

an example. A regular academic student's progress record might be accessible with only teaching faculty and trainee teachers of that particular course. The same condition can be expressed using predicate (Refer Fig. 13.3).

$$((Teacher \, AND \, CE \, dept) \, OR \, (Trainee \, teacher \, AND \, (CE \, dept \, OR \, IT \, dept)).$$

Here we will consider all variables mentioned above by the name of attributes. The predicate above is nothing but access policy of the academic condition. As an example, we have used the simplest access policy but in practical used its complex, and can own quite a huge amount of attributes. So access policy can deny a user from accessing information, can restrict it from using data or allows it depending on fulfillment of condition. It will ensure security of data transmitted or stored over the network.

2.3   ABE Working

As per Sahai and Waters [5], ABE technique is made up of four basic algorithms, mainly Setup, algorithm for Generation of Keys, algorithm for Encryption and Decryption. Process is shown in Fig. 13.4.

Whole process will have a data owner, centralized authority for key generation, and recipient. Data owner is the source of data and have authority to send data, so he is also known as sender of the data. Plaintext is encrypted by data owner with the help of encryption key for defining the access policy of data. Receiver can access data after the decryption of data with the help of decryption key and after satisfying access policy of the data. Key generation authority is responsible for the generation of public and private keys as per attribute group. Access rights are granted as per satisfaction of attributes. Refer Fig. 13.4 for the basic model of ABE.

Algorithm

1. Setup(d):

   (a) G_1, G_2 Bilinear groups. Here prime order is p
   (b) g is the originator or source of G_1
   (c) $e : G_1 \times G_1 \rightarrow G_2$ is Bilinear map.
   (d) d=threshold value.

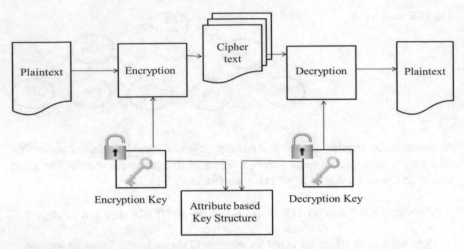

Fig. 13.4 Encryption of symmetric key using ABE

(e) From $t_1, \ldots, t_n$ are randomly chosen

(f) PK is Public key here. $PK = \left(T_1 = g_1^t, \ldots, T_n = g_n^t, Y = e(g, g)^y\right)$ where $y \in \mathbb{Z}_q$ and MK is master key, $(MK = (t_1, \ldots t_n, y))$.

2. Key_Gen(A_U, PK, MK).

(a) Authority generates Private key for users U.

(b) polynomial q is chosen randomly

   $q(0) = y$ where degree$=d - 1$.

(c) D is Private key,

   $D_i = g^{q(i)/t}{}_i$, where $i \in A_U$.

3. Encrypt (A_{CT}, PK, M):

(a) Message $M \in G_2$ is encrypted with set of attributes A_{CT}

(b) Randomly chosen, $s \in \mathbb{Z}_q$

(c) Encrypted data is published as

   $CT = A_{CT}, E = MY^s = e(g, g)^{ys}, E_i = g_i^{ts}$ where $i \in A_{CT}$

4. Decrypt (CT, PK, D):

(a) CT is decrypted with private key D.

(b) d is randomly chosen from $i \in A_U \cap A_{CT}$ to solve.

   $e(E_i, D_i) = e(g, g)^{q(i)s}$, if $|A_U \cap A_{CT}| \geq d$

   and $Y^s = e(g, g)^{q(0)s}) = e(g, g)^{ys}$ with the Lagrange coefficient

$$M = E / Y^s$$

Example:

For example, for class of expressive attributes {*CSE, Prof, IT, M − Tech Scholar*} (refer Fig. 13.3) of an encoded data, threshold value is 3. The encrypted data can be only decrypted if the private key satisfies at least three or more than three number of attributes in the encrypted data. So, receiver may have private key with any three number of attributes to achieve decryption and obtain the original data [10].

The basic ABE model for the decryption of coded text decryption key will only work for decoding of the message if at least d attributes satisfy criteria. For the KP-ABE or CP-ABE scheme, the decryption criteria are that the access structure belongs to a secret key, or cipher text must be satisfied by the attribute set.

3 ABE Categories

The FIBE technique introduced by Sahai and Waters [5] was inspired by IBE scheme [4] and became popular with its descriptive nature. It is considered as rich and strong expression wise. In this scheme, a set of cipher texts are characterized with a group of attributes α and threshold value *d* and another collection of attributes *á* both are adjoined with the secret key of the user. In the basic ABE model for the decryption of coded text, decryption key will only work for decoding if at least d attributes satisfy criteria of overlapping between coded text and private key. There is no flexible policy for access control to support various operations in this scheme. In the FIBE threshold is the access policy which stated as constant in the setup phase. So it does not satisfy the high-demanding need of real-time applications. To answer this limitation of ABE, more advanced schemes are introduced.

ABE technique works by using attributes, so attributes are a vital part of this technique. Public key which is used for encryption purposes is generated by using attributes. Not only that they are also used as an access policy. Depending on the access policy, researches can be divided into two main types. The two main types [8] observed are Key Policy-ABE (KP-ABE) and Ciphertext Policy–ABE (CP-ABE) schemes. The first concept of KP-ABE was proposed by Goyal et al. [8]. Bethencourt et al. [9] suggested the first CP-ABE technique. This section will give a brief outline of the mentioned techniques.

3.1 Key Policy-ABE (KP-ABE)

The first most concept of KP-ABE was proposed by Goyal et al. [8]. Data is encrypted using a collection of attributes, and access policy is built using the secret key of the user. To decrypt the cipher text using the decryption algorithm, the attribute set of the cipher text must be able to satisfy access policy in the secret key of the user. Refer Fig. 13.5 for complete understanding of process. Key generation algorithm of KP-ABE is different from the Key_Gen algorithm of basic ABE. As

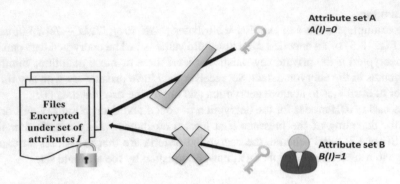

Fig. 13.5 KP-ABE access control

Key production technique is different, so the decryption technique is also changed than basic ABE. Attribute set of cipher text is used to run the decryption algorithm here.

Algorithm for KP-ABE

1. Setup (d):

   (a) G_1, G_2 Bilinear groups. Here prime order is p.
   (b) g is the originator or source of G_1.
   (c) $e : G_1 \times G_1 \rightarrow G_2$ is Bilinear map.
   (d) $d=$ threshold value.
   (e) From t_1, ….., t_n are randomly chosen.
   (f) PK is Public key here. ($PK = T_1 = g_1^t, \ldots, T_n = g_n^t, Y = e(g, g)^y$) where $y \in \mathbb{Z}_q$ and MK is master key, $MK = (t_1, \ldots, t_n, y)$.

2. Key_Gen (A_{U-KP}, PK, MK):

   (a) Authority generates Private Key for users U.
   (b) Authority generates private key components in the access design for every node x.
   (c) Polynomial q_x with $d - 1$ degree is chosen randomly such that

$$q_x(0) = q_{parent(x)}\,(index(x)).$$

   where $x's$ parent node is $parent(x)$ and $index(x)$ is linked with node x.

   (d) D is Private key.

$$D_x = g^{q_x(0)/t}{}_i, \text{ Where } i = leaf\,node.$$

3. Encrypt (A_{CT}, PK, M):

   (a) Message $M \in G_2$ is encrypted with set of attributes A_{CT}
   (b) Randomly chosen, $s \in \mathbb{Z}_q$
   (c) Encrypted data is published as

$$CT = \left(A_{CT}, E = MY^s = e(\mathrm{g}, \mathrm{g})^{ys}, E_i = \mathrm{g}_i'^s A\right) \text{ where } i \in A_{CT}$$

4. Decrypt (CT, PK, D):

By using recursion technique decryption can be achieved.

(a) CT is decrypted with private key D.
(b) If $i=$ leaf node and is included in the access design of the user's private key.

$$e\left(E_i, D_x\right) = e(\mathrm{g}, \mathrm{g})^{s.q}{}_x^{(0)}$$

Also if $i\neq$leaf node, the decrypt node function will be called, and calculate $e(\mathrm{g}, \mathrm{g})^{s.q}{}_x^{(0)}$ by utilizing Lagrange coefficient.

$Y^s = e(\mathrm{g}, \mathrm{g})^{q(0)s} = e(\mathrm{g}, \mathrm{g})^{ys}$ will be evaluated only if the access design in user's secret key is satisfied by the attributes of encrypted data.

The message $M = E/Y^s$ is retrieved.

Example:

For example, the attributes of the encoded data are {*CSE* $\wedge$ *M. tech scholar*}, and the users' secret key with access policy are {*CSE* $\wedge$ (*M. tech scholar* $\vee$ *Prof*)}. Message can be retrieved by the decryption only when the access policy of recipient's private key is fulfilled by the attribute accompanied with the encrypted data.

3.2   Ciphertext Policy-ABE (CP-ABE)

CP-ABE is the most famous type of ABE technique introduced by Bethencourt et al. [9]. There are four basic pillars of this system. Working of CP-ABE is same as of KP-ABE, but only difference is CP-ABE access structure is in encrypted data. Refer Fig. 13.6 for more details.

1. Centralized authority of attributes:

   • Responsible for production of private keys for the purpose of decryption of data. It also produce public key as well as master key.

2. Data Source:

   • As a source key role of a data owner is to decide an access policy to define access of users on data. It will also provide encryption of information using the same policy.

3. Data user:

   • Can decode data using access policy specified by source and using its own secret key. Data can be retrieved using private key and proper match of attributes.

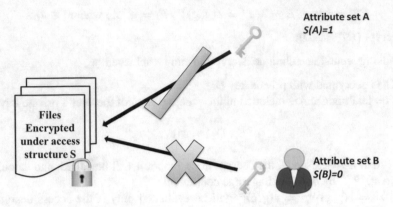

Fig. 13.6 (CP-ABE) Access Control

4. Centralized Server.

   • Management of smooth overall operations.

Algorithm of CP-ABE The following algorithm defines the complete working process of CP-ABE [23].

1. Setup:

   (a) G_0 Bilinear group. Here prime order is p
   (b) g is the originator or source of G_0
   (c) Exponents α, β are chosen randomly from $\mathbb{Z}_q$
   (d) PK is Public key here. $PK = \left(G_0, g, h = g^\beta, f = g^{\frac{1}{\beta}}, e(g, g)^\alpha\right)$ and MK is master key, $MK = (\beta, g^\alpha)$.

2. Key_Gen (A_U, MK):

   (a) Authority generates Private Key for users U.
   (b) Randomly choose $s \in \mathbb{Z}_q$ and s_j for every attribute j
   (c) Polynomial q with $d - 1$ degree is chosen randomly such that

$$q(0) = y.$$

   (d) D is Private key where,
        $D = (DK = g^{(\alpha + s)/\beta}$, where $j \in A_U$

   and $D_j = g^s. H(j)^s{}_j, D^*{}_j = g^s{}_j$ is output.
3. Encrypt $(A_{CT - CP}, PK, M)$:

   (a) Randomly chosen, $y \in \mathbb{Z}_q$,

$$q_r(0) = y.$$

(b) Root node$=r$ and $I=$set of leaf nodes in A_{CT-CP}
(c) Encrypted data is published as

$$CT = A_{CT-CP}, \acute{C} \triangleq Me(g, g)^{\alpha y}, C = h^y$$

where $C_i = g^{q(0)}{}_i$ and $C_{*i}C_i = C^*{}_I = H(att(i)^q)_i{}^{(0)}$

5. Decrypt (CT, D):

(a) CT is decrypted with private key D.
(b) $x=$ leaf node and $k = att(x), k \in A_U$
(c) Decrypt node (CT, D, x)

$$(e(D_k, C_x)) \div \left(e\left(D_k^*, C_x^*\right)\right) = e(g, g)^{sq}{}_x^{(0)}$$

By utilizing Lagrange coefficient ($CT, D, s) = e(g, g)^{ys}$.

$$\acute{C}/\left((e(C, DK)) \div (e(g, g)^{ys}\right) = M$$

Example:
Source of data willing to send confidential information to data users in the system. This huge data is deposited in the cloud. Instead of providing direct access to the data in the cloud, owners of data will give access to the users willing to access data only when they have suitable identifications which are nothing but the correct group of attributes. Access policy is given to the private information. To retrieve information, access policy should match the attribute set of the data user. If this condition satisfies, then only data users may have access to the information.

For example, private keys with access $design = \{CSE \wedge (Prof \vee M.\ Tech\ scholar)\}$ and attributes set of the encrypted $data = \{CSE \wedge M.\ Tech\ scholar\}$. To decrypt ciphertext in order to obtain the plaintext information access design of encoded data satisfied by selected attributes set in recipient private key.

4 Difference between CP-ABE and KP-ABE

Table 13.1 gives basic difference between two policies KP-ABE and CP-ABE (Table 13.1).

Table 13.1 Difference between KP-ABE and CP-ABE with respect to access policy and attribute association

Key-policy ABE	Ciphertext-policy ABE
Ciphertext ↑ Attribute association	Ciphertext ↑ Policy association
Private key ↑ Policy association	Private key ↑ Attribute association

5   Literature Review of KP-ABE and CP-ABE

Goyal et al. [8] proposed concept of more generalized key-policy attribute-based encryption. It is more efficient in terms of encrypted data sharing. This scheme is proven secured under the DBDH supposition. The name KP-ABE is given to this scheme as there is a tree access structure with which every private key is linked. This will define the nature of cipher texts which can be decoded by this private key, which are characterized by a collection of expressive attributes. Cipher text can be decoded only when the access group of attributes can match the access structure of the private key. This is a powerful encryption technique with real-time application. But this scheme will lead to the drawback that the sender cannot decide who can retrieve the message. He can only go for choosing attributes.

Ostrovsky et al. [11] introduced a system in which private keys can be characterized with positive as well as negative attributes. Here the access policy they used is more complex. Though having all these advantages, it suffers from the drawback that it increases the size of private key and cipher text. Which will also create encoding/decoding overheads. Lewko et al. [12] gave a solution to this drawback. This scheme attained user reversal and developed efficient KP-ABE technique.

Most of the techniques of KP-ABE suffer from the size issue. Size of cipher text increases with the number of attributes. This is the most common problem whose effective solution is provided by Attrapadung et al. [13]. Technique proposed by him possesses fixed cipher text size.

Goyal et al. [8] proposed the concept of ciphertext policy–ABE technique, but it does not suggest any actual constructed framework. In this scheme, a private key will be connected with a random number of attributes, and coded text will be connected with access structure. For successful decryption of coded text, attributes should satisfy the access structure of it.

The construction of the first CP-ABE scheme was proposed by Bethencourt et al. [9] in 2007. This scheme provides features of KP-ABE [8]. It also provides feature access control policies which are flexible.

A comparatively secured technique is suggested by Cheung and Newport [14]. As an access policy AND gate is supported by this technique not only for the positive attributes but also for negative attributes. The missing attributes from AND gate are indicated by the do not care component. If compared with Bethencourt et al. [9], this technique proves its security. But efficiency decreases if equated with Bethencourt et al. [9] because of increase in number of attributes size of privacy key and cipher text also.

By considering Cheung and Newport's Scheme [14] as base, Nishide et al. [15] upgraded the efficiency and Emura et al. [16] attained the secret policies. Nishide et al. [15] suggested a technique for multivalued attributes using AND gates. Using similar access policy, Emura et al. [16] proposed improved technique which has the special feature that they achieve constant length of coded text.

Goyal et al. [17] and Liang et al. [18] proposed a technique by using tree structure mainly bounded one. In this scheme, Goyal et al. [12] introduced a bounded technique of CP-ABE which gives a comprehensive transition method which is able to make a transformation of KP-ABE method to a CP-ABE method. It faces a limitation that the sender is limited to make use of the access tree only where the depth of the access tree is fixed at the time of setup. Liang et al. [18] enhanced the bounded technique of CP-ABE [17] by enhancing encoding/decoding process in the algorithm.

Ibraimi et al. [19] suggested approach to eliminate the border constrictions in [17, 18]. In this technique, they make use of *or* and *and* nodes to represent the access tree which outlines the privacy preserving policy. If contrasted with Cheung and Newport's [14], it is found that it needed a smaller number of overheads in encoding, decoding, and other phases.

Later Waters [20] suggested a new procedure for understanding CP-ABE under a solid and non-collaborative environment. Access control is expressed by using the attributes over the matrix M in the system. But if compared deeply, this technique gives comparatively similar performance as Bethencourt et al.'s [9].

Lewko et al. [21]. recommended an ABE method that attains adaptive safety inspired from the technique of Later Waters [20]. But when this method is compared with Later Waters [20], practically it shows less efficiency.

J. Zhang and Z. F. Zhang [22] suggested a new technique. This method works on AND gates though too without making use of bilinear pairings. This method gives proven solutions to the problems related to security. Where most of the other techniques work using bilinear pairing, this technique gives the probability of solving tough security problems without making use of bilinear pairing.

6   Comparison of ABE, KP-ABE, CP-ABE

As we can observe by the above analysis, all three schemes ABE, KP-ABE, and CP-ABE are different in application, complexity measures, flexibility, etc. We can draw a few conclusions from the above study.

The basic model of ABE technique supports only threshold policy. So this model is appropriate for those applicants who require simple policy. And other models like KP-ABE, CP-ABE do support the complex approaches, and are suitable for their use in data distribution which requires more attention.

In KP-ABE systems, the private key of a user is used for construction of access policy. That is the reason data sources cannot decide the third-party individual who can decode the data. When compared CP-ABE with KP-ABE, CP-ABE techniques are more feasible to use in real-time applications.

Let us define it in general form. The KP-ABE techniques are more query-oriented applications, e.g., they are used in TV systems, access to databases, broadcasting having specific targets. While CP-ABE techniques are used for those applications which need access control, e.g., access to different social networking sites.

Comparison of KP-ABE and CP-ABE with respect to performance (Table 13.2).

Table 13.2 Parameter-wise comparison of KP-ABE and CP-ABE

S. No.	Factors	KP-ABE	CP-ABE
1.	Proficiency	Ordinary, but high in case of broadcasting use	Ordinary
2.	Fine-grained access control	Low	Ordinary
3.	Resistance to collusion	Good	Good
4.	Computational overhead	Good	Ordinary

7 The Security Model

Here to prove security of technique the chosen set model is used against the selected plaintext attack [5]. Competitor and an opponent played a chosen ID game.

7.1 Fuzzy Model for ABE

Initial phase.

The competitor announces his identity I, using which he is willing to challenge.

Setup:

Setup algorithm is run by the opponent, he tells the competitor public parameters.

Stage 1:

For the purpose of private keys, competitors can issue queries.

Challenge: The competitors submit M_1 and M_2 off the same length where M_1 and M_2 are two messages. Random coin is flipped by the opponent for choosing value b. It encrypts message M_b with identity I, Coded text is transferred to competitor.

Stage 2:

Stage 1 is repeated.

Guess:

Benefit of Competitor in this game,

$$Benf\ C = |\ pr\ [\dot{b} = b] - 1/2|$$

Definition: In the security model of fuzzy selective-ID, technique is safe if all polynomial-time oppositions have minor benefit majorly in the above game.

CPA security of basic ABE technique is proved by Sahai and Waters [5]. DBDH assumption is used for basic ABE technique and KP-ABE technique [8, 11] also.

CP-ABE systems are more complicated. It is difficult to prove security of CP-ABE techniques as these are complicated one to handle. To prove its security under standard assumption, the main focus of research is on access structure. Depending on access structure, the three main types of research can be observed: Tree, AND gate, and LSSS matrix.

8 Conclusion

Data security is the most crucial aspect in every field in today's world. The need of confidentiality of data demands fine-grained access control. With the evolution in technology, various schemes of encryption were introduced. This chapter gives a brief overview about various encryption techniques like PKE, PKI, and IBE. It mainly studies ABE with its algorithm process and features of it including two main schemes of Key Policy-ABE (KP-ABE) and Ciphertext Policy-ABE (CP-ABE) followed by their working models and algorithms. It also enlightened on some popular techniques in literature review. We have also focused on comparison of ABE, CP-ABE, and KP-ABE with each other. Finally this chapter ended with security models of these schemes.

ABE technique is one of the most successful encryption techniques which ensures data security and high-level access control. Still many problems can be addressed and studied as a future works. Almost all ABE techniques use bilinear pairing as construction method, but high computational complexity of bilinear pairing reduces efficiency of algorithm to some extent. That is why if amount of bilinear pairing operations can be reduced to obtain improved scheme. Improving the construction technique of CP-ABE systems is also important research topic that can be studied extensively. Attempt to make access structure complex added extra redundancy to it. So to prove its security is difficult task. Hence optimization is necessary. Key misuse and key duplicating problem are major headache which can be effectively addressed by accountability. But existing techniques are proved to be safe in the selective model. So there is a need to provide solution to this problem.

References

1. Rivest, R. L., Shamir, A., & Adleman, L. (1978). A method for obtaining digital signatures and public-key cryptosystems. *Communications of the Association for Computing Machinery, 21*(2), 120–126.
2. Pang, L., Li, H., & Wang, Y. (2013). NMIBAS: A novel multi-receiver ID-based anonymous signcryption with decryption fairness. *Computing and Informatics, 32*(3), 441–460.
3. Pang, L., Li, H., & Pei, Q. (2012). Improved multicast key management of Chinese wireless local area network security standard. *IET Communications, 6*(9), 1126–1130.
4. Shamir, A. (1985). Identity-based cryptosystems and signature schemes. In *Advances in Cryptology: Proceedings of (CRYPTO '84), vol. 196 of Lecture Notes in Computer Science* (pp. 47–53). Berlin: Springer.
5. Sahai, A., & Waters, B. (2005). Fuzzy identity-based encryption. In *Advances in Cryptology— EUROCRYPT 20e05, vol. 3494 of Lecture Notes in Computer Science* (pp. 457–473). Berlin: Springer.
6. Lee, C., Chung, P., & Hwang, M. (2013). A survey on attribute-based encryption schemes of access control in cloud environments. *International Journal of Network Security, 15*(4), 231–240.

7. Goyal, V., Jain, A., Pandey, O., Sahai, A. "Bounded Ciphertext policy attribute based encryption". In: Aceto, L., Damgard, I. Goldberg, L.A., Ingolfsdottir, A., Walukiewicz, I. (eds.) ICALP 2008.
8. Goyal, V., Pandey, O., Sahai, A., & Waters, B. (2006). Attribute- based encryption for fine-grained access control of encrypted data. In *Proceedings of the 13th ACM conference on computer and communications security(CCS'06)* (pp. 89–98).
9. Bethencourt, J., Sahai, A., & Waters, B. (2007). Ciphertext-policy attribute-based encryption. In *Proceedings of the IEEE Symposium on Security and Privacy (SP'07)* (pp. 321–334).
10. Delerablée, C., & Pointcheval, D. (2008). Dynamic threshold public-key encryption. In D. Wagner (Ed.), *Advances in cryptology—CRYPTO 2008. Lecture Notes in Computer Science* (pp. 317–334). https://doi.org/10.1007/978-3-540-85174-5_18.
11. Ostrovsky, R., Sahai, A., & Waters, B. (2007). Attribute- based encryption with non-monotonic access structures. In *Proceedings of the 14th ACM Conference on Computer and Communications Security(CCS'07)* (pp. 195–203).
12. Lewko, A., Sanais, A., & Waters, B. (2010). Revocation systems with very small private keys. In *Proceedings of the IEEE Symposium on Security and Privacy (SP '10)* (pp. 273–285). Oakland.
13. Attrapadung, N., Libert, B., & de Panafieu, E. (2011). Expressive key- policy attribute-based encryption with constant-size cipher- texts. In *Public Key Cryptography—PKC2011* (Vol. 6571, pp. 90–108). Springer.
14. Cheungand, L., & Newport, C. (2007). Provably secure ciphertext policy ABE. In *Proceedings of the 14th ACM Conference on Computer and Communications Security(CCS'07)* (pp. 456–465).
15. Nishide, T., Yoneyama, K., & Ohta, K. (2008). Attribute-based encryption with partially hidden encryptor-specified access structures. In *Applied Cryptography and Network Security (ACNS 2008)* (pp. 111–129). Berlin: Springer.
16. Emura, K., Miyaji, A., Omote, K., Nomura, A., & Soshi, M. (2010). A cipher text-policy attribute-based encryption scheme with constant cipher text length. *International Journal of Applied Cryptography, 2*(1), 46–59.
17. Goyal, V., Jain, A., Pandey, O., & Sahai, A. (2008). Bounded cipher text policy attribute based encryption. In *Automata, Languages and Programming: Part II, vol. 5126 of Lecture Notes in Computer Science* (pp. 579–591). Berlin: Springer.
18. Liang, X., Cao, Z., Lin, H., & Xing, D. (2009). Provably secure and efficient bounded cipher text policy attribute based encryption. In *Proceedings of the 4th International Symposium on ACM Symposium on Information, Computer and Communications Security(ASIACCS'09)* (pp. 343–352).
19. Ibraimi, L., Tang, Q., Hartel, P., & Jonker, W. (2009). Efficient and provable secure ciphertext-policy attribute-based encryption schemes. In *Information Security Practice and Experience (ISPE 2009)* (pp. 1–12). Berlin: Springer.
20. Waters, B. (2011). Ciphertext-policy attribute-based encryption: An expressive, efficient, and provably secure realization. In *Public key cryptography (PKC '11)* (pp. 53–70). Berlin: Springer.
21. Lewko, A., Okamoto, T., Sahai, A., & Waters, B. (2010). Fully secure functional encryption: attribute-based encryption and (hierarchical) inner product encryption. In *Advances in Cryptology: EUROCRYPT 2010, vol. 6110 of Lecture Notes in Computer Science* (pp. 62–91). Berlin: Springer.
22. Zhang, J., & Zhang, Z. F. (2012). A ciphertext policy attribute- based encryption scheme without pairings. In *Information Security and Cryptology(ISC'12)* (pp. 324–340). Berlin: Springer.
23. Cheung, L., & Newport, C. (2007). Provably secure ciphertext policy ABE. In *Proceedings of the 14th ACM Conference on Computer and Communications Security—CCS 07* (pp. 456–465). https://doi.org/10.1145/1315245.1315302.

Chapter 14
Digital Signatures

Pinkimani Goswami, Madan Mohan Singh, and Khandakar Tahidur Rahman

Abstract A digital signature is a cryptographic protocol that ensures the authenticity of a message. In this book chapter, we will discuss some algorithms for digital signature. We will also discuss some algorithms for the blind signature scheme, undeniable signature schemes, short signature schemes, and Hierarchical identity-based signature schemes. We will also discuss Signcryption.

Keywords Signature schemes · Blind signature schemes · Undeniable signature schemes · Short signature scheme · HIBS schemes · Signcryption

1 Introduction

A digital signature is a procedure of signing a message stored in electronic form. It is a cryptographic protocol that shows the authenticity of a message. That is the digital signature schemes give the grantee to the receiver that the message was sent by the claimed sender. Therefore, the digital signature schemes provide a way to detect forgery or tampering. The four main security services of a security system are as follows [1]:

1. Confidentiality, i.e., information is kept secret from all but the authorized parties.
2. Message authentication, i.e., the sender of a message is authentic.
3. Integrity, i.e., message has not modified during communication.
4. Non-repudiation, i.e., the sender of a message cannot deny the creation of the message.

P. Goswami (✉) · K. T. Rahman
Department of Mathematics, University of Science & Technology, Meghalaya, India

M. M. Singh
Department of Basic Sciences and Social Sciences, North-Eastern Hill University, Shillong, Meghalaya, India

© Springer Nature Switzerland AG 2021
K. A. B. Ahmad et al. (eds.), *Functional Encryption*, EAI/Springer Innovations in Communication and Computing, https://doi.org/10.1007/978-3-030-60890-3_14

243

Note that the message authentication codes (MACs), which are based on symmetric key cryptography, provide the authenticity and integrity of a message. But MACs cannot prevent the verifier from creating forgeries as they (the signer and the verifier) share the same key. Also, the signer can deny the creation of the message as they are sharing the same key, i.e., it does not achieve non-repudiation. Another disadvantage of MACs is that one cannot prove the authenticity of a message without sharing the secret key. One approach to remove these disadvantages is by replacing the symmetric key cryptography into asymmetric key cryptography. A signature scheme is a security scheme that is based on asymmetric key cryptography. A digital signature scheme achieved not only authenticity and integrality of a message but also non-repudiation. A signature scheme consists of three algorithms:

1. Key generation algorithm: It is a PPT algorithm. For a given security parameter ℓ, the algorithm produces a private key (called signing key) K_{pr} and the corresponding secret key (called verifying key) K_{pub}.
2. Signature Generation: It may be a probabilistic algorithm. For a given message M and secret key K_{pr}, the algorithm produces a signature s on M.
3. Signature Verification: It is a deterministic algorithm. For a message M, a signature s and public key K_{pub}, the algorithm answer "accept" if (M, s) is a valid signature.

As the verification is done by the signer's public key, so anyone can prove the authenticity of the message. Also, no one can create forgeries as the sender used his secret key to sign a message. Note that the sender can get public key from a trusted authority such as a government agency.

1.1 Security Models for Signature Schemes

In this section, we discuss the security models for a signature scheme. Here we list possible goals and attack models of an attacker. Note that a signature scheme cannot be unconditionally secure, and hence the main goal is to construct a signature scheme which is computationally or provably secure [2].

The following are the main goals of an attacker:

- Total break: Adversary can determine the sender public key and therefore he can create a signature of any message.
- Selective forgery: Given a message M, an adversary can determine s such that (M, s) is a valid signature from the sender.
- Existential forgery: Adversary is able to create s for some M such that (M, s) is a valid signature from the sender.

The following are the attack models commonly considered:

- Key-only attack: Adversary only has the sender's public key, i.e., he has accessed the verification algorithm.

- Know-message attack: Adversary only has a list of messages signed by the sender.
- Chosen-message attack: Adversary chooses messages M_1, M_2, $\cdots$ and request the sender to sign. Therefore, he has the corresponding signatures of each message which he has chosen.

One can see [2–4], for security analysis of signature schemes.

1.2 Signature Schemes with Hash Functions

In practical implementation, asymmetric key algorithms are often too inefficient to sign a document. One approach to sign a long message M is to break the message into small parts and use the signature algorithm to each of these parts separately. But it is not a secure approach as the adversary may interchange the position of the parts of the message or may get other information. Also, it will be time-consuming for a long message. One solution to this problem is somehow "compress" the message M prior to signing and hence the hash function comes into picture. Note that a hash function H is a function $H : \{0, 1\}^* \rightarrow \{0, 1\}^r$, where a bit string of arbitrary length is converted to a bit string of length $r \in \mathbb{N}$. So, using a cryptographic hash function, a message with arbitrary length is converted to a fixed length, which is called a message digest. A hash function should have the following properties to prevent the attacks on the signature scheme:

1. Pre-image resistance (or one-wayness): To prevent the existential forgery on the signature scheme using a key-only attack.
2. Second pre-image resistance (or weak-collision resistance): To prevent the existential forgery on the signature scheme using a known-message attack.
3. Collision resistance: To prevent the existential forgery on the signature scheme using a chosen message attack (Fig. 14.1).

1.3 Organization of the Chapter

The main aim of this chapter is to provide a brief idea of signature schemes. Section 2 introduced some well-known signature schemes based on the integer factorization problem (IFP) and discrete logarithm problem (DLP). Section 3 deals with blind signature schemes, where we described the three important blind signature schemes. Section 4 introduced undeniable signature schemes. In Sect. 5, we will discuss Short signature schemes. Section 6 deals with hierarchical identity-based signature schemes, and some other types of signature schemes are briefly mentioned in Sect. 7. Another cryptographic primitive, namely signcryption, is introduced in Sect. 8. The chapter concludes in Sect. 9.

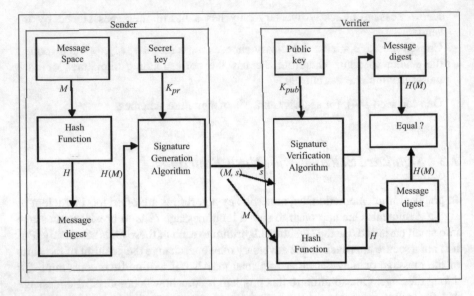

Fig. 14.1 Protocol for signature scheme with hash function

2 Signature Schemes Based on IFP and DLP

In this section, we will discuss some of the signature scheme based on IFP and DLP. We will consider RSA signature scheme, which is based on IFP, ElGamal signature scheme, and some of its variants, which are based on DLP.

2.1 RSA Signature Scheme

We will start with the RSA signature scheme. In 1978 [5], Rivest et al. proposed this scheme. The key generation algorithm of the RSA signature scheme is same as the RSA cryptosystem. The public key of the sender is (p, q, d), and the private key is (n, e), where $ed \equiv 1 \pmod{\phi(n)}$. A hash function H is considered which is public for all.

Signature generation	Signature verification
Consider a message M	1. Compute $m = H(M)$
1. Find the massage digest $m = H(M)$	2. Compute $\overline{m} = s^e \bmod n$
2. Compute $s = m^d \bmod n$	Accept the signature if $m = \overline{m}$.
s is the signature on the message M	

Let us consider one small example to illustrate the RSA signature scheme.

Example: Suppose $n = 18923 = 127 \times 149$ and $e = 1261$. Then $\phi(18923) = 18648$ and $d = 5797$. Suppose the signer wishes to sign a message $m = H(M) = 8990$. She computes $s = 8990^{5797} \bmod 18923 = 7212$. Therefore, 7212 is the signature on 8990.

Anyone can verify the signature by checking $7212^{1261} \bmod 18923 = 8990$.

Hence, the signature is valid.

Efficiency: Both the signature generation and verification algorithm needed one modular exponentiation. The size of a signature is around 1024 bits for a 1024-bit modulo. It is recommended to use e like 3, 257, and 65,537 to speed up the verification algorithm.

Security: The RSA signature scheme is based on IFP. If a secure hash function is used, then the scheme is secure against existential forgery attack.

2.2 ElGamal Signature Scheme

In 1985 [6], ElGamal described a signature scheme named as ElGamal Signature scheme, which is based on DLP. The key generation algorithm is the same as the ElGamal cryptosystem. The secret key of the sender is a, and the public key of the sender is (p, g, A), where $A = g^a \pmod{p}$. A hash function $H : \{0, 1\}^* \to \mathbb{Z}_p$ is considered, which is public to all. The signature generation and verification algorithms of the ElGamal signature scheme are as follows:

Signature generation	Signature verification
For a message M	For the signed message $(M, (s, t))$
1. Compute $m = H(M)$.	1. Compute $m_1 = g^{H(m)} \pmod{p}$
2. Select a random integer k such that	$m_2 = A^s s^t \pmod{p}$.
$\gcd(k, p - 1) = 1$.	Accept the signature
3. Compute $s = g^k \pmod{p}$.	If $m_1 \equiv m_2 \pmod{p}$
$t = k^{-1}(H(m) - as) \pmod{p - 1}$	
(s, t) is the signature on M.	

Let us consider one small example to illustrate the ElGamal signature scheme.

Example: Suppose $p = 31847$, $\alpha = 5$, $a = 7459$, then $\beta = \alpha^a \bmod p = 25703$. Suppose the singer wishes to sign the message $m = H(M) = 8990$. Suppose she chooses $k = 1165$ randomly. She computes

$$s = 5^{1165} \bmod 31847 = 23972 \bmod 31847$$

and $t = (8990 - 7456 \times 23972)1165^{-1} \bmod 31846 = 31396 \bmod 31846$. Therefore the signature of the message 8990 is (23972, 31396).

Anyone can verify the signature by checking

$$5^{8990} \bmod 31847 = 10262$$

and $25703^{23972} \times 23972^{31396} \bmod 31847 = 10262$

Hence, the signature is valid.

Security: The security of ElGamal signature is based on DLP. If a secure hash function is used in ElGamal signature scheme, then it is secure against existential forgery. However, the security of the ElGamal signature scheme can be broken if k is not secret or if k is reuse. If k is not secret and $(s, p - 1) = 1$, then from

$$t = k^{-1} (m - as) \bmod p - 1$$

one can find the secret key a of the signer. Again, suppose the signer used the same k to signed messages m_1 and m_2. Let (s, t_1) and (s, t_2) be the signatures on m_1 and m_2, respectively, where $s = g^k \bmod p$, $t_1 = k^{-1}(m_1 - as) \bmod p - 1$, and $t_2 = k^{-1}(m_2 - as) \bmod p - 1$. Then from t_1 and t_2, one can compute

$$k = (m_1 - m_2) (t_1 - t_2)^{-1} \bmod p - 1$$

and hence can find the secret key of the signer. A detailed description on the security of ElGamal signature scheme is found in [2].

2.3 The Schnorr Signature Algorithm

In 1989 [7], Schnorr proposed a new signature scheme named as the Schnorr signature scheme. It is a modification of the ElGamal signature scheme. For a message of length 1024 bit, the length of a Schnorr signature is around 320 bits, whereas an ElGamal signature is around 2048 bits. The key generation algorithm is similar to the ElGamal Signature scheme. But here instead of considering a cyclic group $\mathbb{Z}_p^*$, of order p, a cyclic subgroup $\langle \alpha \rangle$ of order q is considered, where $q \mid p - 1$. Basically, for p of size 1024 bits, the size of q is 160 bits. We can find α by computing $\alpha = g^{\frac{p-1}{q}} \bmod p$, where g is a generator of $\mathbb{Z}_p^*$. The secret key of the signer is a where $1 \leq a \leq q - 1$ and the corresponding public key is (p, q, α, β), $\beta = \alpha^a \bmod p$. A secure hash function $H : \{0, 1\}^* \to \mathbb{Z}_q$ is considered, which is public for all. The signature generation and verification algorithms are as follows:

Signature Generation	Signature Verification
Choose a message M	For the signed message $(M, (r, s))$
1. Choose a secret random integer k such that $1 \leq k \leq q - 1$	1. Compute $t = H(M \parallel \alpha^s \beta^{-r} \bmod p)$
2. Compute $r = H(M \parallel^1 \alpha^k \bmod p)$	Accept the signature if $t = r$.
$s = (k + ar) \bmod q$	
$(M, (r, s))$ is the signed message.	

Here we are considering one small example to illustrate the scheme.

Example: Suppose $q = 15923$ and $p = 2q + 1 = 31847$. 5 is a primitive root modulo 31847. So $\alpha = 5^2 \bmod 31847 = 25$ such that $\text{ord}(\alpha) = 15923$. Suppose, $a = 101$. Then $\beta = 25^{101} \bmod 31847 = 30484$. Suppose the singer wishes to sign the message $M = 8990$. Suppose she chooses $k = 27$ randomly. She computes

$$r = H\left(M \parallel 25^{27} \bmod 31847\right) = H\left(M \parallel 12401 \bmod 31847\right) = 3438 \quad \text{(say)}$$

and $s = (27 + 101 \times 3438) \bmod 15923 = 12882 \bmod 15923$.

Therefore, the signature of the message 8990 is $(3438, 12882)$.

Anyone can verify the signature by checking

$$25^{12882} \times 30484^{-3438} \bmod 31847 = 12401 \bmod 31847$$

and $t = H(M \parallel 12401 \bmod 31847) = 3438$

Hence, the signature is valid.

Security: The Schnorr signature scheme is based on DLP. However, one can attacks the Schnorr signature scheme in the following approach too:

1. Given a signature (r, s) on a message m, if one can find m' such that $H(m \parallel \alpha^s \beta^{-r}) = H(m' \parallel \alpha^s \beta^{-r})$, then (r, s) is a valid signature for m'. To resist this attack, the hash function should be second pre-image resistant.
2. One can choose any (r, s) and compute $\alpha^s \beta^{-r}$ and then try to find m such that $H(m \parallel \alpha^s \beta^{-r}) = r$. To resist this attack, the hash function should be pre-image resistant.

Efficiency: The signature algorithm needed one exponentiation, one hash function evaluation, and one computation modulo q, Whereas the verification algorithm needed to performs a multi-exponentiation $\alpha^s \beta^{-r}$ and one hash function. The signature algorithm is faster than verifying algorithm.

We refer [8] for a detailed description of the Schonrr signature scheme.

2.4 Nuberg-Rueppel Digital Signature Algorithm

The Nuberg-Rueppple (N-R) signature Scheme [9] is another modification of the ElGamal signature scheme, which is based on DLP. The size of the signature is same as Schnorr signature. The scheme is described as follows:

Key Generation
1. Choose a large p prime and q such that $q \mid (p - 1)$.
2. Choose $\alpha \in \mathbb{Z}_p^*$ such that $\text{ord}(\alpha) = q$.
3. Choose a such that $2 \le a \le q - 1$.
4. Compute $A = \alpha^a \bmod p$.
5. Choose a function $F : G \to \mathbb{Z}_q$ defined as $F(m) = m \bmod q$.

The Public key is (p, q, α, A, F), and the corresponding secret key is a.

Signature generation	Signature verification
Choose a message M	Compute $u = g^s A^r \bmod p$
Choose a random integer k such that	$t = (H(M) + F(u \bmod p)) \bmod q$
$2 \le k \le q - 1$	If $t = r$, then the signature is a valid
Compute	signature.
$r = (H(M) + F(g^k \bmod p)) \bmod q$	
$s = (k - ar) \bmod q$	
$(M, (r, s))$ is the signed message.	

Here we are considering one small example to illustrate the scheme.

Example: Suppose $p = 31847$, then $q = 15923$. Let $g = 25$ and $a = 101$. Then $A = 5905$. Suppose the singer wishes to sign the message $m = H(M) = 8990$. Suppose she chooses $k = 1165$ randomly. She computes

$$r = (8990 + 9516) \bmod 15923 = 2583 \bmod 15923$$

and $s = (8990 - 101 \times 2583) \bmod 15923 = 10973 \bmod 15923$. Therefore, the signature of the message 8990 is $(2583, 10973)$.

Anyone can verify the signature by checking

$$u = 25^{10973} \times 5905^{2583} \bmod 31847 = 9516$$

and $t = (8990 + 9516) \bmod 15923 = 2583$.

Hence, the signature is valid.

The security and efficiency of N-R signature scheme is similar with the Schnorr signature scheme.

2.5 Digital Signature Standard (DSS)

The DSS [10] is another variant of ElGamal signature scheme, which is published in 1994. It is a modification of ElGamal signature scheme. The key generation algorithm is same with ElGamal signature scheme.

Key Generation
1. Choose a large prime p (1024 bits).
2. Choose another prime q (160 bits) such that $q \mid p - 1$.
3. Find α of order q.
4. Choose a such that $0 \le \alpha \le q - 1$.
5. Compute $\beta = \alpha^a \bmod p$.

The public key is (p, q, α, β), and the corresponding secret key is a.

Signature generation	Signature verification
Choose a message M	Compute $u = s^{-1} \bmod q$
Choose a random integer k such that	$t_1 = uH(M) \bmod q$
$2 \le k \le q - 1$	$t_2 = ur \bmod q$
Compute $r = (\alpha^k \bmod p) \bmod q$	$t = (\alpha^{t_1} \beta^{t_2} \bmod p) \bmod q$
$s = (H(M) + ar)k^{-1} \bmod q$	The signature is a valid signature if
$(M, (r, s))$ is the signed message.	$t = r$.

Let us discuss the following example to illustrate DSS.

Example: Suppose $p = 31847$, $q = 15923$, $\alpha = 25$, and $a = 101$, then $\beta = 25^{101} \bmod 31847 = 30484$. Suppose the singer wishes to sign the message $m = H(M) = 8990$. Suppose she chooses $k = 1165$ randomly. She computes

$$r = \left(25^{1165} \bmod 31847\right) \bmod 15923 = 9516 \bmod 15923$$

and $s = (8990 + 101 \times 9516)1165^{-1} \bmod 15923 = 327 \bmod 15923$. Therefore, the signature of the message 8990 is (23972, 31396).

Anyone can verify the signature by checking

$$\left(25^{9182} 30484^{3389} \bmod 31847\right) \bmod 15923 = 9516 \bmod 15923$$

Hence, the signature is valid.

Security: It is based on DLP over a prime field and its subgroups. DSS is secure under random oracle model. Like ElGamal signature scheme, the signer should not use same k for different messages. The hash function used for DSS is SHA-1 [11].

Efficiency: The DSS signature generation required one modular exponentiation, where verification algorithm required two modular exponentiations. It is as fast as Schnorr signature scheme and N-R signature scheme. For a 1024-bit modulo, the DSS signature is 320 bits long.

3   Blind Signature Scheme

In a Blind signature scheme, the signer and the user (who create the message) are different parties. In this scheme, the message is hiding from the signer. In Crypto'82 [12], Chaum introduced the concept of the blind signature scheme. The security of this scheme is based on IFP. In 1994 [13], Camenisch et al. proposed two blind signature schemes, which are based on DLP. Among these ones is based on modified RSA and the other is based on N-R Signature scheme. An ElGamal-based blind signature scheme is proposed in [14]. In [15], Pointcheval et al. described a blind signature scheme based on the Schnorr's signature scheme (called Schnorr's blind signature scheme). In the same paper, he proposed the Okamoto-Schnorr blind signature scheme, which is based on Okamoto Scheme [16]. The scheme is secured under the random oracle model. Numerous blind signature schemes have been proposed [17–21].

A blind signature scheme comprises of three algorithms:

1. Key generation: It is a PPT algorithm, that on given 1^ℓ, for a security parameter ℓ, the signer produces a pair of keys (K_{pub}, K_{pr}), where K_{pub} is a public key, and K_{pr} is the corresponding private key.
2. Blind signature generation: It is a PPT algorithm. The user and signer execute the following steps:

   - Blinding: The user who wants a signature in a message M, first blind that message (say m') and sends the blind message to the signer.
   - Signature generation: Given a blind message m' and the private key K_{pr}, the signer evaluate the signature s' and sends to the user.
   - Unblinding: The user unblind the signature on the blind message and produces the signature s of the original message M.

3. Signature verification: It is a deterministic polynomial time algorithm. For the message M, the signature s, and the public key K_{pub} of the singer, the verification algorithm yield "true" or "false."

The two main security requirement of a blind signature scheme are *blindness* (i.e., the signer should not access the message), and *one-more forgery*, i.e., the user should not create more signature form the signature provided by the signer [22]. The security of a blind signature scheme is studied in [23–26]. Blind signatures have various applications such as in electronic payment systems [12] and electronic voting systems [27].

3.1 Chaum's RSA Blind Signature Scheme

This is the first blind signature scheme, and it is based on the IFP. The Key Generation algorithm is the same as the RSA key generation algorithm. The secret key of the signer is (p, q, d) and corresponding public key is (n, e), where $ed \equiv 1 \bmod \phi(n)$.

Let us consider one small example to illustrate the Chaum's RSA blind signature scheme.

Example: Suppose $n = 18923 = 127 \times 149$ and $e = 1261$. Then $\phi(18923) = 18648$ and $d = 5797$. Suppose the user needs a signature on a message $m = H(M) = 8990$. The user first blind the message by executing the following steps:

Suppose $k = 101$, then $m' = 8990 \times 101^{1261} \bmod 18923 = 8703 \bmod 18923$.

The signer generates the signature $s' = 9338 \bmod 18923$ on $m' = 8703$.

The user computes $s = 7212 \bmod 18923$. So, 7212 is the signature on 8990.

Blind signature generation	Signature verification
Blinding: Suppose $m \in \mathbb{Z}_n$ be a message. 1. Choose a random $k \in \mathbb{Z}_n$ such that $(k, n) = 1$. 2. Compute $m' = mk^e \bmod n$ The user sends the blind message m' to the signer. *Signature Generation:* For the message m', the signer 1. Compute $s' = (m')^d \bmod n$ s' is the signature on m'. *Unblinding:* Using s', the user 1. Compute $s = k^{-1}s' \bmod n$ s is the signature on m.	For a message m and signature s 1. Compute $\overline{m} = s^e \bmod n$ 2. Accept the signature if $m = \overline{m}$.

Anyone can verify the signature by checking $7212^{1261} \bmod 18923 = 8990$. Hence, the signature is valid.

3.2 Schnorr Blind Signature Protocol

This signature scheme is based on DLP. In this scheme the message M is never sent to the signer, so blinding and unblinding algorithm is not required. The hash function $H : \{0, 1\}^* \rightarrow \mathbb{Z}_q$ is public for all.

Key Generation
1. Choose two large primes p and q such that $q \mid (p - 1)$.
2. Choose an element $g \in \mathbb{Z}_p^*$ such that $\text{ord}(g) = q$.
3. Choose d such that $2 \leq d \leq q - 1$ and compute $A = g^d \bmod p$.
    Here the signing key d and corresponding verification key is (q, g, A).

Signature Generation
Suppose M is the message to be signed. The user asks the signer to initiate a communication. The signer has to execute the following steps:

1. Choose a random integer $\overline{d} \in [2,\ q - 1]$.
2. Compute $\overline{u} = g^{\overline{d}} \bmod p$.

   The signer sends $\overline{u}$ to the user.
   The user has to execute the following steps:

1. Select two random numbers $\alpha, \beta \in \mathbb{Z}_q$.
2. Compute $u = \overline{u}g^\alpha A^\beta \bmod p$
3. Compute $s = H(M \parallel u)$ and $\overline{s} = (s - \beta) \bmod q$.

   The user sends $\overline{s}$ to the signer.
   The signer computes $\overline{t} = \overline{d} - d\overline{s} \bmod q$ and sends $\overline{t}$ to the user.

The user computes $t = \left(\bar{t} + \alpha\right) \bmod q$. The blind signature on M is (s, t).

Signature Verification
1. Compute $t_1 = H(M \parallel g^t A^s)$
2. If $t_1 = s$, the signature is a valid signature.
   Correctness
1. Compute $g^t A^s = g^{\bar{t}+\alpha} A^{\bar{s}+\beta} = u$, so $t_1 = H(M \parallel u) = s$.

3.3 Okomoto-Schnorr Blind Signature Protocol

This scheme is the modification of the Schnorr blind signature, which is based on DLP.

Key Generation
1. Choose two large primes p and q such that $q \mid (p - 1)$.
2. Choose two elements $g_1, g_2 \in \mathbb{Z}_p^*$ such that $\operatorname{ord}(g_1) = \operatorname{ord}(g_2) = q$.
3. Compute $A = g_1^{d_1} g_2^{d_2} \bmod q$, where $2 \leq d_1, d_2 \leq q - 1$.

Here the secret key is (d_1, d_2), and the public key is (q, g_1, g_2, A).

Signature Generation
The user asks the signer to initiate a communication. The signer executes the following steps:

1. Chose random numbers $\overline{d_1}, \overline{d_2} \in \{2, 3, \cdots, q - 1\}$.
2. Compute $\bar{u} = g_1^{\overline{d_1}} g_2^{\overline{d_2}} \bmod q$ and send $\bar{u}$ to the user.

The user executes the following steps:

1. Select three random integers $\alpha, \beta, \gamma \in \mathbb{Z}_q$.
2. Compute $u = \bar{u} \, g_1^{\alpha} g_2^{\beta} A^{\gamma} \bmod p$
3. Compute $s_1 = H(M \parallel u)$ and $\overline{s_1} = (s - \gamma) \bmod q$.
4. Send $\overline{s_1}$ to the signer.

The signer executes the following steps:

1. Compute $\overline{s_2} = \left(\overline{d_1} - d_1 \overline{s_1}\right) \bmod q$ and $\overline{s_3} = \left(\overline{d_2} - d_2 \overline{s_1}\right) \bmod q$.
2. Sends $\overline{s_2}$ and $\overline{s_3}$ to the user.

The user executes the following steps:

1. Compute $s_2 = (\overline{s_2} + \alpha) \bmod q$ and $s_3 = (\overline{s_3} + \beta) \bmod q$.

The blind signature on M is (s_1, s_2, s_3).

Signature Verification
1. Compute $t_1 = H\left(M \parallel g_1^{s_2} g_2^{s_3} y^{s_1}\right)$.
2. If $t_1 = s_1$, then the signature is a valid signature.
   Correctness

$$g_1^{s_2} g_2^{s_3} y^{s_1} = g_1^{\overline{s_2}+\alpha} g_2^{\overline{s_3}+\beta} y^{\overline{s_1}+\gamma} = u$$

So, $t_1 = H(M \parallel u) = s_1$ and hence signature is verified.

4   Undeniable Signature Schemes

The signature schemes which we have discussed in the last two sections have the property that anyone who has the knowledge of signer public key and message-signature pair can verify the signature. However, this is not suitable for many other situations such as licensing software, where a software vendor might want to sign on their products such that only the paying customer can verify the validity of these signatures. The undeniable signature scheme is useful in such scenario. In an undeniable signature scheme, the verification algorithm depends on the active participation of the signer. In this type of signature scheme, a verifier can not only verify signature but also check whether the signature is forged or the signer is trying to deny his signature.

In 1989 [28], Chaum et al. introduced the concept of the undeniable signature scheme. It is based on the DLP over a prime field. Various undeniable signature schemes have been proposed since its introduction. The most of the undeniable signature scheme is based on DLP (e.g. [29–32]). In 1991 [33], Boyen et al. suggested a research direction for construction of undeniable signature schemes based on RSA. In 1997 [34], Gennaro et al. proposed the first RSA-based undeniable signature scheme. After that various undeniable signature scheme was proposed which are based on RSA or related problems (e.g., [35–37]). Some other undeniable signature schemes have been proposed based on some other intractable problems like [38–41]. The various notion of security had been discussed in [42]. For the recent work on the undeniable signature scheme, one can check [43–46].

An undeniable signature scheme comprises of four algorithms:

1. Key generation: It is a PPT algorithm, where for given security parameter ℓ, the signer generates a pair of keys($K_{\text{pub}}, K_{\text{pr}}$), where K_{pr} is the secret key and K_{pub} is the corresponding public key of the signer.
2. Signature generation: It is a PPT algorithm, where for a given message m and the signer private key K_{pr}, which produces a signature s for the message m.
3. Signature verification: For a given message M, signature s and public key K_{pub} of the signer, the verifier executes the following steps:

   - Challenge: The verifier sends a challenge (say z) to the signer.

- Response: Given the challenge z, the message M, and his private key K_{pr}, the signer evaluates a response (say r) and sends it to the verifier.
- Verification: Given a massage m, the signature s, the signer public key K_{pub} and response r, the signature verification algorithm returns an answer "true" or "false" depending on s is a valid signature or not. If the verification return "false," the verifier executes the following steps:

4. Denial protocol: Here the verifier will check whether the signature s is forged or the signer is trying to deny his valid signature by sending challenges and checking responses of the signer.

4.1 Chaum-Van Antwerper Undeniable Signature Scheme

In this section, we will consider the Chaum-Van Antwerper (CvA) undeniable signature scheme proposed in [28]. We have considered the algorithm from [3, 22].

Key Generation
1. Choose a large prime p of the form $p = 2q + 1$, where q is prime.
2. Choose $g \in \mathbb{Z}_p^*$ such that $\text{ord}(g) = q$.
3. Choose $a \in \mathbb{Z}_q$ such that $\gcd(a, q) = 1$
4. Compute $A = g^a \bmod q$
5. Select a hash function $H : \{0, 1\}^* \to \mathbb{Z}_p$, which is public to all.
   The Public key (p, q, g, A) and the corresponding secret key a.
   Signature Generation: For a given message M and public key (p, q, g, A).

1. Compute $m = H(M)$
2. Compute $s = m^a \bmod p$
   (M, s) is the signed message on M.

Signature Verification
- Challenge: The verifier selects two random integers $u, v \in \mathbb{Z}_q$ and compute $z = s^u A^v \bmod p$. The verifier sends the challenge z to the signer.

  - Response: Given the challenge z, the signer evaluates $w = z^{a^{-1} \bmod q} \bmod p$ and sends the response w to the verifier.

- Verification: If $w = m^u g^v \bmod p$, then the verifier accepts the signature. Otherwise, he rejects it.

Denial Protocol
Suppose u and w are the challenge and response, respectively, generated in the verification algorithm. If the verification fails, then there are two possibilities either (i) the signature is forged or (ii) the signer trying to deny the signature. In that case, the verifier performs a second verification as follows:

- Challenge: The verifier

(i) selects two random integers u', $v' \in \mathbb{Z}_q$,
(ii) computes $z' = s^{u'} A^{v'} \bmod p$.

The verifier sends the challenge z' to the signer.
- Response: Given the challenge z', the signer evaluates

$$w' = z'^{a^{-1} \bmod q} \bmod p$$

The signer sends the respond w' to the verifier.
- Verification: If $\left(wg^{-v}\right)^{u'} = \left(w'g^{-v'}\right)^{u}$, then the verifier concludes that the signature is forged. Otherwise, he concludes that the signer is trying to deny the signature.

Correctness.

The correctness of the denial protocol is based on the following two lemmas [3]:

Lemma 4.1.1:

1. For any $z \in \mathbb{Z}_p^*$, there are q pairs (u, v) such that $z \equiv s^u A^v \bmod p$.
2. If $s \not\equiv m^a \bmod p$ and if $z \in \mathbb{Z}_p^*$, then for each $w \in \mathbb{Z}_p^*$ there is exactly one pair (u, v) such that $w \equiv m^u g^v \bmod p$ and $z \equiv s^u A^v \bmod p$.

Lemma 4.1.2:

If $s = m^a \bmod p$ and w is fixed with $w \not\equiv m^u g^v \bmod p$, then for every $w' \in \mathbb{Z}_p^*$ there is exactly one pair $(u', v') \in \mathbb{Z}_q \times \mathbb{Z}_q$ such that

$$\left(wg^{-v}\right)^{u'} \equiv \left(w'g^{-v'}\right)^{u} \bmod p.$$

From the lemma 4.1.1, the signer knows that for a given z, the verifier has q possible choices of pairs $(u, v) \in \mathbb{Z}_q \times \mathbb{Z}_q$. Also, for any two distinct pairs (u_1, v_1) and (u_2, v_2), the value of w is different. So, the probability to choose the correct (u, v) by the signer is $\frac{1}{q}$, which is very small for large q.

Suppose the signature is a valid signature but the signer wants to deny the signature, then he has to respond w' for challenge z' such that $\left(wg^{-v}\right)^{u'} = \left(w'g^{-v'}\right)^{u}$, but the verification fails. By Lemma 4.1.1, there are q pairs (u', v') that gives z'. By Lemma 4.1.2, for every w', there is exactly one pair (u', v') such that $\left(wg^{-v}\right)' = \left(w'g^{-v'}\right)^{u}$. The signer has no idea which pair (u', v') was chosen by the verifier and hence no idea which w' to choose. The best he can do is to select w' randomly and hence he has $\frac{1}{q}$ chances to obtain the right w', which is very small for large q. Therefore, if $\left(wg^{-v}\right)^{u^{-1}} = \left(w'g^{-v'}\right)^{u}$, then the verifier accepts that the signature is invalid; otherwise, he knows with probability at least $1 - \frac{1}{q}$ that the signer trying to deny the signature.

Example: Suppose $p = 31847 = 2 \times 15923 + 1$ and $g = 25$. Suppose $a = 101$. Then $A = 25^{101} \bmod 31847 = 30484$. Suppose the singer wants to sign the message $M = 8990$. She computes $s = 8990^{101} \bmod 31847 = 15457$. Therefore, the signature of the message 8990 is 15457.

To verify the message, verifier chooses two random values $u = 39$ and $v = 7932$ and sends the challenge $z = 15457^{39} 30484^{7932} \bmod 31847 = 17205 \bmod 31847$ to the signer. Signer responds with $w = 17205^{101^{-1} \bmod 15923} \bmod 31847 = 21486$. The verifier checks the response and verifies the signature by checking

$$8990^{39} 25^{7932} \bmod 31847 = 21486$$

Hence, the signature is valid.

4.2 RSA-Based Undeniable Signature Schemes

In 1997 [34], Gennaro et al. proposed an undeniable signature scheme (called GKR scheme) based on RSA. We have considered the scheme from [3, 22]. The scheme is defined as follows:

Key Generation
1. Choose two large random primes p and q such that both $p' = \frac{p-1}{2}$ and $q' = \frac{q-1}{2}$ are also primes.
2. Compute $n = pq$
3. Select two integers e and d such that $ed \equiv 1 \bmod \phi(n)$.
4. Choose $g \in \mathbb{Z}_n^*$ and compute $A \equiv g^d \bmod n$.
   Public key (n, g, A) and secret key (e, d).

Signature Generation
1. Choose a message M.
2. Compute $m = H(M)$
3. Compute $s = m^d \bmod n$
   (M, s) is the signed message on M.

Signature Verification
- Challenge: The verifier

    (i) computes $m = H(M)$
    (ii) selects two random integers $i, j \in \{1, 2, \cdots, n\}$,
    (iii) computes $u = s^{2i} A^j \bmod n$.

   The verifier sends the challenge u to the signer.
- Response: Given the challenge u, the signer evaluates.

    (i) $w = u^e \bmod n$ and sends it to the verifier.

- Verification: The verifier evaluates

$$w' = m^{2i} g^j \bmod n$$

(i) If $w = w'$, then the verifier concludes that the signature is a valid signature.

- **Denial protocol**: Suppose u and w are the challenge and response, respectively, generated in the verification algorithm. The verification fails, then there are two possibilities either (i) the signature is forged or (ii) the signer is trying to deny the signature. In that case, the verifier performs a second verification as follows:
- Challenge: The verifier

  (i) selects two random integers $i \in \{4, 8, 12, \cdots, 4k\}$ and $j \in \{1, 2, \cdots, n\}$,
  (ii) computes $w_1 = m^i g^j \bmod n$ and $w_2 = s^i A^j \bmod n$.

  The verifier sends the challenge (w_1, w_2) to the signer.
- Response: Given the challenge (w_1, w_2), the signer evaluates

  (i) $i' \in \{4, 8, 12, \cdots, 4k\}$ such that

$$w_1 w_2^{-e} = \left(ms^{-e}\right)^{i'} \bmod n$$

  The respond i' sends to the verifier.
- Verification: If $i = i'$, then the verifier concludes that the signature is forged. Otherwise, he concludes that the signer is trying to deny the signature.

Correctness

In order to see how this protocol prevents the signer from denying a valid signature, first let us consider the case that (M, s) is a valid signature on M. In that case,

$$\begin{aligned} w_2^e &\equiv \left(s^i A^j\right)^e \bmod n \\ &\equiv m^i g^j \bmod n \\ &\equiv w_1 \bmod n \end{aligned}$$

Again,

$$s^e \equiv m \bmod n$$

Therefore, for any $i' \in \{4, 8, \cdots, 4k\}$

$$w_1 w_2^e = \left(ms^{-e}\right)^{i'} \equiv 1 \bmod n$$

Thus, the signer can guess that the secret value i was chosen by the verifier, and the guess is correct with a probability of $\frac{1}{k}$. On the other hand, if (M, s) is a forged signature, then $w_1 w_2^e \equiv (ms^{-e})^{i'} \equiv 1 \bmod n$ holds only for a single i', i.e., $i' = i$. Therefore, the verifier concludes that the signature is forged.

Note that there exists at least one i' such that $w_1 w_2^e \equiv (ms^{-e})^{i'} \equiv 1 \bmod n$ holds. Therefore if the signer could not find one such i', then it implies that the verifier sent at least one of the value of w_1 or w_2 wrongly.

In order to reduce the probability of successful cheating, it is suggested to repeat the protocol a few times instead of increasing k. If $k = 1024$, the signer can successfully cheat in eight executions of the denial protocol with a probability of 2^{-80} [22].

5 The Short Signature Scheme

The shortest possible signature is needed when a human is asked to manually key in a signature. Note that the two most commonly used signature schemes are RSA and DSS. But the size of the signature is too large to be keyed. For example, an RSA signature scheme and a standard DSS or ECDSS (elliptic curve DSS) produce a signature of length 1024 bit and 320 bit, respectively, for a 1024-bit modulus.

Various approaches have been made to shorten the signature of a signature scheme with the same level of security. In 2000 [47], Naccache et al. proposed a variant of the DSS, where the signature is approximately 240 bits long. Another variant of the DSS, which is secure in a random oracle model, is proposed in [48]. But the length of the signature generated by this scheme is the same as in [47]. Signature with message recovery is an another method for minimizing the length of a signature (e.g., [5, 49]). By this method, one can minimize the length of the signature for a long message, but for short messages, the length is the same with the DSS signature [50]. In 2001 [51], Boneh et al. introduced the idea of a short signature scheme by using bilinear pairing. Their signature scheme (called BLS short signature scheme or BLS signature scheme) generates a signature of approximately 160-bit-long with a similar level of security to 320-bit DSS signature. The security of the BLS signature scheme is based on the difficulty of solving CDH assumption over elliptic curves, and it is proved that the scheme prevents the existential forgery under a chosen message attack in the random oracle model [51]. In 2003 [52], Choon et al. presented a variant of BLS (called identity-based BLS) signature scheme where the user's public ID was used as a public key. Goh et al. [53] also proposed a signature scheme based on CDH assumption. In 2004 [54], Boneh et al. proposed another short signature scheme where the length of the signatures is almost same with the BLS signature scheme, but the security does not need a random oracle model. Their scheme is based on the strong Diffie-Hellman (SDH) assumption. Numerous short signature schemes have been proposed, and most of them are based on either the DLP and its variants or the problem for Bilinear pairing (e.g., [55–58]). In 2010 [59], Yu et al. proposed a short signature scheme which is based on IFP. Their scheme is secure under the strong RSA subgroup assumption, and it produces a signature of length 420 bit.

In the next two sections, we will discuss the BLS signature scheme and the Boneh-Boyen signature scheme.

5.1 The Boneh-Lynn-Shacham Short Signature Scheme

We will recall some definitions before presenting the BLS short signature scheme.

5.1.1 Bilinear Pairing [51]

Let q be a prime number and let G_1, G_2, and G_T be cyclic groups of order q. Let G_1, G_2 be two additive groups and G_T be a multiplicative group. A mapping $e : G_1 \times G_2 \to G_T$ is called a bilinear map if it satisfies the following conditions:

1. Bilinearity: $e(aP, bQ) = e(P, Q)^{ab}$, $\forall P \in G_1$, $\forall Q \in G_1$ & $a, b \in \mathbb{Z}_q^*$.
2. Non-degeneracy: $e(P, Q) \neq 1$, i.e., if P, Q are the generators of G_1 and G_2, respectively, then $e(P, Q)$ is a generator of G_T.
3. Computable: There exists an efficient algorithm to compute $e(P, Q)$ $\forall P \in G_1$, $\forall Q \in G_2$.

Note that if $G_1 = G_2$, then the pairing is called symmetric, otherwise it is said be to asymmetric pairing. The modified Weil pairing and Tate pairing are some examples of cryptographic bilinear maps.

5.1.2 Computation Assumptions [51]

Definition 1 [11]: For given (P, aP, bP) for some $a, b \in \mathbb{Z}_q^*$, the computational Diffie-Hellman (CDH) problem asked to find abP.

Definition 2 [11]: The CDH assumption states that for every PPT algorithm $\mathcal{A}$, $\mathrm{Succ}_{\mathcal{A},G_1}^{\mathrm{CDH}} = \Pr\left[\mathcal{A}(P, aP, bP, abP) = 1 : a, b \in \mathbb{Z}_q^*\right]$ is negligible.

Definition 3 [11]: For given (P, aP, bP, cP) for some a, b, $c \in \mathbb{Z}_q^*$, the Decisional Diffie-Hellman (DDH problem) asked to check $c = ab \bmod q$ or not.

Definition 4 [11]: The DDH assumption states that for every PPT algorithm $\mathcal{A}$, $\mathrm{Succ}_{\mathcal{A},G_1}^{\mathrm{CDH}}$ is negligible, where

$$\mathrm{Succ}_{\mathcal{A},G_1}^{\mathrm{CDH}} = \left| \Pr\left[\mathcal{A}(P, aP, bP, cP) = 1\right] - \Pr[\mathcal{A}(P, aP, bP, abP) = 1] \cdot a, b \in \mathbb{Z}_q^* \right|.$$

5.1.3 The Boneh-Lynn-Shacham Short Signature Scheme

Let G_1, G_2, and G_T be cyclic groups of order q and $e : G_1 \times G_2 \to G_T$ be a cryptographic bilinear map. Let P and Q be the generators of G_1 and G_2, receptively. Consider a hash function $H : \{0,1\}^* \to G_1$, which is public for all. The BLS short signature scheme works as follows:

Key generation	Signature generation	Signature verification
Choose a random number $x \in \mathbb{Z}_q^*$ Compute $A = xP \in G_2$ The public key is (P,A) and the secret key is x.	Given a secret key x and a message $m \in \{0,1\}^*$, Compute $s = xH(m) \in G_1$. (m,s) is the signed message on m.	Given a public key (P,A), a message m and a signature s, if $e(Q,s) = e(A, H(m))$, then the signature is a valid signature, otherwise the signature is invalid

The signature s is an element of G_1. Hence to construct a short signature, the elements of the group G_1 must have a short representation. In [51], Bonehet et al. constructed such group over elliptic curve E modulo q, where $q = 3^l$, $l \geq 1$ (i.e., G_1 and G_2 are the subgroups of the group of points on E). In that case, instead of taking s as a signature, one can store the x-coordinate of s (say σ) as signature. In the verification, one can consider a point s having σ as x-coordinate. Since there are two points having σ as x-coordinate, so it is possible picked $-s$ too. Therefore, a signature is considered as a valid signature even if $e(s, Q)^{-1} = e(A, H(m))$. For details, see [51].

Efficiency: The key generation algorithm of the BLS signature scheme required one scalar multiplication in G_1. Again, it required one hash operation and one scalar multiplication in G_1 to generate a signature. The verification algorithm required one hash operation and two pairing computation.

Security: The BLS short signature scheme is secure against existential forgery under adaptive chosen message attack in the random oracle model assuming the CDH problem is hard in G_1 [51]. Note that G_1 is a GAP group, i.e., there is an efficient PPT algorithm to solve DDH problem, but hard to solve CDH problem in G_1. One can see [51] for the security analysis of the BLS signature scheme.

5.2 The Boneh-Boyen Short Signature Scheme

In 2004 [54], Boneh et al. described a short signature scheme based on pairing. The length of the signature generated by their scheme is the same as [51], but it is more efficient. Also, the security of their scheme is based on the SDH problem without using a random oracle model. Boneh et al. present two versions of their signature scheme, a basic scheme and a full scheme. Before presenting the schemes, let us recall some definitions.

5.2.1   Computational Assumption

The following definitions are considered from [54, 60].

Definition 1: Let G_1, G_2 be two cyclic groups (not necessarily distinct) of prime order p. Let g_1, g_2 be the generator of G_1 and G_2, respectively. For a randomly chosen element $x \in \mathbb{Z}_p^*$ and a given $(q+2)$-tuple $\left(g_1, g_2, g_2^x, \cdots, g_2^{x^q}\right) \in G_1 \times G_2^{q+1}$, the strong Diffie-Hellman (q-SDH) problem in (G_1, G_2) is asked to find a pair $\left(c, \; g_1^{\frac{1}{x+c}}\right)$ where $c \in \mathbb{Z}_p^*$.

Note that an algorithm $\mathcal{A}$ has advantage ϵ in solving q-SDH in (G_1, G_2) if

$$\Pr\left[\mathcal{A}\left(g_1, g_2, g_2^x, g_2^{x^2}, \cdots, g_2^{x^q}\right) = \left(c, g_1^{\frac{1}{x+c}}\right)\right] \geq \varepsilon$$

where the probability is over the random choice of x in $\mathbb{Z}_p^*$ and the random bits consumed by $\mathcal{A}$.

Definition 2: If there is no t- time algorithm has the advantage at least ε in solving the q-SDH problem in (G_1, G_2), then we called (q, t, ε)-SDH (or q-SDH) assumption holds in (G_1, G_2).

5.2.2   The Boneh-Boyen Short Signature Scheme

The Basic Signature Scheme

Let (G_1, G_2) be bilinear groups of prime order p. Let g_1 be a generator of G_1 and g_2 is a generator of G_2.

Key Generation
1. Choose a random number $x \in \mathbb{Z}_p^*$
2. Compute $A = g_2^x \in G_2$

The public key is (g_1, g_2, A) and the secret key is x.

Signature Generation: Suppose $m \in \mathbb{Z}_p^*$ be the messages to be signed.

1. Compute $s = g_1^{\frac{1}{x+m} \bmod p} \in G_1$. If $x + m \equiv 0 \bmod p$, then $s = 1$.

(m, s) is the signed message on m.

Signature Verification
For the signed message (m, s), verify that

$$e\left(s, A \cdot g_2^m\right) = e\left(g_1, g_2\right)$$

The signature is a valid signature if equality holds. Again, for $s = 1$, if $A \cdot g_2^m = 1$, then the signature is a valid signature.

The Full Signature Scheme

Let (G_1, G_2) be bilinear groups of prime order p. Let g_1 and g_2 be the generators of G_1 and G_2, respectively. Suppose that a message m is considered from $\mathbb{Z}_p^*$. The domain can be extended to all of $\{0, 1\}^*$ using a collision-resistant hash function $H : \{0, 1\}^* \rightarrow \mathbb{Z}_p$ [54].

Key Generation
1. Choose two random numbers $x, y \in \mathbb{Z}_p^*$.
2. Compute $A = g_2^x \in G_2$ and $B = g_2^y \in G_2$.

The public key is (g_1, g_2, A, B), and the secret key is (x, y).
Signature Generation: For a given message $m \in \mathbb{Z}_p^*$,

1. Choose a random number $r \in \mathbb{Z}_p^*$.
2. Compute $s = g_1^{\frac{1}{x+m+yr} \bmod p} \in G_1$. If $x + m + yr \equiv 0 \bmod p$, then choose a different random r.

$(m, (s, r))$ is the signed message on m.
Signature Verification
For the signed message (m, s), verify that

$$e\left(s, A \cdot g_2^m \cdot B^r\right) = e(g_1, g_2)$$

The signature is a valid signature if the equality holds, otherwise the signature is invalid.

Efficiency: The key generation and signature generation algorithm required the same times with BLS signatures. The verification algorithm requires only one pairing and one multi-exponentiation, and therefore it is faster than the BLS signature scheme. One can compute the value $e(g_1, g_2)$ at initialization time. A signature produced by the Boneh-Boyen signature scheme contains two elements (s, r), where the length of each of the elements has approximately $\log_2 p$ bits, therefore the size of the signature is approximately $2\log_2 p$ [54].

Security: The scheme is secure against existential forgery under chosen message attacks provided that the q-SDH assumption holds in (G_1, G_2) [54]. The basic signature scheme is secure against existential forgery under a weak chosen message attack [54]. For details, see [54]. In [61], it is proved that the forging the Boneh-Boyen signature is equivalent to solving the q-SDH problem.

6 Hierarchical Identity-Based Signature Scheme

In 1984 [62], Shamir introduced the notion of identity-based signature (IBS) scheme. In an IBS scheme, a public key can be extracted from user ID, and a corresponding secret key can be assessed by the public key generator (PKG).

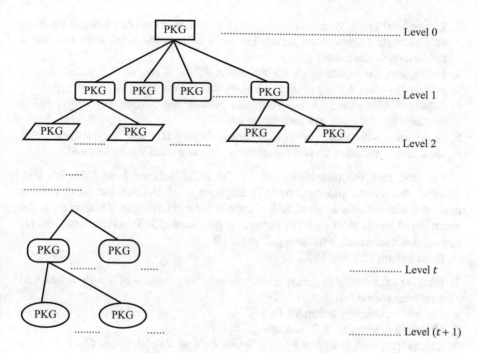

A PKG at Level i generates the private key of a PKG at Level $(i + 1)$ and an entity at Level i is identified by ID-tuple $(ID_1, ID_2, ...,ID_i)$.

Fig. 14.2 Protocol for HIBS scheme

Numerous IBS schemes have been presented (e.g., [63–68]). Since a single PKG is used in IBS schemes, so they are impractical for a large organization. The hierarchical identity-based signature (HIBS) scheme has removed this problem by introducing multiple PKGs, which are arranged in a tree structure (Fig. 14.2). Each PKG can generate secret keys for its next lower-level PKGs. Hence it diminishes the workload of the root PKG and is very useful for an enormous group.

In 2002 [69], Gentry et al. presented the first HIBS scheme, which is based on the difficulty of solving Bilinear Diffie-Hellman (BDH) problem. In 2004 [70], Chow et al. proposed a HIBS scheme, which is provably secure under a random oracle model. Various HIBS schemes have been proposed since it is introduced along with their security (e.g., [19, 71–77]).

A HIBS scheme comprises of five algorithms [69]:

1. Root setup: For a security parameter ℓ, the root PKG generates $(msk, \ param)$, where the secret key of the root PKG is msk and $param$ is the system parameters, which is publicly available. The description of message space $\mathcal{M}$ and the signature space $\mathcal{S}$ is also contained in a system parameter.

2. Lower-level setup: With the knowledge of system parameters, a lower-level PKG can generate a lower-level secret. For each extraction, it can also generate a random one-time secret.
3. Extraction: On inputting an ID-tuple of a PKG, it returns the corresponding private key SK_{ID} for any of its children by using *param* and its private key.
4. Signature Generation: A signer inputs *param*, the private key SK_{ID}, and a message M, and generates a signature $s \in \mathcal{S}$.
5. Signature Verification: For a given signer identity ID-tuple, a message M, and signature s, a verifier accepts the signature if s is a valid signature on M.

One can read the security notion for the HIBS scheme from [69]. We now consider the scheme proposed by [69]. Suppose level$_t$ denotes the set of entities at level t, where level$_0$ = {rootPKG}. Suppose for a given security parameter ℓ, the setup algorithm produces a BDH parameter generator $\mathcal{IG}$. Two hash functions H_1 and H_2 are considered, which are public for all.

Root Setup: The root PKG

1. For given ℓ, runs $\mathcal{IG}$ to generate two prime order groups G_1, G_2 with order q and a bilinear map $e : G_1 \times G_2 \to G_2$.
2. Choose an arbitrary generator $P_0 \in G_1$.
3. Choose a random $s_0 \in \mathbb{Z}_q$ and sets $Q_0 = s_0 P_0$.
4. Choose two hash functions $H_1 : \{0, 1\}^* \to G_1$ and $H_2 : \{0, 1\}^* \to G_1$.

The signature scheme space is $\mathcal{S} = G_1^{t+1}$, where t denotes the level of signer. The system parameters *param* $= (G_1, G_2, e, P_0, Q_0, H_1, H_2)$ and $s_0 \in \mathbb{Z}_q$ is the secret key of root PKG.

Lower-level setup: Entity $E_t \in$ Level$_t$ picks an integer $s_t \in \mathbb{Z}_q$, which it keeps secret.

Extraction: Let E_t be an entity in Level$_t$ with ID-tuples $(ID_1, ID_2, \cdots, ID_t)$ where $(ID_1, ID_2, \cdots, ID_i)$ for $1 \le i \le t$ is the ID-tuple of E_t's ancestor at level Level$_i$. Suppose, S_0 denotes the identity element of G_1. Then E_t's parent:

1. computes $P_t = H_1(ID_1, ID_2, \cdots, ID_t) \in G_1$
2. sets E_t's secret point S_t to be $S_{t-1} + s_{t-1} P_t = \sum_{i=1}^t s_i P_i$ and
3. also gives E_t the values of $Q_i = s_i P_0$, $1 \le i \le t - 1$.

Signing: To sign a message M with ID-tuple $(ID_1, ID_2, \cdots, ID_t)$, the signer at level$_t$ executes the following steps:

1. Compute $P_M = H_2(ID_1, ID_2, \cdots, ID_t, M) \in G_1$
2. Compute $Sig(ID - tuple, M) = S_t + s_t P_M$
3. Send $Sig(ID - tuple, M)$ and $Q_i = s_i P_0$, $1 \le i < t$.

$[Sig, Q_1, \cdots, Q_t] \in \mathcal{S}$ is the signature for $(ID - tuple, M)$.

Verification: For a signature $[Sig, Q_1, \cdots, Q_t] \in \mathcal{S}$ on a message($ID - tuple$, M), if $e(P_0, Sig) = e(Q_0, P_1) e(Q_0, P_M) \Pi_i^t e(Q_{i-1}, P_i)$, then the signature is a valid signature.

Note that the size of the signature is proportional to the depth of the signer in the hierarchy [69]. In the same paper, the author discussed some methods to reduce the length of the signature.

7 Other Signature Schemes

There are some other notions of signature schemes, which are beyond the scope of this chapter. We will briefly discuss some of them for interested readers.

One-time signature (OTS): Lamport [78] proposed the first OTS scheme. The OTS scheme can be used to sign one message per key pair. The signature can be verified an arbitrary number of times. It can be constructed from one-way function and the signature generation and verification algorithm is fast.

Online/off-line signature: Even et al. [79] proposed the concept of online/off-line signature scheme. Here the signature generation algorithm performs two faces: online and off-line face. The first face is off-line and when message to be signed is known, then the online face is executed. It is useful for smart card application.

Group signature: In 1991 [80], Chaum et al. proposed the first Group signature scheme. In a group signature scheme, any member of the groups can sign messages in place of the group, and the receiver can verify the signature, but cannot discover which member of the group has signed. If required, the trusted authority can identify the signer with or without the help of group members.

Ring signature scheme: Rivest et al. [81] introduce the notion of ring signature scheme. This is similar with the group signature scheme. But in this scheme, there is no way to identify the signer.

Threshold signature scheme (TSS): Here, the secret key is assigned among n groups with or without the help of trusted authority by running an interactive protocol among all groups. In a (t, n)-threshold signature scheme, any t or more signer can sign a message in place of a group of n singers. It does not reveal the identification of group members who has signed the message.

Multisignature scheme: Here, a group of signers can together sign a message such that a verifier is convinced that each of the subgroup engaged in signing. This is more compact than a collection of signatures from all signers.

Unique signature scheme: Goldwasser et al. [82] introduced the notion of unique signature scheme, which is also called as invariant signature scheme. It is a building block of constructing a verifiable random function.

Proxy signature scheme: Mambo et al. [83], proposed the notion of proxy signature scheme. Here, a proxy signer is allowed to sign a message on behalf of an original signer within a given context.

8 Signcryption

In 1997 [84], Zheng proposed the idea of a new protocol in asymmetric key cryptography called signcryption. The main idea of this protocol is to transmit a message in a secure and authenticated way with more efficiency. That is a signcryption scheme must satisfy the properties of encryption and digital signature scheme. Note that a sign-and-encrypt scheme also ensures the confidentiality and authenticity of a message, where a sender first signed the message and the signed message is encrypted (called sign-and-encrypt alg.), whereas the receiver first decrypts the message and then verify (called decrypt-and-verify alg.). The main motivation of a signcryption scheme is to construct a scheme that has the same properties with a sign-and-encrypt scheme but with lower computational cost compared to sign-and-encrypt scheme. In a signcryption scheme, both the encryption and signature generation algorithms are performed together in a well-organized way.

A signcryption scheme comprises of three algorithms:

1. Key generation: It is a randomized algorithm that for given 1^k, where k is a security parameter produces a pair of key (K_{pub}, K_{pr}), K_{pub} is the public key and K_{pr} is the corresponding private key. Both the sender and the recipient of a message must execute the key generation algorithm before they can communicate. Suppose the key pair for sender is $\left(K_{pub_S}, K_{pr_S}\right)$ and the receiver is $\left(K_{pub_R}, K_{pr_R}\right)$.
2. Signcryption Algorithm: It is a randomized algorithm that for given a message M, the sender's secret key K_{pr_S}, and the receiver's public key K_{pub_R} produce a signcrypted message C.
3. Unsigncryption Algorithm: It is a deterministic algorithm that for the given signcrypted message C, sender's public key K_{pub_S} and the receiver's secret key K_{pr_R} produce the message M.

Signcryption is based on a shortened DSS. SDSS1 and SDSS2 are two different shortened digital signature schemes which are obtained by applying the shortening method. Both SDSS1 and SDSS2 schemes are described in Table 14.1. We considered the following table from [85].

Note that both SDSS1 and SDSS2 schemes produce a signature of length $|H(\cdot)| + |q|$, whereas the signature generated by DSS is $2|q|$. Now, we describe Signcryption scheme with Shortened signature [85] as follows:

Key Generation
1. Choose a prime p of large size.
2. Choose a prime factor q of $p - 1$.
3. Choose $g \in \mathbb{Z}_p$ such that ord $(g) = q$.
4. Choose a one-way hash function H whose output has at least 128 bits.
5. Choose a keyed hash function KH.
6. The (E, D) is the encryption and decryption algorithms of a symmetric cipher.

Table 14.1 SDSS1 and SDSS2 Schemes

Algorithms	SDSS1	SDSS2
Key generation	1. Choose two large primes p and q such that $q \mid (p-1)$. 2. Choose $g \in \mathbb{Z}_p^*$ such that $\text{ord}(g) = q$ 3. Choose a one-way hash function H 4. Choose $e_a \in \{1, \cdots, q-1\}$ and compute $d_a = g^{e_a} \bmod p$ Public key (p, q, g, e_a, H) and the corresponding private key d_a.	
Signature generation	For a message M, Choose $d \in \{1, \cdots, q-1\}$ Compute $r = H\left(g^{d'} \bmod p, M\right)$ $s = d(r + d_a)^{-1} \bmod q$	For a message M, Choose $d \in \{1, \cdots, q-1\}$ Compute $r = H\left(g^{d'} \bmod p, M\right)$ $s = d(1 + d_a r)^{-1} \bmod q$
Signature verification	For a signature (r, s) and message M, compute $k = (e_a g^r)^s \bmod p$ Verify if $r = H(k, M)$	For a signature (r, s) and message M, compute $k = \left(g e_a^r\right)^s \bmod p$ Verify if $r = H(k, M)$

The sender executes the following steps to construct a key pair.

1. Choose a random integer $d_a \in [1, \ p-1]$.
2. Compute $e_a \equiv g^{d_a} \bmod p$.

e_a is the sender public key and d_a is the sender corresponding secret key. The receiver executes the following steps to a construct key pair.

1. Choose a random integer $d_b \in [1, \ p-1]$.
2. Compute $e_b \equiv g^{d_b} \bmod p$.

e_b is the receiver public key and d_b is the corresponding secret key.
Signcryption Algorithm:

1. Choose a random integer $d \in [1, \ q-1]$.

2. $(k_1, k_2) = H\left(e_b^d \bmod p\right)$

3. $c = E_{k_1}(m)$

4. $r = KH_{k_2}(m)$

5. $s = \begin{cases} d(r + d_a)^{-1} \bmod q & \text{if SDSS1 is used} \\ d(1 + r d_a)^{-1} \bmod q & \text{if SDSS2 is used} \end{cases}$

The signcrypted message is (c, r, s)

Unsigncryption Algorithm:

1. Recover $(k_1, k_2) = \begin{cases} H\left((e_a g^r)^{sd_b} \bmod p\right) & \text{if SDSS1 is used} \\ H\left((g e_a^r)^{sd_b} \bmod p\right) & \text{if SDSS2 is used} \end{cases}$

2. $m = D_{k_1}(c)$

3. If $KH_{k_2}(m) = r$, then accept m.

Note that there are two signcryption schemes described together, one is named SCS1 when SDSS1 is used and the other is called SCS2 [85]. The signcryption algorithm is more efficient than sign-and-encrypt scheme in the sense that it required only one modular exponentiation, whereas in sign-and-encrypt scheme, at least one or more modular exponentiations are required separately for each signature generation and encryption algorithm. The unsigncryption algorithm required two modular exponentiations, whereas decrypt-and-verify algorithm required at least one or more modular exponentiations for each decryption and verification algorithm. The advantages of signcryption over sign-and-encrypt scheme are thoroughly discussed in [85]. For the completeness of this section, we have mentioned Table 14.2, which shows the advantage of SCS1 and SCS2 over RSA-based sign-and-encrypt schemes and DLP-based sign-and-encrypt schemes in terms of average computational cost and communication overhead. Table 14.2 is the combination of Tables 4 and 5 of [85].

A signcryption scheme is considered to be secure if it satisfies the conditions: unforgeability, non-repudiation, and confidentiality. In [85], it is proved that if the keyed hash function (KH) behaves like a random function, then both the SCS1 and SCS2 are unforgeable against adaptive attacks. The detailed description of the security of signcryption can be found in [85]. Some works on signcryption scheme can be found in [86–95], etc.

9 Conclusion

Signature scheme is a cryptographic primitive, which provides authenticity of a message. It is categorized as signature scheme with message recovery and signature scheme with appendix [22]. The signature schemes with hash function are considered in the second category. In this chapter, we have studied the signature schemes with hash function. We have discussed some example of signature scheme whose security is based on IFP and DLP like RSA signature scheme, ElGamal signature scheme, and some of its variants.

We have also presented some special signature schemes. Blind signature schemes, where the singer and the user are two different parties and the signer does not know about the message. In this chapter, we have described Chaum's RSA blind signature scheme, Schnorr blind signature scheme, and Okomoto-Schnorr blind signature scheme.

Table 14.2 Advantage of SCS1 and SCS2 in terms of comp. Cost and comm. Overhead

| Length of security parameters | | | Adv. over sign-and-encrypt scheme based on | | | |
| | | | RSA with small e | | Schnorr signature and ElGamal encryption scheme | |
p	q	KH	Adv. in average comp. cost	Adv. in average comm. overhead	Adv. in average comp. cost	Adv. in average comm. overhead
512	144	72	0%	78.9%	58%	70.3%
1024	160	80	32.3%	88.3%	58%	81.0%
1536	176	88	50.3%	91.4%	58%	85.3%
2048	192	96	59.4%	93.0%	58%	87.7%
4096	256	128	72.9%	95.0%	58%	91.0%
8192	320	160	83.1%	97.0%	58%	94.0%
10240	320	160	86.5%	98.0%	58%	96.0%

In an undeniable signature scheme, the verification algorithm depends on the active participation of the signer. It also can check whether a signature signed by the signer is forged or the signer is trying to deny his signature. In this chapter, we have explained two undeniable signature schemes, namely Chaum-Van Antwerper undeniable signature scheme and RSA-based undeniable scheme.

The RSA and DSS are the two most frequently used signature schemes, which produce a signature of large size. But in some situations, one needed a short signature. The concept of short signature scheme is introduced by Boneh et al. We discussed two short signature schemes: Boneh-Lynn-Schacham short signature scheme and Boneh-Boyen short signature scheme.

We also provide a short introduction to hierarchical identity-based signature scheme. We have briefly mentioned some other signature schemes like one-time signature, online/off-line signature, group signature, ring signature, threshold signature, multisignature, unique signature, and proxy signature scheme.

Signcryption is a cryptographic protocol which provide authenticity and confidentiality of a message with more efficiency than the sign-and-encrypt scheme. Zheng proposed the idea of signcryption. We have introduced a short introduction to signcryption.

References

1. H. Tolgay. Cryptography II and secure communication.https://www.hakantolgay.com/files/Cryptography_Day_II.pptx
2. Stinson, D. R. (2013). *Cryptography theory and practice, discrete mathematics and its applications*. Chapman & Hall/CRC, Taylor & Francis Group.
3. Buchmann, J. A. (2004). *Introduction to cryptography. Undergraduate texts in mathematics*. New York: Springer.
4. Pointcheval, D., & Stern, J. (1996). Security proofs for signature schemes. In U. Maurer (Ed.), *Advances in Cryptology - EUROCRYPT 1996.LNCS 1070, (pp. 387–398)*. Berlin, Heidelberg: Springer.
5. Rivest, R. L., Shamir, A., & Adleman, L. (1978). A method for obtaining digital signatures and public key cryptosystem. *Communications of ACM, 21*(2), 120–126.
6. ElGamal, T. (1985). A public-key cryptosystem and a signature scheme based on discrete logarithms. In G. R. Blakley & D. Chaum (Eds.), *Advances in Cryptology - CRYPTO 1984.LNCS 196, (pp. 10–18)*. Berlin: Springer.
7. Schnorr, C. P. (1991). Efficient signature generation by smart cards. *Journal of Cryptology, 4*, 161–174.
8. Galbraith, S. D. (2012). *Mathematics of public key cryptography*. Cambridge University Press.
9. Nyberg, K., & Rueppel, R. (1995). Message recovery for signature schemes based on the discrete logarithm problem. In A. De Santis (Ed.), *Advances in Cryptology - EUROCRYPT 1994.LNCS 950* (pp. 182–193). Springer.
10. Digital Signature Standard. Federal Information Processing Standards Publications 186, NIST; 1994.
11. Secure Hash Function (SHA-1). Fderal Information Processing Standards Publications 180–1, NIST; 1995.
12. Chaum, D. (1983). Blind signatures for untraceable payments. In D. Chaum, R. L. Rivest, & A. T. Sherman (Eds.), *Advances in Cryptology (pp. 199–203)*. Boston, MA: Springer.

13. Camenisch, J., Piveteau, M., & Stadler, M. (1994). Blind signatures based on the discrete logarithm problem. In A. De Santis (Ed.), *Advances in Cryptology -EUROCRYPT 1994.LNCS 950, (pp. 428–432)*. Berlin, Heidelberg: Springer.
14. Mohammed, F., Emarah, A. E., & El-Shennawy, K. (2000). A blind signature scheme based on ElGamal signature. In *IEEE/AFCEA EUROCOMM 2000 Information Systems for Enhanced Public Safty and Security* (pp. 51–53).
15. Pointcheval, D., & Stern, J. (1996). Provably secure blind signature schemes. In K. Kim & T. Matsumoto (Eds.), *Advances in Cryptology - ASIACRYPT 1996.LNCS 1163, (pp. 252–265)*. Berlin, Heidelberg: Springer.
16. Okamoto, T. (1993). Provably secure and practical identification schemes and corresponding signature schemes. In E. F. Brickell (Ed.), *Advances in Cryptology - CRYPTO 1992.LNCS 740, (pp. 31–53)*. Berlin: Springer.
17. Fan, C.-I., & Lei, C. L. (1996). Efficient blind signature scheme based on quadratic residues. *Electronics Letters, 32*(9), 811–813.
18. Gao, W., Hu, Y., & Liu, M. (2017). Identity-based blind signature from lattices. *Wuhan University Journal of Natural Sciences, 22*, 355–360.
19. Ruckert, M. (2010). Lattice-based blind signatures. In M. Abe (Ed.), *Advances in Cryptology - ASIACRYPT 2010.LNCS 6477, (pp. 413–430)*. Berlin, Heidelberg: Springer.
20. Wang, F., Hu, Y. P., & Wang, C. X. (2010). A lattice-based blind signature scheme. *Geometrics and Information Science of Wuhan University, 35*(5), 550–553.
21. Zhang, F., & Kim, K. (2003). Efficient ID-based blind signature and proxy signature from bilinear pairings. In *8th Australasian Conference on Information Security and Privacy - ACISP 2003* (pp. 312–323). Springer-Verlag.
22. Das, A., & Madhavan, C. E. V. (2009). *Public-key cryptography : Theory and practice*. Delhi: Pearson Education.
23. Juels, A., Luby, M., & Ostrovsky, R. (1997). Security of blind digital signatures. In B. S. Kaliski (Ed.), *Advances in Cryptography - CRYPTO 1997.LNCS 1294, (pp. 150–164)*. Berlin, Heidelberg: Springer.
24. D. Pointcheval. Strengthened security for blind signature. In K. Nyberg (Ed.), Advances in Cryptology - EUROCRYPT 1998.LNCS 1403, (pp. 391–405), Springer, Berlin, Heidelberg, 1998.
25. Pointcheval, D., & Stern, J. (2000). Security arguments for digitl signatures and blind signatures. *Journal of Cryptology, 13*(3), 361–396.
26. Schroder, D., & Unruh, D. (2012). Security of blind signatures revisited. In M. Fischlin, J. Buchman, & M. Manulis (Eds.), *Public Key Cryptography - PKC 2012.LNCS 7293, (pp. 662–679)*. Berlin, Heidelberg: Springer.
27. Fujioka, A., Okamoto, T., & Ohta, K. (1992). A practical secret voting scheme for large scale elections. In J. Seberry & Y. Zheng (Eds.), *Advances in Cryptology - AUSCRYPT 1992.LNCS 718, (pp. 244–251)*. Berlin, Heidelberg: Springer.
28. Chaum, D., & van Antwerpen, H. (1990). Undeniable signatures. In G. Brassard (Ed.), *Adavances in Cryptology - CRYPTO 1989.LNCS 435, (pp. 212–216)*. New York, NY: Springer.
29. Camenisch, J., & Michels, M. (2000). Confirmer signature schemes secure against adaptive adversaries. In B. Preneel (Ed.), *Advances in Cryptology - EUROCRYPT 2000.LNCS 1807, pp. 243–258*. Berlin, Heidelberg: Springer.
30. Chaum, D. (1990). Zero-knowledge undeniable signatures (extended abstract). In I. B. Damgard (Ed.), *Advances in Cryptology - EUROCRYPT 1990.LNCS 473, (pp. 458–464)*. Berlin, Heidelberg: Springer.
31. Chaum, D., van Heijst, E., & Pfitzmann, B. (1992). Cryptographically strong undeniable signatures, unconditionally secure for the signer. In J. Feigenbaum (Ed.), *Advances in Cryptology - CRYPTO 1991.LNCS 576, (pp. 470–484)*. Berlin, Heidelberg: Springer.
32. Michels, M., & Stadler, M. (1997). Efficient convertible undeniable signature schemes (extended abstract). In *Selected Area in Cryptography - SAC 1997* (pp. 231–244). Ottawa.

33. Boyar, J., Chaum, D., Damgard, I., & Pedersen, T. (1991). Convertible undeniable signature. In A. J. Menezes & S. Vanstone (Eds.), *Advances in Cryptology - CRYPTO 1990.LNCS 537, (pp. 189–205)*. Berlin, Heidelberg: Springer.
34. Gennaro, R., Krawczyk, H., & Rabin, T. (1997). RSA-based undeniable signatures. In B. S. Kaliski (Ed.), *Advances in Cryptology - CRYPTO 1997.LNSC 1294, (pp. 132–149)*. Berlin, Heidelberg: Springer.
35. Galbraith, S. D., & Mao, W. (2003). Invisibility and anonymity of undeniable and confirmer signatures. In H. Joye (Ed.), *Topics in Cryptology - CT-RSA 2003.LNSC 2612, (pp. 80–97)*. Berlin: Springer.
36. Galbraith, S. D., Mao, W., & Paterson, K. G. (2002). RSA-based undeniable signatures for general moduli. In B. Preneel (Ed.), *Topics in Cryptology - CT-RSA 2002.LNCS 2271, pp. 200–217*. Berlin, Heidelberg: Springer.
37. Miyazaki, T. (2000). An improved scheme of the Gennaro-Krawczyk-Rabin undeniable signature system based on RSA. In D. Won (Ed.), *Information Security and Cryptology - ICISC 2000.LNCS 2015, (pp. 135–149)*. Berlin, Heidelberg: Springer.
38. Aguilar-Melchor, C., Bettaieb, S., Gaborit, P., & Schrek, J. (2013). A code-based undeniable signature scheme. In M. Stam (Ed.), *14th IMA International Conference on Cryptography and Coding IMACC 2013.LNCS 8308* (pp. 99–119). Berlin: Springer.
39. Libert, B., & Quisquater, J.-J. (2004). Identity based undiniable signatures. In T. Okamoto (Ed.), *Topic in Cryptology - CT-RSA 2004.LNCS 2964, pp. 112–125*. Berlin, Heidelberg: Springer.
40. Monnerat, J., & Vaudenay, S. (2004). Generic homomorphic undeniable signatures. In P. J. Lee (Ed.), *Advances in Cryptology - ASIACRYPT 2004.LNCS 3329, (pp. 354–371)*. Berlin, Heidelberg: Springer.
41. J. Monnerat, and S. Vaudenay. Undeniable signatures based on characters: How to sign with one bit. In F. Bao, R. Deng, & J. Zhou (Ed.), Public key cryptography - PKC 2004.LNCS 2947, pp. 69–85. Springer, Berlin, Heidelberg, 2004.
42. Desmedt, Y., & Yung, M. (1991). Weakness of undeiable signature schemes. In D. Davies (Ed.), *Advances in Cryptology - EUROCRYPT 1991.LNCS* (Vol. 547, pp. 205–220). Berlin, Heidelberg: Springer.
43. Behnia, R., Heng, S. H., & Gan, C. S. (2015). An efficient certificateless undeniable signature scheme. *International Journal of Computer Mathematics, 92*(7), 1313–1328.
44. Huang, Q., & Wong, D. S. (2013). Short and efficient convertible undeniable signature schemes without random oracles. *Theoritical Computer Sciences, 476*, 67–83.
45. K. Kurosawa, and S-H. Heng. 3-Move undeniable signature scheme. In R. Cramer (Ed.), Advances in cryptology - EUROCRYPT 2005, LNCS 3494, (pp. 181–197), Springer, *Berlin, Heidelberg* 2005.
46. Zhu, H. Universal Undeniable Signatures. In *IACR Cryptology ePrint Archive, Report 2004/005*. http://eprint.iacr.org/2004/005.
47. Naccache, D., & Stern, J. (2001). Signing on a postcard. In Y. Frankel (Ed.), *Financial Cryptography - FC 2000.LNCS 1962, (pp. 121–135)*. Berlin, Heidelberg: Springer.
48. I. Mirnov, A short signature scheme as secure as DSA. Preprint, 2001.
49. Nyberg, K., & Ruepple, R. A. (1993). A new signature scheme based on DSA giving message recovery. In *1st ACM Conference on Communication and Computer Security - CCS 1993* (pp. 58–61). USA.
50. Tso, R., Gu, C., Okamoto, T., & Okamoto, E. (2007). Efficient ID-based digital signatures with message recovery. In F. Bao, S. Ling, T. Okamoto, H. Wang, & C. Xing (Eds.), *Cryptology and Network Security, CANS 2007.LNCS 4856, (pp. 47–59)*. Berlin, Heidelberg: Springer.
51. Boneh, D., & Lynn, B. (2001). andH. Shacham. Short signature from the Weil pairing. In C. Boyd (Ed.), *Advances in Cryptology - ASIACRYPT 2001.LNSC 2248, (pp. 514–532)*. Berlin, Heidelberg: Springer.
52. Choon, J. C., & Hee Cheon, J. (2003). An identity-based signature from gap Diffie-Hellman groups. In Y. Desmedt (Ed.), *Public Key Cryptography - PKC 2003.LNCS 2567, (pp. 18–30)*. Berlin, Heidelberg: Springer.

53. Goh, E., & Jarecki, S. (2003). A signature scheme as secure as the Diffie-Hellman problem. In E. Biham (Ed.), *Advances in Cryptology - EUROCRYPT 2003.LNCS 2656, (pp. 401–415)*. Berlin: Springer.

54. Boneh, D., & Boyen, X. (2004). Short signatures without random oracles. In C. Cachin & J. L. Camenisch (Eds.), *Advances in Cryptology - EUROCRYPT 2004.LNCS 3027* (pp. 56–73). Berlin: Springer.

55. Akleylek, S., Kirlar, B. B., Sever, O., & Yuce, Z. (2011). Short signature scheme from bilinear pairings. *Journal of Telecomunication and Information Technology, 1-13*.

56. Boneh, D., Boyen, X., & Shacham, H. (2004). Short group signatures. In M. Franklin (Ed.), *Advances in Cryptology - CRYPTO 2004.LNCS 3125, (pp. 41–55)*. Berlin: Springer.

57. Ng, T., Tan, S., & Chin, J. (2018). A Variant of BLS Signature Scheme with Tight Security Reduction. In J. Hu, I. Khalil, Z. Tari, & S. Wen (Eds.), *Mobile Networks and Management MONAMI 2017.Lecture Notes of the Institue for Computer Sciences, Social Informatics and Telecommunications Engineering 235* (pp. 150–163). Cham: Springer.

58. Zhang, F., Safavi-Naini, R., & Susilo, W. (2004). An efficient signature scheme from bilinear pairing and its applications. In F. Bao, R. Deng, & J. Zhou (Eds.), *Public Key Cryptography - PKC 2004.LNSC 2947, (pp. 277–290)*. Berlin, Heidelberg: Springer.

59. Yu, P., & Xue, R. (2011). A short signature scheme from the RSA-family. In M. Burmester, G. Tsudik, S. Magliveras, & I. Ilic (Eds.), *Information Security. ISC 2010.LNSC 6531, (pp. 307–318)*. Berlin, Heidelberg: Springer.

60. Tanaka, N., & Saito, T. On the q-Strong Diffie-Hellman problem. In *IACR Cryptology ePrint Archive, Report 2010/215*. http://eprint.iacr.org/2010/215.

61. Jao, D., & Yoshida, K. (2009). Boneh-Boyen signatures and the strong Diffie-Hellman problem. In H. Shacham & B. Waters (Eds.), *Pairing-Based Cryptography- Pairing 2009.LNCS 5671, (pp. 1–16)*. Berlin: Springer.

62. Shamir, A. (1985). Identity-based cryptosystems and signature schemes. In G. R. Blakley & D. Chaum (Eds.), *Advances in Cryptology - CRYPTO 1984.LNCS 196, (pp. 47–53)*. Berlin, Heidelberg: Springer.

63. Barreto, P., Libert, B., McCullagh, N., & Quisquater, J. (2005). Efficient and provably-secure identity-based signatures and signcryption for bilinear maps. In B. Roy (Ed.), *Advances in Cryptology - ASIACRYPT 2005.LNCS 3788* (pp. 515–532). Berlin: Springer.

64. Chai, Z., Cao, Z., & Dong, X. (2007). Identity-based signature scheme based on quadratic residues. *Sci China Ser F, 50*, 373–380. https://doi.org/10.1007/s11432-007-0038-1.

65. Hess, F. (2003). Efficient identity-based signature schemes based on pairings. In K. Nyberg & H. Heys (Eds.), *Selected Areas in Cryptography - SAC 2002.LNCS 2595, (pp. 310–324)*. Berlin: Springer.

66. Huang, X., Susilo, W., Mu, Y., & Zhang, F. (2008). Short designated verifier signature scheme and its identity-based variant. *Internatonal Journal of Network Security, 6*(1), 82–93.

67. Paterson, K., & Schuldt, J. (2006). Efficient identity-based signatures secure in the standard model. In L. Batten & R. Safavi-Naini (Eds.), *The 11th Australasian Conference on Information Security and Privacy - ACISP 2006.LNCS 4058, (pp. 207–222)*. Berlin, Heidelberg: Springer.

68. Xiong, H., Qin, Z., & Li, F. (2010). Identity-based threshold signature secure in the standard model. *International Journal of Network Security, 10*(1), 75–80.

69. Gentry, C., & Silverberg, A. (2002). Hierarchical ID-based cryptography. In Y. Zheng (Ed.), *Advances in Cryptology - ASIACRYPT 2002.LNCS 2501, (pp. 548–566)*. Berlin, Heidelberg: Springer.

70. Chow, S., Hui, L., Yiu, S., & Chow, K. (2004). Secure hierarchical identity based signature and its application. In J. Lopez, S. Qing, & E. Okamoto (Eds.), *Information and Communications Security ICICS 2004.LNCS 3269, (pp. 480–494)*. Berlin, Heidelberg: Springer.

71. Au, M., Liu, J., Yuen, T., & Wong, D. Practical hierarchical identity based encryption and signatue schemes without random oracles. In *IACR Cryptology ePrint Archive, Report 2006/368*. http://eprint.iacr.org/2006/368.

72. Au, M., Liu, J., Yuen, T., & Wong, D. Efficient hierarchical identity based signature in the standard model. In *IACR Cryptology ePrint Archive, Report 2007/068*. http://eprint.iacr.org/2007/068.

73. Tian, M., Huang, L., & Yang, W. (2012). A new hierarchical identity-based signature scheme from lattices in the standard model. *International Journal of Network Security, 14*(6), 310–315.

74. Ye, F., Qian, Y., & Hu, R. (2015). HIBaSS: Hierarchical identity-based signature scheme for AMI downlink transmission. *Security and Communication Networks, 8*(16), 5262–5277.

75. Yuen, T. H., & Wei, V. K. Constant-size hierarchical identity-based signature/ signcryption without random oracles. In *IACR Cryptology ePrint Archive, Report 2005/412*. http://eprint.iacr.org/2005/412.

76. L. Zhang, Y. Hu, andQ. Wu,"New constraction of short hierarchical ID-based signature in the standard model" Fundamenta Informaticae, *vol. 90, no.* 1–2, 191–201, 2009.

77. Zhang, L.-Y., & Wu, Q. (2010). Adaptively secure hierarchical identity-based signature in the standard model. *The Journal of China Universities of Posts and Telecommunications, 17*(6), 95–100.

78. Lamport, L. (1979). Constructing digital signatures from a one-way function. In *SRI International: Technical Report CSL-98*.

79. Even, S., Goldreich, O., & Micali, S. (1990). On-line/ Off-line digital signatures. In G. Brassard (Ed.), *Advances in Cryptology - CRYPTO 1989.LNCS 435, pp. 263–277*. New York, NY: Springer.

80. Chaum, D., & van Heyst, E. (1991). Group signatures. In D. Davies (Ed.), *Advances in Cryptology- EUROCRYPT 1991.LNCS 547, (pp. 257–265)*. Berlin, Heidelberg: Springer.

81. Rivest, R. L., Shamir, A., & Tauman, Y. (2001). How to leak a secret. In C. Boyed (Ed.), *Advances in Cryptology - ASIACRYPT 2001.LNCS 2248, (pp. 552–565)*. Berlin, Heidelberg: Springer.

82. Goldwasser, S., & Ostrovsky, R. (1993). Invariant signatures and non-interactive zero-knowledge proofs are equivalent (extended abstract). In E. F. Brickell (Ed.), *Advances in Cryptology - CRYPTO 1992.LNCS 740, (pp. 228–245)*. Berlin, Heidelberg: Springer.

83. Mambo, M., Usuda, K., & Okamoto, E. (1996). Proxy signature: Delegation of the power to sign messages. *IEICE Transactions on Fundamentals of Electronics, Communications and Computer Sciences, E79-A*(9), 1338–1354.

84. Zheng, Y. (1997). Digital signcryption or how to achive cost (signature & encryption)<<cost(signature)+cost(encryption). In B. S. Kaliski (Ed.), *Advances in Cryptology - CRYPTO 1997.LNCS 1294, (pp. 169–179)*. Berlin, Heidelberg: Springer.

85. Zheng, Y. (1998). Signcryption and Its Applications in Efficient Public key Solutions. In *Proceedings of the 1st International Workshop on Information Security - ISW 1997* (pp. 291–312).

86. Baek, J., Steinfeld, R., & Zheng, Y. (2002). Formal proofs for the security of signcryption. In D. Naccache & P. Paillier (Eds.), *Public Key Cryptography - PKC 2002.LNCS 2274* (pp. 80–98). Berlin: Springer.

87. F. Bao, and R-H. Deng. A Signcryption scheme with signature directly verifiable by public key. In H. Imai, & Y. Zheng (Ed.), Public key Cryptography - PKC 1998.LNCS 1431, pp. 55–59. Springer, Berlin, Heidelberg 1998.

88. Dodis, Y., Freedman, M., Jarecki, S., & Walfish, S. Optimal Signcryption from any trapdoor permutation. In *IACR Cryptology ePrint Archive, Report 2004/020*. http://eprint.iacr.org/2004/020.

89. Li, F., & Takagi, T. (2013). Secure identity-based Sincryption in the standard model. *Mathematical and Computer Modelling, 57*, 2685–2694.

90. J. Malone-Lee. Identity-based signcryption. IACR Cryptology ePrint Archive, Report 2002/098. http://eprint.iacr.org./2002/098.

91. Petersen, H., & Michels, M. (1998). Cryptoanalysis and improvement of signcryption schemes. *IEE Proceedings - Computers and Digital Techniques, 145*(2), 149–151.

92. Shin, J. B., Lee, K., & Shim, K. (2003). New DSA-verifiable signcryption schemes. In P. J. Lee & C. Lim (Eds.), *Information Security and Cryptology - ICISC 2002.LNCS 2587, pp. 35–47*. Berlin, Heidelberg: Springer.

93. Toorani, M., & Beheshti, A. A. (2010). *An elliptic curve-based signcryption scheme with forward secrecy*. arXive. doi:arXiv:1005.1856.
94. Ullah, I., Amin, N., Khan, J., Rehan, M., Naeem, M., Khattak, H., & Ali, H. (2019). A novel provable secured signcryption scheme PSSS: A hyper-elliptic curve-based approach. *Mathematics, 7*, 1–16.
95. Verma, V., & Gupta, D. (2016). An efficient signcryption algorithm using bilinear mapping. In *3rd International Conference on Computing for Sustainable Global development. IEEE.*
96. Menezes, A., Okamoto, T., & Venstone, S. (1993). Reducing elliptic curve logarithms to logarithms in a finite field. *IEEE Transactions on Information Theory, 39*(5), 1639–1164.
97. Ruckert, M. (2010). Strongly unforgeable signatures and hierarchical identity-based signatures from lattices without random oracles. In N. Sendrier (Ed.), *Post-Quantum Cryptography - PQCrypto 20s10.LNCS 6061, (pp. 182–200).* Berlin, Heidelberg: Springer.
98. Wu, Q., & Zhang, L. (2013). New efficient hierarchical identity-based signature. *Journal of Computers, 8*(3), 803–810.

Chapter 15
QUIET: Quatro-Inverse Exponential Cipher Technique

Harshit Bhatia, Rahul Johari, and Kalpana Gupta

Abstract Majority of the available traditional symmetric key cipher techniques rely on a single key encoding scheme to secure the data before transmitting it over insecure channel. Here, in this chapter a new symmetric shared key cipher technique "QUIET: Quatro-Inverse Exponential Cipher Technique" has been proposed. The technique makes use of multiple keys such as User's Personal keys, his geographical location along with dynamic session key (256 bit to 512 bit), all of which play a very significant role in the process of hiding user's critical information before transmitting it on the network. The newly introduced cryptographic technique is symmetric key technique that deals with encrypting and decrypting the text message by using a set of four keys blended with a set of mathematical operations on those keys. The simulation of the proposed QUIET, Quatro-Inverse Exponential Cipher Technique, was successfully performed in Java programming language and the results, when compared with Triplicative Cipher technique, have been positive and encouraging.

Keywords Cryptography · Symmetric · Encryption · Decryption · Triplicative · Exponential

H. Bhatia (✉)
REVAL India Private Limited, Gurugram, India

R. Johari
SWINGER (Security, Wireless IoT Network Group of Engineering and Research) Lab, USICT, GGSIP University, Sector-16C, Delhi, India
e-mail: swinger@ipu.ac.in

K. Gupta
Centre for Development of Advanced Computing (C-DAC), Noida, India

© Springer Nature Switzerland AG 2021
K. A. B. Ahmad et al. (eds.), *Functional Encryption*, EAI/Springer Innovations in Communication and Computing, https://doi.org/10.1007/978-3-030-60890-3_15

1   Introduction

The world has been exponentially progressing toward a digital era, with most of the day-to-day activities involving humongous data propagating through the networks(s) every second. This has led to an ever-increasing demand of securing user's or client information shared over the web. As the saying goes "Data never Sleeps," thereby meaning that every day, on $24 \times 7 \times 365$ basis, user's data is continuously processed on the web by Banks, Insurance Companies, Educational Institutions, Hospitals, Hotels, etc. This increase in network traffic calls for better and safer protection by designing and developing efficient techniques that can secure user's data before it is sent out onto the network. The science of cryptography comes to the rescue by securing the data against such attacks [1, 2]. In lieu of this, this chapter aims at proposing a symmetric key technique that can be deployed in the real-world to secure the sensitive data from invulnerabilities of the attackers that aim to steal information. This technique makes use of multiple keys on the plaintext to encrypt the data before it can be sent to the receiver over an unsecure network.

For the sake of simplicity and clarity, the rest of the paper is organized as follows: Sect. 2 describes the Problem Statement; Sect. 3 describes the proposed cryptosystem; Sect. 4 describes the Methodology Adopted; Sect. 5 describes Key Selection; Sect. 6 describes Key Sharing; Sect. 7 describes Encoding and Decoding; Sect. 8 describes Algorithm Formulated; Sect. 9 describes Encryption Simulation; Sect. 10 describes Encryption Flow Chart; Sect. 11 describes Results; Sect. 12 describes Conclusion and Future Work followed by References.

2   Problem Statement

With growing commercial transaction on the Web, the security of the user's data is of paramount importance. Today the customer performs online shopping, perform online banking transactions, uses digital wallet for online and offline payments, pays the premium of his insurance policies and performs n number of transactions where his personal information as well as credit and debit card information is always available on the web. As security expert, it is a big challenge to protect customer information. To protect and safeguard the customer data, Quatro-Inverse Exponential Cipher technique has been proposed in this chapter. It is a symmetric cipher technique that uses rotational dynamic keys and exponential mathematical operations to encrypt a plaintext message over a wide array of networks and computer devices. The keys are chosen to be different with every message transmission to make the network secure. The transmission of the chosen keys between the sender and the receiver is done by using the enhanced Diffie-Hellman key exchange technique with digital signatures. Even though the Quatro-Inverse Exponential Cipher technique uses complex exponential operations, yet the technique is able to perform well when encrypting the plain text.

3 Proposed Cryptosystem

The cryptosystem that is introduced in this chapter is an extension of a previously proposed system [3] in which the encryption and decryption mechanism used a set of three private keys. The technique introduced here makes use of four private keys with basic mathematical operations to encrypt the plaintext before being sent to the receiver and uses the inverse mathematical operations with the same set of four keys to decrypt the cipher text received by the receiver to finally obtain the intended plain text message. The technique makes use of four private keys and exponential mathematical operations and hence is named as "QUIET: Quatro-Inverse Exponential Cipher Technique." Increasing the number of keys from a single key to four disjoint keys increases the security of the cryptosystem. The basic idea is to preserve the privacy of the data while outsourcing the data and making use of well-defined yet hard to decipher keys for securing the data. The "QUIET: Quatro-Inverse Exponential Cipher Technique" is a lightweight cryptographic technique that can easily be plugged in any network or onto any device to secure the data that is being transmitted on an otherwise insecure network.

The cryptosystem makes use of lightweight basic mathematical operations. These mathematical operations like addition, subtraction, multiplication, division and Exclusive OR (XOR) that can directly be implemented on the hardware and hence are fast and consume smaller CPU cycles for computations. Therefore, the cryptosystem is fast and lightweight, and owing to this fact, the technique can even be plugged into a portable handheld device with limited battery power.

4 Methodology Adopted

The technique is a symmetric key cipher technique implying that the same set of keys will be used in the process of encrypting the plain text message as well as during the decryption of the ciphertext message. These set of keys are private and are shared only between the sender and the receiver via a secure channel with no one else having access to them. The Quatro-Inverse Exponential Cipher Technique is a lightweight symmetric cipher technique that uses pre-defined mathematical operations along with the set of four unique keys to secure the plaintext message.

5 Key Selection

The cryptosystem has four private keys that are secret and only known to the sender and the receiver. A strong cryptosystem is highly dependent on the set of chosen keys for the process and keeping this view in mind the Quatro-Inverse Exponential Cipher Technique makes use of a combination of strong private keys that are long and hard to guess, and are based on inputs that are provided by the sender. The keys are strategically extracted from personal information of the user which are both private as well as hard to guess by the hacker. The set of four keys is given as $K_1, K_2, K_3,$ and K_4:

K1 = Passport Number XOR current latitude location
K2 = Social Security Number AND current longitude location
K3 = Sequential sum of digits*current Epoch Timestamp
K4 = Unique session key re-generated for every session

The user input is limited to two private information—passport number and social security number. These two sets of input values are private and known to the user, however, to limit the "social engineering attacks," the cryptosystem introduces variables that are not known to the user like the dynamic session key, current latitude and longitude and the chosen current timestamp. Since this information is concealed from the end user, it is difficult for the attacker to guess this information. The "social engineering cryptography attacks" are aimed at getting the information from the people instead of the traditional crypt attacks. It has been identified that the humans are one of the weakest links of even a string cryptosystem [4]. Therefore, by not allowing the users to choose their keys, the technique chooses the secure keys itself and encrypts the plaintext using these chosen keys. Authors in [5] analyzed various social attacks and conducted them on the social networking websites to further conclude that humans are a weak link to the security as they may reveal secret information to the attacker. Hence, removing the human element from the equation adds to the strength of the cryptographic technique.

The only restriction exists that the K_1 key must be a co-prime with the number of characters in the character set. The keys are chosen to be the private information of the sender which would not be readily available to the attacker and the introduction of the variable factor like the current timestamp and location coordinates with this information make up for a hard to guess keys and thus increasing the security of the cryptosystem.

The fourth key is a unique SHA256 hash key that is generated for every session. Each transmission of the cipher text from sender to receiver uses a different hash key. This hash key is called as a dynamic session key because it is a rotational key which is generated for every session and then immediately discarded once the transfer is successful and then re-generated for next transmission. Making the session key unique per transmission greatly increases the security of the technique. This is because any interceptor would not be able to guess the pattern of the cipher text in cryptosystem since each transmission has a randomly generated session key. This key serves two purposes, it ensures that the keys are always different per transmission, thereby making the cryptosystem free from redundancies, and helps the receiver to validate the identity of the sender that is sending the message.

The third key makes use of the fourth dynamic key to calculate the sum of its digits. However, the sum of digits of dynamic session key is chosen according to the iteration (or character position). If the character position is even, then the even position digits are chosen from the dynamic session key (fourth key) and their sum is calculated. For odd positioned characters, the odd positioned digits are chosen from the dynamic session key, and their sum is calculated in similar fashion. This obtained sum is then multiplied with the current epoch timestamp to obtain the third key. The product will always be a unique number since the timestamp will always be different. Making use of two unique keys, the security of the technique is increased by making it extremely difficult for the attacker to recognize the pattern in the encryption keys.

6   Key Sharing

The four private keys that are generated by the cryptosystem need to be shared between the sender and the receiver so that only the two parties have the information of these private keys and no one else knows about them. One of the most pressing question with symmetric key ciphers that arises is how the common keys are shared between the sender and the receiver. Since, symmetric key ciphers employ the use of same keys for both encryption on the sender's end as well as decryption of the cipher text on the receiver's end. The eco-cipher technique uses the Diffie-Hellman Key exchange algorithm for the transfer of the keys from sender to receiver [6]. Since Diffie-Hellman Key exchange is susceptible to the man in the middle attack, hence the key exchange algorithm is bundled together with the digital signatures to authenticate the two users and only upon the verification of identity, the key exchange happens.

The signing authority generates the pair of public and private keys which are used in the Digital Signature Algorithm (DSA) [7]. A hash function generates a message digest which is further used to create the actual digital signature on both the receiver and the sender end. This signature is sent along with the keys, and each end user can verify the authenticity of where the keys came from with the help of this digital signature.

7   Encoding and Decoding Operation

The "QUIET: Quatro-Inverse Exponential Cipher Technique" allows all characters to be inserted as the plain text. However, since the technique involves mathematical operations as an exponential value hence the plaintext domain is restricted to 26 alphabets and 10 numerical characters where the alphabets are encoded as "A" = 1, "B" = 2, "C" = 3, and so on till "Z" = 26 comprising a total of 26 characters. If the character set is assumed to be large such as ASCII values, then the exponential value increases and the computations become heavier. Hence, for systems with fast processing speeds the character set can easily be increased to include more characters. However, for systems with limited computation power and limited battery, like the portable handheld devices, the encoding scheme of limiting the character set to A–Z is much suited.

7.1   Encryption Operation

The encryption operation involves taking the plaintext and the four keys as the input and returns a cipher text that can be then transmitted to the receiver's end. If the encryption operation is depicted as a function, it can be denoted as follows:

$$C(y) \leftarrow f(p, k)$$

The above function dictates that the cipher function C(y) is obtained by the encryption function on the input set of plaintext "p" and keys "k."

The encryption operations involve simple mathematical operations that are performed on the chosen set of four keys in a pre-defined order which are briefly described below with the help of equations. The plaintext is denoted by P_t, and the four keys are denoted by K_1, K_2, K_3, K_4. The intermediary encryption steps are given by E_n, and final cipher text is denoted by C_t. The "N" denotes the number of characters in the assumed character set. For the basic version of the technique, value of N for alphabet is 26 and for numerals is 10.

$$E_1 = (P_t * K_1) \bmod N$$
$$E_2 = E_1 * K_2$$
$$E_3 = E_2 + K_3$$
$$E_4 = E_3 \text{XOR } K_4$$
$$C_t = e^{E}4$$

$$\therefore C = e^{[\{(P*E1)\bmod N*K2\}+K3 \, XOR \, K4]}$$

The above mathematical operation can be better illustrated as a small example. It has been assumed that size of N is 26 in examples, and for simplicity of mathematical calculations the values for key are assumed to be small.

$$P_t = \textbf{ENCRYPT}$$
$$N = 26$$
$$K_1 = 5$$
$$K_2 = 7$$
$$K_3 = 13$$
$$K_4 = 8$$

Table 15.1 depicts the example with the above plaintext and keys:

Table 15.1 Encryption Table

P_t	E_1	E_2	E_3	E_4	C_t
E (5)	(5*5) mod 26 = 25	25*7 = 175	175 + 13 = 188	188 ⊕ 8 = 180	$e^{180} = 1.489384200781383E78$
N (14)	(14*5) mod 26 = 18	18*7 = 126	126 + 13 = 139	139 ⊕ 8 = 131	$e^{131} = 7.808671073519151E56$
C (3)	(3*5) mod 26 = 15	15*7 = 105	105 + 13 = 118	118 ⊕ 8 = 126	$e^{126} = 5.261441182666386E54$
R (18)	(18*5) mod 26 = 12	12*7 = 84	84 + 13 = 97	97 ⊕ 8 = 105	$e^{105} = 3.989519570547216E45$
Y (25)	(25*5) mod 26 = 21	21*7 = 147	147 + 13 = 160	160 ⊕ 8 = 168	$e^{168} = 9.151092805295634E72$
P (16)	(16*5) mod 26 = 2	2*7 = 14	14 + 13 = 27	27 ⊕ 8 = 19	$e^{19} = 1.784823009631872 5E8$
T (20)	(20*5) mod 26 = 22	22*7 = 154	154 + 13 = 167	167 ⊕ 8 = 175	$e^{175} = 1.003539180614329 5E76$

The sender sends the string of numbers as the cipher text to the receiver as follows:

1.4893842007818383E78   7.808671073519151E56   5.261441182666386E54
3.989519570547216E45   9.151092805295634E72   1.7848230096318725E8
1.0035391806143295E76

7.2   Cipher Text Decoding: Decryption Operation

The "QUIET: Quatro-Inverse Exponential Cipher Technique" is a symmetric key cipher and hence it would use the same set of four private keys for decryption on receiver's end as the keys that sender used to encrypt the plaintext message. The decryption function can be seen as an inverse encryption function that would yield the plaintext message. It can be represented as a mathematical function that accepts the cipher text and keys as input and yields the plaintext. The inverse on function denotes that decryption is inverse of encryption function.

$$P(y) \leftarrow f^{-1}(c, k)$$

The inverse operations point to the special pair of operations which counter the effect of each other and when used on same set of numbers will cancel each other out reversing each other's effect. They are extremely helpful in symmetric cryptography wherein one operation (for instance, addition) is used to encrypt the number to result in a different number and then the other counterpart operation (for instance, subtraction) when used on the second number will cancel out the first operation and result in the first number. This counterpart operation is used in decrypting the cipher text to get the intended plaintext message. The decryption operation for Quatro-Inverse Exponential Cipher Technique is a series of mathematical operations, and they can be best described as mathematical equations for each step. Each intermediary decryption step is denoted by D_n, and the keys are denoted as K_1^{-1}, K_2, K_3, K_4. The mathematical equations are described as follows:

$$D_1 = \log_e (C_t)$$
$$D_2 = D_1 \text{XOR } K_4$$
$$D_3 = D_2 - K_3$$
$$D_4 = D_3/K_2$$
$$P_t = \left(D_4 * K_1^{-1}\right) \bmod N$$

$$\therefore P = \left[\left(\log_e(C) \, XOR \, K4 - K3\right)/K2\right] * K1^{-1} \bmod N$$

Continuing along the same example that was used in the encryption operation, we try to depict the decryption operation that happens on the receiver's end to obtain the plaintext message.

The decryption operation needs an Inverse Modulo of the key K_1. This key is the multiplicative inverse and is denoted as K_1^{-1}. In order to find the modular multiplicative inverse, the Extended Euclidean algorithm is used.

The modular multiplicative inverse of an integer K_1 must satisfy the following condition:

$$\left(K_1 * K_1^{-1}\right) \equiv 1 \bmod N$$

and, gcd $(K_1, N) = 1$ (K_1 and N should be co-primes)

The modular multiplicative inverse of 5 mod 26 is 21, hence $K_1^{-1} = 21$ (Table 15.2).

The plain text that is obtained at the receiver's end is "ENCRYPT" which is same as the original plaintext message that was transmitted by the sender.

8 Algorithm Formulated

8.1 Key Generation Algorithm

The Quatro-Inverse Exponential Cipher Technique is a symmetric key technique with four private keys. These keys are generated by the cryptosystem automatically based on the input of the user. The key selection and sharing algorithm are described in the steps that follow:

Step 1: start.
Step 2: accept Passport Number from user → P.
Step 3: accept Social security Number from user → S.
Step 3: current longitude → Lx.
Step 4: current latitude → Ly.
Step 5: current timestamp → T.
Step 6: P XOR Lx → K1.
Step 7: S AND Ly → K2.
Step 8: Dynamic session key (variable: To be kept as 512 bits for networks with better computing power, and for networks with limited computation power the size is limited to 256 bits) → K4.
Step 9: IF character position IS even THEN.
Step 10: Calculate sum of even digits of K4 → Sum.
Step 11: ELSE.
Step 12: Calculate sum of odd digits of K4 → Sum.
*Step 13: Sum * T → K3.*
Step 14: Generate a SHA256 hash for digital signature.

Table 15.2 Decryption Table

C_t	D_1	D_2	D_3	D_4	P_t
1.4893842007818383E78	Log(c) = 180	180 ⊕ 8 = 188	188−13 = 175	175/7 = 25	(25 * 21) mod 26 = 5 (E)
7.8086710735191151E56	Log(c) = 131	131 ⊕ 8 = 139	139−13 = 126	126/7 = 18	(18 * 21) mod 26 = 14 (N)
5.2614411826638654E54	Log(c) = 126	126 ⊕ 8 = 118	118−13 = 105	105/7 = 15	(15 * 21) mod 26 = 3 (C)
3.9895195705472164E45	Log(c) = 105	105 ⊕ 8 = 97	97−13 = 84	84/7 = 12	(12 * 21) mod 26 = 18 (R)
9.1510928052956344E72	Log(c) = 168	168 ⊕ 8 = 160	160−13 = 147	147/7 = 21	(21 * 21) mod 26 = 25 (Y)
1.7848230096318725E8	Log(c) = 19	19 ⊕ 8 = 27	27−13 = 14	14/7 = 2	(2 * 21) mod 26 = 16 (P)
1.0035391806143295E76	Log(c) = 175	175 ⊕ 8 = 167	167−13 = 154	154/7 = 22	(22 * 21) mod 26 = 20 (T)

Step 15: Transfer digitally signed keys using Diffie-Hellman key exchange algo-
*   rithm.*
Step 14: end.

8.2   Encryption Algorithm

The Quatro-Inverse Exponential Cipher Technique uses the four private keys
generated in the previous section to encrypt the plain text and create a cipher text
which is then sent over to the receiver. The steps of the encryption algorithm are
described in the text that follows:

Step 1: start.
Step 2: accept first key from generation algorithm → K1.
Step 3: accept second key from generation algorithm → K2.
Step 3: accept third key from generation algorithm → K3.
Step 4: accept fourth key from generation algorithm → K4.
Step 5: accept plaintext from user → PT.
Step 6: Number of characters in encoder → N.
Step 7: Encode (PT) → P.
Step 8: Multiply (P, K1) mod N → E1.
Step 9: Multiply (E1, K2) → E2.
Step 10: Addition (E2, K3) → E3.
Step 11: XOR (E3, K4) → E4.
Step 12: exp (E4) → CT.
Step 13: return final Cipher Text as stream of numbers.
Step 14: end.

9   Encryption Simulation

The Quatro-Inverse Exponential Cipher Technique was simulated on a Java platform
as a standalone application. The simulation was carried out on a smaller set of input
keys, and the key generation mechanism was not incorporated in the simulation
runs to simplify the runs. However, it is always possible to incorporate the same by
using the key generation algorithms provided in the previous section. The simulation
environment is detailed in Table 15.3:

The simulation runs of the encryption and decryption are briefly described below
on three different text messages with different key combinations.

Simulation 1: Refer Tables 15 4 and 15.5.

Table 15.3 Simulation
environment specifications

Specifications	Java
O.S. used	Windows 10 Enterprise
Device model	Dell latitude 7400
Processor	Intel i7-8665U
RAM	32 GB
Development IDE	IntelliJ IDEA
IDE version	Build #IC-193.6911.18
Compile SDK	JDK 11
Language	Java

$$P_t = \textbf{CIPHER}$$
$$N = 26$$
$$K_1 = 11 \text{ and } K_1^{-1} = 19$$
$$K_2 = 23$$
$$K_3 = 9$$
$$K_4 = 17$$

Cipher text is as follows: 1.633308100216833E81 1.1373425541353215E221
2.000349215698554E196   2.045494911349825E110   1.8112390828890233E41
2.2182652975385555E156

The plain text that is obtained at the receiver's end is "CIPHER" which is same
as the original plaintext message that was transmitted by the sender.

Simulation 2: Refer Tables 15.6 and 15.7.

$$P_t = \textbf{INVERSE}$$
$$N = 26$$
$$K_1 = 17 \text{ and } K_1^{-1} = 23$$
$$K_2 = 12$$
$$K_3 = 19$$
$$K_4 = 5$$

Cipher text is as follows: 8.818602191274965E125 2.515438670919167E30
4.675374784632515E61   3.637970947608805E42   6.0975343934414735E113
2.5526681395254553E63 3.637970947608805E42

The plain text that is obtained at the receiver's end is "INVERSE" which is same
as the original plaintext message that was transmitted by the sender.

Simulation 3: Refer Tables 15.8 and 15.9.

Table 15.4 Encryption Table

P_t	E_1	E_2	E_3	E_4	C_t
C (3)	(3*11) mod 26 = 7	7*23 = 161	161 + 9 = 170	170 ⊕ 17 = 187	e^{187} = 1.6333081002168336E81
I (9)	(9*11) mod 26 = 21	21*23 = 483	483 + 9 = 492	492 ⊕ 17 = 509	e^{509} = 1.1373425541353215E221
P (16)	(16*11) mod 26 = 20	20*23 = 460	460 + 9 = 469	469 ⊕ 17 = 452	e^{453} = 2.0003492156985554E196
H (8)	(8*11) mod 26 = 10	10*23 = 230	230 + 9 = 239	239 ⊕ 17 = 254	e^{254} = 2.0454949113498256E110
E (5)	(5*11) mod 26 = 3	3*23 = 69	69 + 9 = 78	188 ⊕ 17 = 95	e^{95} = 1.8112390828890233E41
R (18)	(18*11) mod 26 = 16	16*23 = 368	368 + 9 = 377	377 ⊕ 17 = 360	e^{360} = 2.2182652975385555E156

Table 15.5 Decryption Table

C_t	D_1	D_2	D_3	D_4	P_t
1.6333081002168833E81	Log(c) = 187	$187 \oplus 17 = 170$	$170 - 9 = 161$	$161/23 = 7$	$(7 * 19) \bmod 26 = 3$ (**C**)
1.1373425541353215E221	Log(c) = 509	$509 \oplus 17 = 492$	$492 - 9 = 483$	$483/7 = 21$	$(21 * 19) \bmod 26 = 9$ (**I**)
2.0003492156985 54E196	Log(c) = 452	$452 \oplus 17 = 469$	$469 - 9 = 460$	$460/7 = 20$	$(20 * 19) \bmod 26 = 16$ (**P**)
2.0454949113 49825E110	Log(c) = 254	$254 \oplus 17 = 239$	$239 - 9 = 230$	$230/7 = 10$	$(10 * 19) \bmod 26 = 8$ (**H**)
1.8112390828890233E41	Log(c) = 95	$95 \oplus 17 = 78$	$78 - 9 = 69$	$69/7 = 3$	$(3 * 19) \bmod 26 = 5$ (**E**)
2.2182652975385555E156	Log(c) = 360	$360 \oplus 17 = 377$	$377 - 9 = 368$	$368/7 = 16$	$(16 * 19) \bmod 26 = 18$ (**R**)

Table 15.6 Encryption Table

P_t	E_1	E_2	E_3	E_4	C_t
I (9)	(9*17) mod 26 = 23	23*12 = 276	276 + 19 = 295	295 $\oplus$ 5 = 290	$e^{290} = 8.8186021912749665E125$
N (14)	(14*17) mod 26 = 4	4*12 = 48	48 + 19 = 67	67 $\oplus$ 5 = 70	$e^{70} = 2.5154386709191667E30$
V (22)	(22*17) mod 26 = 10	10*12 = 120	120 + 19 = 139	139 $\oplus$ 5 = 142	$e^{142} = 4.6753747846325515E61$
E (5)	(5*17) mod 26 = 7	7*12 = 84	84 + 19 = 103	103 $\oplus$ 5 = 98	$e^{98} = 3.6379709476088805E42$
R (18)	(18*17) mod 26 = 20	20*12 = 240	240 + 19 = 259	259 $\oplus$ 5 = 262	$e^{262} = 6.0975343934414735E113$
S (19)	(19*17) mod 26 = 11	11*12 = 132	132 + 19 = 151	151 $\oplus$ 5 = 146	$e^{146} = 2.5526681395254553E63$
E (5)	(5*17) mod 26 = 7	7*12 = 84	84 + 19 = 103	103 $\oplus$ 5 = 98	$e^{98} = 3.6379709476088805E42$

Table 15.7 Decryption Table

C_t	D_1	D_2	D_3	D_4	P_t
8.8186021912749465E125	Log(c) = 290	290 ⊕ 5 = 295	295−19 = 276	276/12 = 23	(23 * 23) mod 26 = 9 (**I**)
2.5154386709919167E30	Log(c) = 70	70 ⊕ 5 = 67	67−19 = 48	48/12 = 4	(4 * 23) mod 26 = 14 (**N**)
4.6753747846632515E61	Log(c) = 142	142 ⊕ 5 = 139	139−19 = 120	120/12 = 10	(10 * 23) mod 26 = 22 (**V**)
3.6379709476088805E42	Log(c) = 98	98 ⊕ 5 = 103	103−19 = 84	84/12 = 7	(7 * 23) mod 26 = 5 (**E**)
6.0975343934414735E113	Log(c) = 262	262 ⊕ 5 = 259	259−19 = 240	240/12 = 20	(20 * 23) mod 26 = 18 (**R**)
2.5526681395254553E63	Log(c) = 146	146 ⊕ 5 = 151	151−19 = 132	132/12 = 11	(11 * 23) mod 26 = 17 (**S**)
3.6379709476088805E42	Log(c) = 98	98 ⊕ 5 = 103	103−19 = 84	84/12 = 7	(7 * 23) mod 26 = 5 (**E**)

Table 15.8 Encryption Table

P_t	E_1	E_2	E_3	E_4	C_t
Q (17)	$(17*5) \bmod 26 = 7$	$7*7 = 49$	$49 + 13 = 62$	$62 \oplus 8 = 54$	$e^{54} = 2.8307533032774694E23$
U (21)	$(21*5) \bmod 26 = 1$	$1*7 = 7$	$7 + 13 = 20$	$20 \oplus 8 = 28$	$e^{28} = 1.4462570642914775E12$
A (1)	$(1*5) \bmod 26 = 5$	$5*7 = 35$	$35 + 13 = 48$	$48 \oplus 8 = 56$	$e^{56} = 2.0916594960129962E24$
T (20)	$(20*5) \bmod 26 = 22$	$22*7 = 154$	$154 + 13 = 167$	$167 \oplus 8 = 175$	$e^{175} = 1.0035391806143295E76$
R (18)	$(18*5) \bmod 26 = 12$	$12*7 = 84$	$84 + 13 = 97$	$97 \oplus 8 = 105$	$e^{105} = 3.9895195705472216E45$
O (15)	$(15*5) \bmod 26 = 23$	$23*7 = 161$	$161 + 13 = 174$	$174 \oplus 8 = 166$	$e^{166} = 1.2384657367292132E72$

Table 15.9 Decryption Table

C_t	D_1	D_2	D_3	D_4	P_t
2.83075330327469E23	Log(c) = 54	54 ⊕ 8 = 62	62−13 = 49	49/7 = 7	(7 * 21) mod 26 = 17 (**Q**)
1.44625706429147E12	Log(c) = 28	28 ⊕ 8 = 20	20−13 = 7	7/7 = 1	(1 * 21) mod 26 = 21 (**U**)
2.09165949601299E24	Log(c) = 56	56 ⊕ 8 = 48	48−13 = 35	35/7 = 5	(5 * 21) mod 26 = 1 (**A**)
1.00353918061432E76	Log(c) = 175	175 ⊕ 8 = 167	167−13 = 154	154/7 = 22	(22 * 21) mod 26 = 20 (**T**)
3.98951957054721E45	Log(c) = 105	105 ⊕ 8 = 97	97−13 = 84	84/7 = 12	(12 * 21) mod 26 = 18 (**R**)
1.23846573672921E72	Log(c) = 166	166 ⊕ 8 = 174	174−13 = 161	161/7 = 23	(23 * 21) mod 26 = 15 (**O**)

```
"D:\Installed\IntelliJ\IntelliJ IDEA Community Edition 2019.3.4\jbr\bin\java.exe" "-javaagent:D:\Installed\IntelliJ\IntelliJ IDEA Community Edition 2019.3.4\lib\idea_rt
.jar=60611:D:\Installed\IntelliJ\IntelliJ IDEA Community Edition 2019.3.4\bin" -Dfile.encoding=UTF-8 -classpath
D:\Personal\JavaPrograms\QuatroInverseCipher\out\production\QuatroInverseCipher com.harshit.quatro.QuatroInverseCipher
****************** SENDER'S END ******************
Enter Plain text(alphanumeric)
CIPHER
Enter First Key
11
Enter Second Key
23
Enter Third Key
9
Enter Fourth Key
17
Encrypting plain text message CIPHER
The cipher text has been generated.
Cipher Text is: 1.633308100216833E81 1.1373425541353215E221 2.000349215698554E196 2.045494911349825E110 1.8112390828890233E41 2.2182652975385555E156

****************** RECEIVER'S END ******************
Cipher text obtained at receiver: 1.633308100216833E81 1.1373425541353215E221 2.000349215698554E196 2.045494911349825E110 1.8112390828890233E41 2.2182652975385555E156
Decrypting received cipher text message at receiver's end.
Plain Text message sent by sender is: CIPHER
```

Fig. 15.1 Simulation results of Quatro-Inverse Exponential Cipher Technique: CIPHER

```
"D:\Installed\IntelliJ\IntelliJ IDEA Community Edition 2019.3.4\jbr\bin\java.exe" "-javaagent:D:\Installed\IntelliJ\IntelliJ IDEA Community Edition 2019.3.4\lib\idea_rt
.jar=60831:D:\Installed\IntelliJ\IntelliJ IDEA Community Edition 2019.3.4\bin" -Dfile.encoding=UTF-8 -classpath
D:\Personal\JavaPrograms\QuatroInverseCipher\out\production\QuatroInverseCipher com.harshit.quatro.QuatroInverseCipher
****************** SENDER'S END ******************
Enter Plain text(alphanumeric)
INVERSE
Enter First Key
17
Enter Second Key
12
Enter Third Key
19
Enter Fourth Key
5
Encrypting plain text message INVERSE
The cipher text has been generated.
Cipher Text is: 8.818602191274965E125 2.515438670919167E30 4.675374784632515E61 3.637970947608805E42 6.097534393441473SE113 2.5526681395254553E63 3.637970947608805E42

****************** RECEIVER'S END ******************
Cipher text obtained at receiver: 8.818602191274965E125 2.515438670919167E30 4.675374784632515E61 3.637970947608805E42 6.097534393441473SE113 2.5526681395254553E63
3.637970947608805E42
Decrypting received cipher text message at receiver's end.
Plain Text message sent by sender is: INVERSE
```

Fig. 15.2 Simulation results of Quatro-Inverse Exponential Cipher Technique: INVERSE

$$P_t = \mathbf{QUATRO}$$
$$N = 26$$
$$K_1 = 5 \text{ and } K_1^{-1} = 21$$
$$K_2 = 7$$
$$K_3 = 13$$
$$K_4 = 8$$

Cipher Text is as follows: 2.830753303274694E23 1.446257064291475E12 2.091659496012996E24   1.0035391806143295E76   3.989519570547216E45 1.2384657367292132E72.

The plain text that is obtained at the receiver's end is "QUATRO" which is same as the original plaintext message that was transmitted by the sender.

The simulation results (Simulations 1, 2, and 3) are detailed in Figs. 15.1, 15.2 and 15.3:

Fig. 15.3 Simulation results of Quatro-Inverse Exponential Cipher Technique: QUATRO

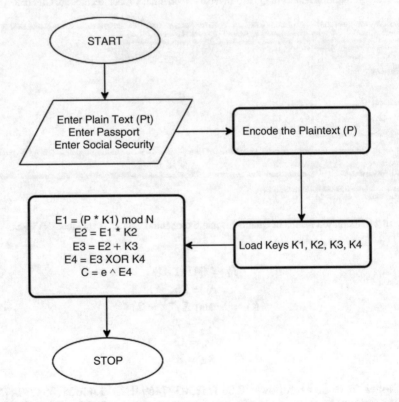

Fig. 15.4 Encryption flowchart

10   Encryption Flowchart

The detailed steps of the encryption process can be denoted as form of a flowchart that is shown in Fig. 15.4.

11   Result

An illustrative comparison graph is drafted to highlight the difference in the running time of Quatro-Inverse Exponential Cipher Technique for three different types of input data, namely for alphabets, numerals, and alpha numerals (Fig. 15.5). Comparison is also drawn between Quatro-Inverse Exponential Cipher Technique and its predecessor cipher technique—Triplicative cipher technique on three different input texts (Fig. 15.6; Table 15.10). The comparison of the two techniques reveals that the Quatro-Inverse Exponential Cipher Technique is able to perform more complex mathematical operations on four keys with a very small fraction of increase in the running time. Increasing the number of keys and making the mathematical operations more complex than the predecessor makes the technique arithmetically stronger as well as secure at the cost of very small increase in running time. The space complexity of the two techniques is same on same input set. For "n" number of characters in plain text, the space complexity for both techniques is Bi-Oh of n-O(n). Furthermore, the Quatro-Inverse Exponential Cipher Technique was simulated in Java, and the running time of the technique is summarized and briefed to portray the output of executing the source code on the mentioned simulation in Java language.

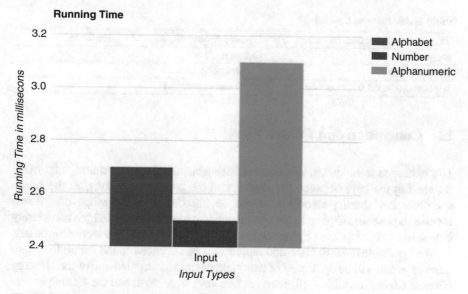

Fig. 15.5   Comparison of Running Time

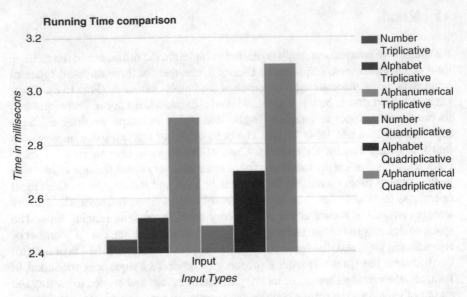

Fig. 15.6 Comparison of the results of Triplicative and Quatro-Inverse Exponential Cipher Technique

Table 15.10 Running time details

Triplicative Cipher time (in ms)	Quatro-Inverse Exponential Cipher (in ms)
Numeric: 2.45	Numeric: 2.5
Alphabet: 2.5	Alphabet: 2.7
Alphanumerical: 2.9	Alphanumerical: 3.1

12   Conclusion and Future Work

The presented technique in this chapter uses a standardized key sharing algorithm for sharing the keys between the sender and the receiver. However, in the future a custom key sharing algorithm would be introduced in the scope of Quatro-Inverse Exponential Cipher Technique which would make the technique completely independent and thus can be plugged into any kind of network and work seamlessly.

The Quatro-Inverse Exponential cipher will be enhanced in the future to accept more keys and will be optimized of handheld devices as an extension of the "Energy efficient Cipher technique" [8] wherein five keys were used, and the technique was further optimized for handheld devices.

Furthermore, the Quatro-Inverse Exponential cipher currently uses basic mathematical operations, and these operations would be replaced with more complex ones for networks and devices with higher computation power to further enhance the security of the presented technique.

References

1. Forouzan, B. A. (2007). *Cryptography and network security*. Mc. Graw-Hill Special Indian Edition.
2. Stallings, W. (2007). *Cryptography and network security-principles and practices* (fourth ed.). Pearson.
3. Johari, R., Bhatia, H., Singh, S., & Chauhan, M. (2016). Triplicative cipher technique. *Procedia Computer Science, 78*, 217–223.
4. Thomason, S., & People, C. D. The Weak Link in Security. *Global Journal of Computer Science and Technology, 13*. ISSN 0975–4172.
5. Orgill, G. L., Romney, G. W., Bailey, M. G., & Orgill, P. M. (2004). The urgency for effective user privacy-education to counter social engineering attacks on secure computer systems. In *Proceedings of the 5th conference on Information technology education* (pp. 177–181).
6. Diffie, W., & Hellman, M. New directions in cryptography. *IEEE Transactions on Information Theory, 22, 6*, 644–654.
7. Kravitz, D. W. US Department of Commerce. In *Digital signature algorithm, U.S. Patent 5,231,668*.
8. Bhatia H, Johari R, Gupta K. E2CT: Energy efficient cipher technique. MICSECS: The 11th Majorov international conference on software engineering and computer systems, vol 2590

Index

A

ABE, *see* Attribute-based encryption (ABE)
Additive inverse
    addition and subtraction property, 7, 8
    inverse
        addition and subtraction property, 7, 8
    modular arithmetic, 7
    and multiplicative inverse, 2
    notation, 158–161
Asymmetric encryption (ASE), 82
Asymmetric key cryptography, 37, 104, 113,
    244
Attribute-based encryption (ABE)
    access policy, 230–231
    basic model
        complexity assumptions, 230
    categories
        CP-ABE, 235–237
        KP-ABE, 233–235
    comparisons, 239–240
    difference, 237
    encryption, 226
    IBE, 228
    literature review, 238–239
    PKI, 227, 228
    public key encryption, 226–227
    security, 225, 240
    working, 231–233
Attributes
    Boneh-Franklin IBE, 138
    FIBE, 206
    public parameters, 162
    See also Attribute-based encryption (ABE)

B

Baby-step giant-step algorithm, 30–31
Bilinear mapping, 20, 26–27, 34–35
Blind signature schemes
    Chaum's RSA, 252–253
    Okomoto-Schnorr protocol, 254–255
    Schnorr protocol, 253–254
BLS short signature/BLS signature scheme,
    260
Boneh-Boyen IBE, 152–156
    algorithm
        decryption, 155
        encryption, 155
        extract, 154, 155
    basic scheme
        additive notation, 158
        decrypting, 160–161
        encryption, 160
        private key extraction, 60
        security, 167–169
        setup of parameters, 158–160
    classification of IBE schemes, 156–157
    full scheme
        decrypting, 167
        encrypting, 166
        private key extraction, 166
        setup of parameters, 165–166
    HIBE
        decryption, 199
        encryption, 198
        extraction of the private key, 198
        setup, 197–198
    limitation, 152

Boneh-Boyen IBE (*cont.*)
   multiplicative notation
      decrypting, 164
      encrypting, 163–164
      extraction of the private key, 162–163
      setup of parameters, 162
   security, 167–169
   setup algorithm, 154
Boneh-Franklin IBE, HIBE
   Boneh-Boyen IBE, 197–199
   decryption, 196–197
   encryption, 195–196
   extraction of the private key, 195
   lower level setup, 195
   root setup, 194–195
Boneh-Franklin scheme
   bilinear Diffie-Hellman problem, 191
   decryption, 141–143
   elliptic curve, 156
   encryption, 141–143
   examples, 146–148
   furtherworks on, 144–146
   HIBE, 194–199
   IBE, 137–139
   private key extraction, 140–142
   proof, 169
   security, 144
   setup of parameters, 139–140, 142

C
Chaum-Van Antwerper (CvA), 256–258
Ciphertext policy-ABE (CP-ABE), 233,
   235–241
Cocks scheme
   correctness of cocks IBE, 130
   decryption, 125–126
   encryption, 123–125
   examples, 126–129
   features, 119
   IBE, 118
   mathematical concepts, 119–120
   parameters, 121–122
   private key extraction, 122–123
   security, 131–133
   working, 118–119
Conjugate element, 39, 40
Convolutional multiplication, 106, 114
CP-ABE, *see* Ciphertext policy-ABE
   (CP-ABE)
Cryptography
   HECC (*see* Hyperelliptic curve
      cryptography (HECC))

   pairing-based (*see* Pairing-based
      cryptography)
   public key, 188–189
   XTR-DH protocol, 50–51
   XTR-DSA signature scheme, 52–53
   XTR-ElGamal encryption scheme, 51
   XTR-Nyberg-Rueppel (NR) signature
      scheme, 52
   XTR version of DH problem and its
      variants, 49–50
Cryptosystem
   ElGamal, 247
   elliptic curve and bilinear mapping-based,
      34–35
   log-based, 34
   LUC, 104
   NTRU, 105–111
   physical faults, 97
   pitfalls, 190
   proposed, 281
   RSA-based, 34, 60
   XTR, 38, 50, 53
CvA, *see* Chaum-Van Antwerper (CvA)
Cyclotomic polynomial, 40

D
Decryption, 110–111, 125–126, 141–143
   algorithm, 155
   encoded version, 11
   FE scheme, 2
   HIBE, 196–197, 199
   hyperelliptic curve, 72–73
   operation, 286–287
   private key, 104
   public key cryptosystem, 104
   Sakai-Kasahara IBE, 177, 180
   table, 292, 294, 296
Diffie-Hellman key exchange, 35, 86–88, 280,
   283, 289
Digital signature algorithm (DSA), 24, 34, 49,
   52–53, 283
Digital signatures
   algorithms, 244
   hash functions, 245
   hierarchical IBS, 264–267
   other schemes, 267
   security
      models, 244–245
      services, 243
   signcryption, 268–270
Digital signature standard (DSS), 250–251,
   260, 272

Discrete logarithm (DL) problem, 30, 31, 34, 38, 60, 83, 95, 138, 245
Divisors
  explanation 1, 62
  explanation 2, 63
  explanation 3, 63
  explanation 4, 63
  explanation 5, 63
  semi-reduced and reduced divisors, 65–67
DL problem, *see* Discrete logarithm (DL) problem
DSA, *see* Digital signature algorithm (DSA)
DSS, *see* Digital signature standard (DSS)

E
ECC, *see* Elliptic curve cryptography (ECC)
EC discrete logarithm problem (ECDLP), 83, 94
ECDLP, *see* EC discrete logarithm problem (ECDLP)
ECSTR, *see* Efficient and compact subgroup trace representation (ECSTR)
Efficient and compact subgroup trace representation (ECSTR), 38
ElGamal scheme, 49, 51, 52, 54
ElGamal signature method, 73–74
Elliptic curve cryptography (ECC)
  data files, 60
  MOV attack, 34
Elliptic curve factorization, 31–32
Encryption
  flowchart, 298
  functional (*see* Functional encryption)
  hyperelliptic curve, 72
  simulation, 289–298
Exponential
  algorithms, 92
  Cipher technique, 280
  mathematical operations, 281
  See also Quatro-inverse exponential cipher technique (QUIET)
Extensions of IBE
  FIBE, 206, 209–211
  IBCPRE, 220–221
  IBKA, 207–209
  LEKS, 213–214
  PEKS, 206, 214–216
  secret session key, 205
  TBEKS, 211–212
  WIBE, 207, 216–217
  WKD-IBE, 218–219

F
FE, *see* Functional encryption (FE)
FIBE, *see* Fuzzy identity-based encryption (FIBE)
Finite field, 24, 61–62, 85–86
  definitions and results, 39–40
  factoring polynomials, 33
  hyperelliptic curves, 68–73
  irreducible polynomials, 33
  properties, 25–26
Functional encryption (FE)
  applications, 6
  challenges, 5–6
  cryptography impact, 98
  data security, 1
  encrypted version, 2
  fully homomorphic encryption, 4–5
  functionality, 3
  obfuscation, 4–5
  scheme, 3–4
  symmetric, 2
Functional field sieve, 31
Functionalities test
  equality test, 27–28
  inequality test, 28
  inner product evaluation, 28
Fuzzy IBE, 157
Fuzzy identity-based encryption (FIBE), 206, 222, 229, 230, 233
  formal model, 210
  security properties, 210–211

G
Greatest common divisor (GCD), 14, 63, 66, 67
  Bezout's algorithm, 15–16
  Euclidean/ Euclid's algorithm, 14–15
  extended Euclid's algorithm, 16–17
Groups (mathematical), 20
Group theory
  axiom, 21
  bilinear mapping, 26–27
  cryptography, 20
  cryptosystem, 33–35
  finite field properties, 25–26
  illustrations, 21–22
  properties of group
    Abelian group, 22–23
    finite field, 24
    Lagrange's theorem, 23
    Schnorr group, 23–24
  science and conceptual variable-based math, 19

H
HECC, *see* Hyperelliptic curve cryptography
    (HECC)
HECC signature algorithm
    ElGamal signature method, 73–74
    security, 76
    signature
        generation, 74–75
        verification, 75–76
HIBE, *see* Hierarchical identity-based
    encryption (HIBE)
HIBS schemes, 265, 266
Hierarchical identity-based encryption (HIBE),
    192–194
    Boneh-Franklin IBE, 194–197
    identity-based encryption, 190–192
    master secret sharing, 199–201
    pitfalls, PKC, 190
    public key cryptography, 188–189
    security, 201–202
Hyperelliptic curve cryptography (HECC)
    computer algorithm-based systems, 60
    divisors, 62–63
    ECC, 60
    encryption and decryption, 72–73
    finite fields
        proof, 68–70
    Jacobian
        Hasse-Weil theory, 64
        instruction, 64
        methodology, 65
        zeta function, 64–65
    mathematical terminologies
        arithmetical closure, 61
        cryptography of hyperelliptic curve, 61
        finite field, 61–62
        interpretation, 62
    Mumford arithmetic, 67–68
    pair generation, 70–72
    semi-reduced and reduced divisors, 65–67
    signature algorithm, 73–76

I
IBCPRE, *see* Identity-based conditional proxy
    re-encryption (IBCPRE)
IBE, *see* Identity-based encryption (IBE)
IBKA, *see* Identity-based key agreement
    (IBKA)
Identity-based conditional proxy re-encryption
    (IBCPRE), 207, 222
    formal model, 220–221
    security requirements, 221
Identity-based encryption (IBE), 118, 138–140

Boneh-Boyen (*see* Boneh-Boyen IBE)
classification
    "commutative blinding", 157
    "exponent inversion", 157
    "full domain hash", 156–157
    "quadratic residuosity", 156
FIBE (*see* Fuzzy identity-based encryption
    (FIBE))
HIBE (*see* Hierarchical identity-based
    encryption (HIBE))
Sakai-Kasahara (*see* Sakai-Kasahara IBE)
See also Boneh-Franklin scheme
Identity-based encryption with wildcard key
    derivation (WKD-IBE)
    formal model, 218–219
    security requirements, 219
Identity-based key agreement (IBKA), 206,
    207
    formal model, 208
    security requirements
        Oracles, 209

J
Jacobian
    hyper ECs, 84
    hyperelliptic curve
        Hasse-Weil theory, 64
        instruction, 64
        methodology, 65
        zeta function, 64–65
    quotient group, 74

K
Key policy-ABE (KP-ABE), 233–235, 237,
    238, 241
Keyword search, 206
    IBCPRE, 220–221
    LEKS, 213–214
    PEKS, 206, 214–216
    WIBE, 207, 216–217
    WKD-IBE, 218–219
KP-ABE, *see* Key policy-ABE (KP-ABE)

L
Lagrange's theorem, 23, 34
Lattice, 104, 112–114
LC, *see* Linear congruence (LC)
LEKS, *see* Linear encryption with keyword
    search (LEKS)
Linear congruence (LC), 11–12

Linear encryption with keyword search
(LEKS), 206, 222
formal model, 213
security requirements, 213–214
Low-Hamming-weight product, 112

M
Master secret
basic scheme—additive notation, 159
PKG, 121
public parameters, 154
sharing, 199–201
user's identity, 190
Mathematical tools
elliptic curve factorization, 31–32
factoring polynomials, 33
functional field sieve, 31
functionalities, 27–28
irreducible polynomials, 33
Paley–Wiener hypothesis, 28
primality test lset 1
Fermat's little theorem, 29
Fermat's strategy, 29
Miller-Rabin method, 29
school method, 29
Solovay–Strassen primality test, 29–30
random square factoring, 32
Matrices, 2, 9–11, 113
Multiplicative
Boneh-Boyen scheme, 156
decrypting, 164
encrypting, 163–164
extraction of the private key, 162–163
finite field, 25, 52
inverse, 2, 16, 17, 287
modular arithmetic, 8–9
multiplication property of modular
arithmetic, 8, 9
polynomial expressions, 16
S-K IBE, 177–180
setup of parameters, 162
subgroup, 38
Mumford representation, 67–68

N
NR scheme, 52
NTRU cryptosystem
algorithm, 107
asymmetric key cryptosystem, 104
complexity, 105
cryptography, 103
decryption, 110–111

encryption, 109
improvements, 113
key generation, 107–109
optimization, 111–112
parameters, 107
public key cryptosystem, 104
security, 112
symbols and notations, 105–107
working method, 111

O
Obfuscation, 2, 4–6

P
Pair generation
divisor order, 70–71
hyperelliptic curve cryptographic
arrangement, 71–72
Pairing-based cryptography
Ate, 94
drawbacks/vulnerabilities, 94–95
Eta, 93
mathematical terms and concepts
ASE, 82
Diffie–Hellman algorithm, 86–88
ECC, 89–90
ECDLP, 83
elliptic curves, 83–84
field, 85
finite field, 85–86
groups, 85
Jacobian of hyper ECs, 84
key escrow, 83
Miller's algorithm, 88–89
public key encryption, 82
RO, 81
SE, 82
subexponential algorithm, 82
turing machine, 83
public key, 81
secret key mechanism, 80
security, 95–97
Tate-Lichtenbaum, 91–92
Weil pairing, 91
PEKS, *see* Public key encryption with keyword
search (PEKS)
PKC, *see* Public key cryptography (PKC)
PKG, *see* Private key generator (PKG)
PKI, *see* Public key infrastructure (PKI)
Polynomial
capacity, 31
coefficient, 201

Polynomial (*cont.*)
   deterministic, 252
   Euclidean method, 66
   expressions, 16
   factoring, 33
   irreducible, 33
   See also NTRU cryptosystem
Prime
   numbers, 12–13, 25, 28–30, 34, 40, 61, 85,
      87, 91, 119–121, 126, 134, 146, 190,
      261
   and relative prime numbers, 12–13
Private key extraction, 134, 140–142, 176, 179,
      181, 208
Private key generator (PKG), 118, 121, 138,
      145, 173, 190–195, 198–201, 266
Public key
   encryption, 82, 226–227
   PEKS, 214–216
   PKC (*see* Public key cryptography (PKC))
   PKI (*see* Public key infrastructure (PKI))
   RSA, 60
   XTR, 50
Public key cryptography (PKC), 60, 81, 138,
      188–189
Public key encryption with keyword search
      (PEKS), 206, 207, 222
   formal model, 215
   security requirements, 215–216
Public key infrastructure (PKI), 151, 152, 227,
      228

Q
Quadratic residuosity problem, 118, 131–133,
      172, 191
Quadratic sieve algorithm (QS), 32
Quatro-inverse exponential cipher technique
      (QUIET)
   algorithm formulated
      encryption algorithm, 289
      key generation algorithm, 287–289
   digital era, 280
   encoding and decoding operation
      cipher text decoding, 286–287
      encryption, 284–286
   key
      selection, 281–283
      sharing, 283
   methodology adopted, 281
   problem statement, 280
   proposed cryptosystem, 281
   result, 299–300

QUIET, *see* Quatro-inverse exponential cipher
      technique (QUIET)

R
Random Oracle (RO), 81, 133, 144, 155–157,
      201, 251, 252, 260, 265
Rivest, Shamir, Adleman (RSA)
   blind signature scheme, 252–253
   cryptosystem, 23
   encryption, 16
   group theory, 34
   signature scheme, 246–247
   undeniable signature schemes, 258–260
RO, *see* Random Oracle (RO)

S
Sakai-Kasahara IBE (S-K IBE)
   additive notation
      decryption, 177
      encryption, 176–177
      extraction of the private key, 176
      setup of parameters, 174–175
   Cocks IBE scheme, 172
   encrypted message, 173
   full scheme
      decryption, 183
      encryption, 182
      extraction of the private key, 181–182
      security, 183–184
      setup of parameters, 181
   motivation, 172
   multiplicative notation
      decryption, 179–180
      encryption, 179–180
      extraction of the private key, 179
      setup of parameters, 178
   operations, 173–174
Schnorr group, 20, 23–24, 34, 35
Security
   Boneh-Boyen IBE, 167–169
   chosen ciphertext security, 133
   fault attacks
      Duursma–Lee, 96
      Miller's algorithm, 97
   fuzzy model for ABE, 240
   HIBE, 201–202
   NTRU cryptosystem, 112
   proof, 133
   quadratic residuosity problem, 131–133
   Sakai-Kasahara IBE, 183–184
   side channel analysis, 97

Semi-reduced and reduced divisors
 accession, 66
 Euclidean polynomial methods, 66
 methodology for accession, 66–67
SE, *see* Symmetric encryption (SE)
Short signature scheme
 Boneh-Boyen, 263–264
 computational assumption, 263
 Boneh-Lynn-Shacham, 262
 bilinear pairing, 261
 computation assumptions, 261
Side channel analysis
 power analysis, 97
 timing attacks, 97
Sieve method, 32
Signature algorithm
 ElGamal signature method, 73–74
 generation, 74–75
 security, 76
 verification, 75–76
Signature schemes
 algorithms, 244
 blind, 251–255
 DSS, 250–251
 ElGamal, 247–248
 encryption schemes, 49
 forgery/tampering, 243
 hash functions, 245
 hierarchical IBS, 264–267
 IFP and DLP, 246–251
 N-R, 249–250
 other schemes, 267
 public key, 145
 RSA, 246–247
 Schnorr algorithm, 248–249
 security
  models, 244–245
  services, 243
 short scheme, 260–264
 shortest possible signature, 260–264
 signcryption, 268–270
 undeniable, 255–260
 XTR-DSA, 52–53
Signcryption, 245, 268–270, 272
S-K IBE, *see* Sakai-Kasahara IBE (S-K IBE)

Symmetric encryption (SE), 2, 51, 52, 82, 206, 226

T
Tate-Lichtenbaum pairing, 91–92
 hyperelliptic, 92
Threshold broadcast encryption with keyword search (TBEKS)
 formal model, 211–212
 security requirements, 212
Trace
 ECSTR, 38
 electromagnetic waves, 97
Triplicative, 299, 300

U
Undeniable signature schemes
 CvA, 256–258
 RSA-based, 258–260

W
Weil pairing, 91
Wildcard identity-based encryption (WIBE)
 formal model, 216–217
 security requirements, 217
Wildcards key derivation (WKD), 207, 218–219, 222
WKD, *see* Wildcards key derivation (WKD)

X
XTR algorithm
 arithmetic operation in $GF(p^2)$, 41–42
 cryptographic applications, 49–53
 definitions and results, finite field, 39–40
 development, 53–54
 DH key exchange protocol, 37
 ECSTR, 38
 group, 40–41
 parameter selection, 47–48
 subgroup selection, 48–49
 translation of arithmetic operation of G to $GF(p^2)$, 42–47

Printed in the United States
by Baker & Taylor Publisher Services